International Conference on Advances in Nuclear Science and Engineering (ICANSE 2024)

IOP Conference Series: Materials Science and Engineering
Volume 1330

Bandung, Indonesia
2 - 4 September 2024

ISBN: 979-8-3313-2472-8
ISSN: 1757-8981

Printed from e-media with permission by:

Curran Associates, Inc.
57 Morehouse Lane
Red Hook, NY 12571

proceedings.com

Some format issues inherent in the e-media version may also appear in this print version.

Institute of Physics
Dirac House, Temple Back
Bristol BS1 6BE UK

Phone: 44 1 17 929 7481
Fax: 44 1 17 920 0979

techtracking@iop.org

Additional copies of this publication are available from:

Curran Associates, Inc.
57 Morehouse Lane
Red Hook, NY 12571 USA
Phone: 845-758-0400
Fax: 845-758-2633
Email: curran@proceedings.com
Web: www.proceedings.com

TABLE OF CONTENTS

IOP Conf. Series: Materials Science and Engineering **1330** (2025) 011001 doi:10.1088/1757-899X/1330/1/011001

PREFACE OF ICANSE 2024

Nuclear science and engineering have become increasingly important, especially with global challenges related to energy, health, and the environment. The International Conference on Advances in Nuclear Science and Engineering (ICANSE) is conducted and carried out as a vital forum for scientists, researchers, students, and practitioners to share the latest ideas, knowledge, and findings in nuclear science and engineering. Indonesia faces a growing demand for clean and reliable energy. Nuclear energy presents a potential solution to meet this demand while reducing greenhouse gas emissions. Utilizations of nuclear energy have been used not only for energy used application but also for non-energy application such as co-generation system for water desalination, hydrogen production, for coal gasification and liquefaction applications and enhanced oil recovery (EOR) applications. In addition, the applications for non-energy applications for health and medicine, pharmaceuticals, nuclear batteries, metal industry, agricultural environment, livestock, and so on based on the radioisotope utilization.

As the regularly seminar, the 7th International Conference on Advances in Nuclear Science and Engineering, (ICANSE) 2024 as the series international conference since 2007, aims at summarizing recent research activities relevant to the advanced development of application in the nuclear science and engineering and facilitate communication among relevant experts. After join symposium between Bandung Institute of Technology (ITB) and Tokyo Institute of Technology (Tokyo Tech) at 2005, the conference of ICANSE was established based on one of two years event 2007, 2009 and 2011. It was organized by Bandung Institute of Technology (ITB), National Nuclear Energy Agency (BATAN) and supported by ministry of education of Indonesia and COE-INES and CRINES of Tokyo Institute of Technology (Tokyo Tech). Participants, invited speaker and contributor speaker are invited from academia in university, research center and governmental organization as well as industry and public who interested in nuclear science and technology. In this 7th ICANSE 2024, as an international conference after long break after covid-19 pandemic, the event is held In conjunction with Annual Meeting and Business Nuclear Society, organized by Bandung Institute of Technology - Department of Physics, Department of Nuclear Science and Engineering (ITB) Faculty of Mathematics and Natural Sciences (FMIPA), Center For Nuclear Science, Technology and Innovation (PSTIN), Bandung Institute of technology. The event is sponsored by LPPM ITB, PT Pertamina (Persero), Indonesia and Rosatom Russia and some supporting organizations Institute for Science and Technology Development (LPIT), Bandung Institute of technology, Indonesian Nuclear Society Association (HIMNI), Indonesian Nuclear Society (HIMNI), Research Organization for Nuclear Energy, National Research and Innovation Agency (BRIN), The theme of the International Conference on Advances in Nuclear Science and Engineering (ICANSE) 2024 is "Optimizing Strategy of Nuclear Sciences and Technology Utilization on Global NZE, Energy Security, Health, and Economy Growth".

In this conference, role of nuclear technology in line with global challenges has been focused which related to energy, health, and the environment, and emphasizes the importance of innovative research and applications in promoting sustainable development. The topics of the conference are listed but not limited to:

- Innovative nuclear energy systems
    - Nuclear utilization systems based on fuel cycle
    - Simultaneous solution for safety, radioactive waste and proliferation problems
    - Advanced Small Reactors without on-site Refueling
    - Small Modular Reactors
    - Generation IV Nuclear Reactors
- Innovative transmutation systems
- Innovative separation and fuel cycles/radioactive wastes
- Nuclear nonproliferation issues
- Innovative energy systems
    - Hydrogen energy system

IOP Conf. Series: Materials Science and Engineering **1330** (2025) 011001 doi:10.1088/1757-899X/1330/1/011001

- o Cogeneration system
- o Thermal energy utilization system
- Material and process for innovative energy systems
- Radiation Physics and Protection
- Biophysics, medical physics and nuclear medicine
- Nuclear Physics and Nuclear Data
- Accelerator and Sub-critical Assembly.
- Theoretical and Computational nuclear physics and particle physics
- Nuclear education, social and public acceptance
- Energy management and planning, Policy and Regulation

Organizer

Chief Organizer : Prof. Dr.Eng Ir. Sidik Permana, S.Si., M.Eng, M.B.A.

Vice Chair : Prof. Abdul Waris, Ph.D

Secretary : Dr. Cici Wulandari

Financial Treasure : Dr. Triati Dewi Kencana Wungu, Dr. Galih R. F. Suwandi, S.Si., M.Si.

Publication Committee : Dr. Syeilendra Pramuditya

Sponsorship Committee : Dr. Eng. Dwi Irwanto

Consumption Committee : Dr. Nur Asiah

Logistic Committee : Dr. Asril Pramutadi Andi Mustari

Event Committee : Dr. Novitrian

Book of Abstract Committee : Dr. Sparisoma Viridi

Proceeding Committee : Dr. Khairul Basar

Documentation Committee : Dede Enan, SAP

Editorial Board

IOP Conf. Series: Materials Science and Engineering **1330** (2025) 011001 doi:10.1088/1757-899X/1330/1/011001

12. Prof. Marco Ricotti (Politecnico di Milano)
13. Prof. Li Fu (INET, Tsinghua University)
14. Prof. Sumer Sahin (Atilim University, Turkey)
15. Prof. H. Mehmet Sahin (Gazi University,Turkey)
16. Prof. Rizal Kurniadi (ITB)
17. Prof. Ir. Wahyu Srigutomo (ITB)
18. Prof. Widayani (ITB)
19. Prof. Idam Arif (ITB)
20. Prof. Siti Nurul Khotimah (ITB)

Chairperson of ICANSE 2024

IOP Conf. Series: Materials Science and Engineering **1330** (2025) 011002 doi:10.1088/1757-899X/1330/1/011002

Peer Review Statement

All papers published in this volume have been reviewed through processes administered by the Editors. Reviews were conducted by expert referees to the professional and scientific standards expected of a proceedings journal published by IOP Publishing.

- **Type of peer review:** Single Anonymous
- **Conference submission management system:** Morressier
- **Number of submissions received:** 31
- **Number of submissions sent for review:** 30
- **Number of submissions accepted:** 30
- **Acceptance Rate (Submissions Accepted / Submissions Received × 100):** 96.8
- **Average number of reviews per paper:** 1.7
- **Total number of reviewers involved:** 15
- **Contact person for queries:**
  Name: sidik permana
  Email: psidik@itb.ac.id
  Affiliation: Institut Teknologi Bandung

Design and testing of the oblique and V-shaped Cu-ITO microelectrode arrays to generate dielectrophoretic force on red blood cells

Edwar Iswardy[1,2,*], Khazanna[3], Elin Yusibani[1], Mursal[1], Kurnia Lahna[1] and , Sri Fitriyani[4]

[1] Department of Physics, Faculty of Mathematics and Natural Sciences, Universitas Syiah Kuala, Banda Aceh 23111, Indonesia
[2] Science, Technology, Engineering, and Mathematics (STEM) Research Center, Universitas Syiah Kuala, Banda Aceh 23111, Indonesia
[3] Master Program in Physics, Department of Physics, Faculty of Mathematics and Natural Sciences, Universitas Syiah Kuala, Banda Aceh 23111, Indonesia
[4] Department of Dental Materials, Faculty of Dentistry, Universitas Syiah Kuala, Banda Aceh 23111, Indonesia

*E-mail: edwariswardy@usk.ac.id

Abstract. Dielectrophoresis (DEP) is a phenomenon in which a force is applied to a particle induced by a gradient of an electric field. The dielectrophoretic technique is popular for manipulating bioparticles, because it requires only a small sample, is label-free, rapid, and inexpensive. Manipulation of biosample can be in the form of monitoring, separation, sorting, capturing, etc., so that the DEP method can be applied as a biosample analysis tool. However, research on the application of the DEP method is still developing on various electrode arrays and bioparticles. In this work, a lab-on-chip device with an oblique and V-shaped 3D microelectrode array has been developed to manipulate red blood cells (RBCs) using the dielectrophoresis (DEP) method. The microelectrodes were fabricated with copper and indium tin oxide films on a glass substrate, while the microchannel was constructed using double-sided tape insulators. RBCs samples were prepared in deionized water and EDTA medium with an electrical conductivity of 1.5 S/m. The test of dielectrophoretic force characteristics on RBCs was carried out by applying an AC signal to the microelectrode, and the phenomenon was observed using a microscope with a CCD camera. The results showed that negative DEP forces were observed at frequencies of 5-7 MHz, 3.5–5 MHz, and 2-4 MHz in the oblique electrode spacing area and in the middle area of the V-shaped electrode curve. While positive DEP forces were observed at frequencies of 8–14 MHz, 6–13 MHz, and 5–11 MHz in the edge area of the oblique electrode and in the inner tip area of the V-shaped electrode curve, respectively at voltages of 5 Vpp, 10 Vpp, and 15 Vpp. The results of this work show the promising potential of lab-on-chip devices with oblique and V-shaped microelectrode arrangements to manipulate particularly RBCs and generally bioparticles so that biosamples can be further analyzed.

Keywords: Oblique and V-shaped microelectrodes, Dielectrophoresis, Red blood cells, Non-uniform electric field

1. Introduction

Whole blood is one of the most commonly used biosamples in the medical field [1]. Red blood cells (RBCs) are a component of blood that is often used as a test sample for diagnosis [2,3]. RBCs contain hemoglobin which is responsible for the process of transporting O_2 and CO_2 gases [4]. Conventionally, manipulation of whole blood like separating blood components is carried out using centrifugation technique [5], however this technique requires large amount whole blood sample. Several new methods, such as fluorescence, magnetic and optical method, have been applied in RBCs analytical devices [6-9]. Even though some of those methods provide also quantitative and qualitative analysis results of RBCs, however in general they require expensive equipment, skillful operators, large room, and more time consuming. Dielectrophoresis is an electrical method that utilizes the effect of particle polarization by a non-uniform electric field that causes particle movement based on spacial gradient of the electric field [1]. This method is popular because it requires a small amount of sample, label-free samples, a short time, can increase sensitivity, and can perform in-situ testing [9, 10]. The generation of non-uniform electric fields to manipulate biosamples can be conducted with various designs of microelectrode arrays that have special functions. By applying the dielectrophoretic method, an oblique electrode array in a microchannel can be used to direct the biosample in continous flow [11,12], while a V-shaped electrode array can be used to capture the biosample in continous flow [13]. The dielectrophoretic force experienced by particles in a non-uniform electric field can be expressed by the equation (1),

$$F_{DEP} = 2\pi r^3 \varepsilon_0 \varepsilon_m \, Re(f_{CM}) \, \nabla E^2 \quad (1)$$

where r, ε_o, ε_m, and ∇E represent respectively particle radius (m), vacuum permittivity (F/m), relative permittivity of the suspension medium (dimensionless), and the electric field gradient (V/m). $Re(f_{CM})$ denotes the real part of the Clausius-Mossotti (CM) factor which shows the degree of particle polarization and can be written in the form of equation (2),

$$f_{CM} = \frac{\varepsilon_p^* - \varepsilon_m^*}{\varepsilon_p^* + 2\varepsilon_m^*} \quad (2)$$

with ε_p^* and ε_m^* being the complex electrical permittivity of particle and medium (both dimensionless), respectively. For every detail, $\varepsilon_p^* = \varepsilon_p - (j\sigma_p/\omega)$ and $\varepsilon_m^* = \varepsilon_m - (j\sigma_m/\omega)$ with ε_p is the permittivity of the particle (dimensionless), ε_m is the permittivity of the medium (dimensionless), σ_m is the electrical conductivity of medium (S/m), σ_p is the electrical conductivity of particle (S/m), j is an imaginary number with $j = \sqrt{-1}$, and ω is angular frequency (rad/s). Application of appropriate parameters is necessary to obtain suitable particles manipulation based on dielectrophoretic force.

In this work, we have developed a lab on chip device containing an 3D oblique and V-shaped microelectrode arrays. The aim of this research is to identify dielectrophoretic forces on RBCs using the designed device which can be used to particularly manipulate RBCs using dielectrophoretic method.

2. Materials and Methods

2.1 Red Blood Cells (RBCs) preparation

Whole blood samples were prepared in Ethylene Diamine Tetra Acetic Acid (EDTA) medium. EDTA medium was utilized as a standard anticoagulant to inhibit coagulation and to preserve morphology of RBCs during experiment. The samples were then diluted with deionized water with a ratio of 1:20 and the electrical conductivity of the test sample in the medium was 1.5 S/m.

2.2 Design and fabrication of microelectrodes and microchannel

The design of oblique and V-shaped microelectrode patterns is as shown in Figure 1a. The width of microelectrodes were 500 μm, the spacing between electrodes were 100 μm, and the spacing between the tips of oblique electrode were 100 μm. The angle of 45° for the oblique and V-shaped microelectrode patterns was elected because it offered optimum sample directing and trapping according to previous report [13]. The copper was chosen as electrode since it is convenient to be patterned by simple lithography technique. While the ITO was utilized due to its electrically conductive and transparent properties, so that the movement of RBCs caused by dielectrophoretic force can be obviously observed by microscope. The fabrication of oblique and V-shaped microelectrodes from copper films was carried out using the lithography method using 1 M $FeCl_3$ solution. While the microchannel was formed using double tape insulators with the width of 5 mm and the height of 30 μm. The bottom of the microchannel contains Cu microelectrodes and the top of the channel contains ITO electrodes. Thus, microelectrode fabrication with a combination of Cu-ITO in the microchannel were obtained (Figure 1b). The final step was to wire the Cu-ITO microelectrode terminals (Figure 1c).

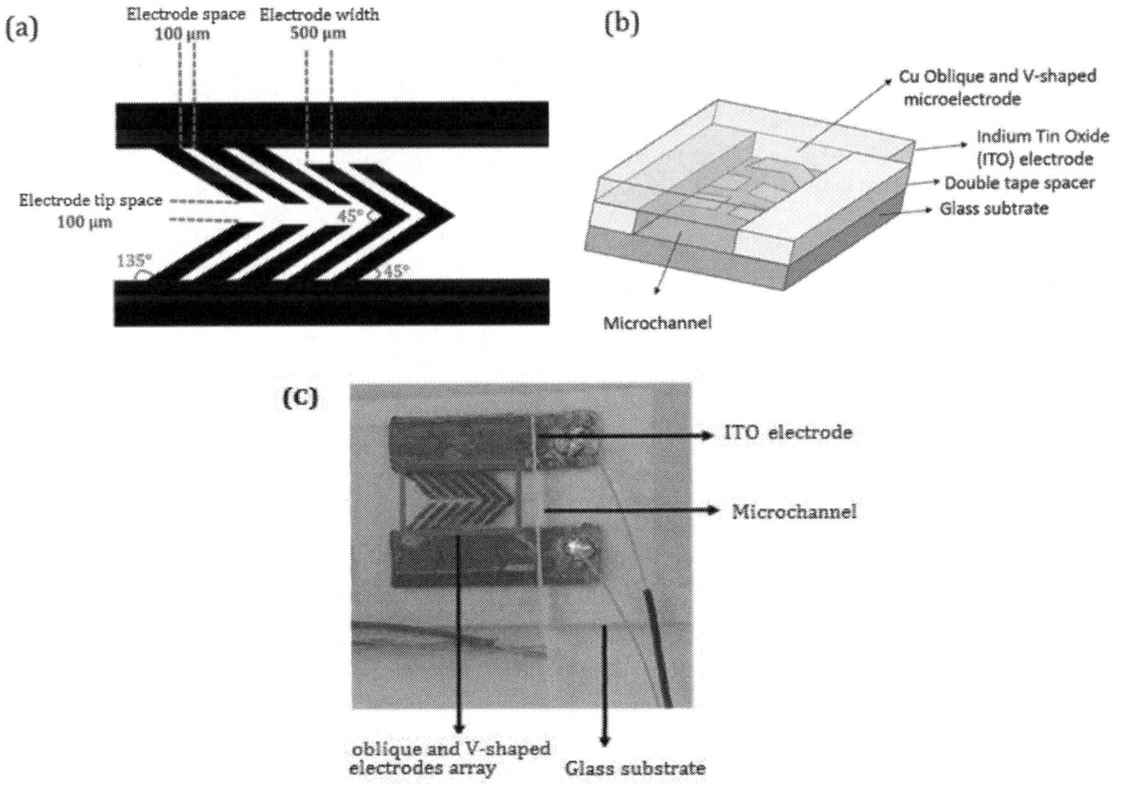

Figure 1. Design and the result of microelectrodes fabrication, (a) Design of Cu oblique dan V-shaped microelectrode arrays, (b) Design of combination Cu-ITO microelectrodes in microchannel, (c) Top view of 3D microelectroda Cu-ITO and wiring

2.3 Instrument set up for dielectrophoresis testing of RBCs

The equipment set up (Figure 2) was employed by connecting the Cu-ITO microelectrode terminals to the function generator. The dielectrophoretic (DEP) force phenomenon was observed using a microscope with a CCD camera (LBA Digimi) that had been connected to a computer. The AC signal on the function generator applied to the microelectrodes was varied in terms of magnitude of the peak-to-peak voltage (Vpp) and frequency to obtain an adequate gradient of the electric field strength that affects the DEP force on RBCs. The voltages were varied from 0 to 20 Vpp, and the frequencies were varied from 0 Hz to 15 MHz, following the previous research [13]. Within those range of voltage and frequency, DEP phenomena occured on RBCs were investigated.

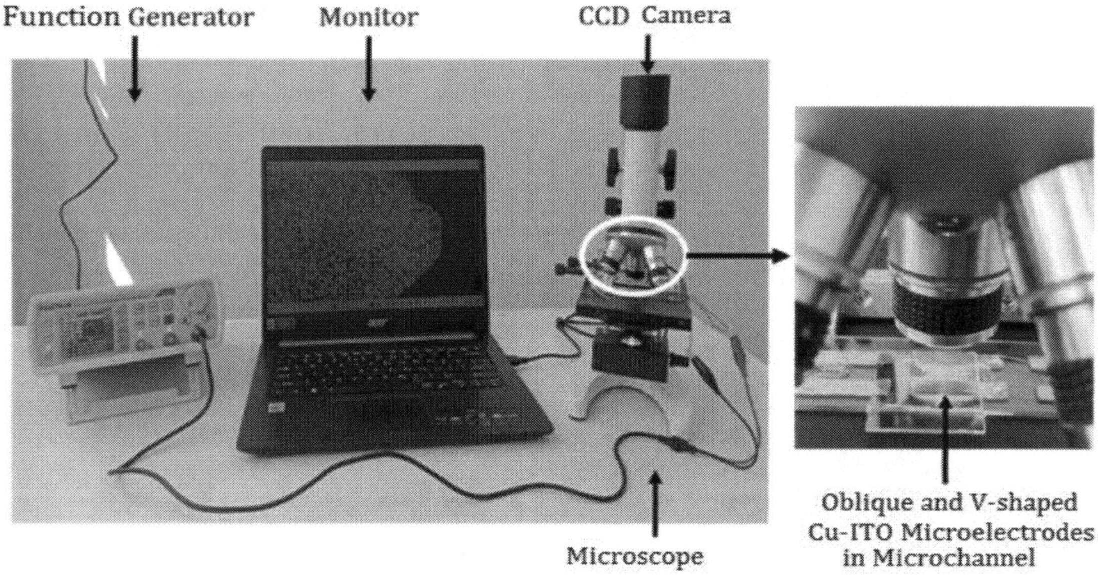

Figure 2. Instrument set up for dielectrophoresis testing of RBCs

3. Results and discussion

3.1 Result of RBCs preparation

RBCs were prepared in deionized water and EDTA medium with an electrical conductivity of 1.5 S/m. Observation of the test samples was carried out to validate the shape and size of the diameter of RBCs. Based on images from an optical microscope with a magnification of 50x, 100x, and 400x and dimensional analysis using Image-J software, it is confirmed that RBCs are biconcave disc-shaped with an average diameter of 7.50 μm (Figure 3) according to reference [14].

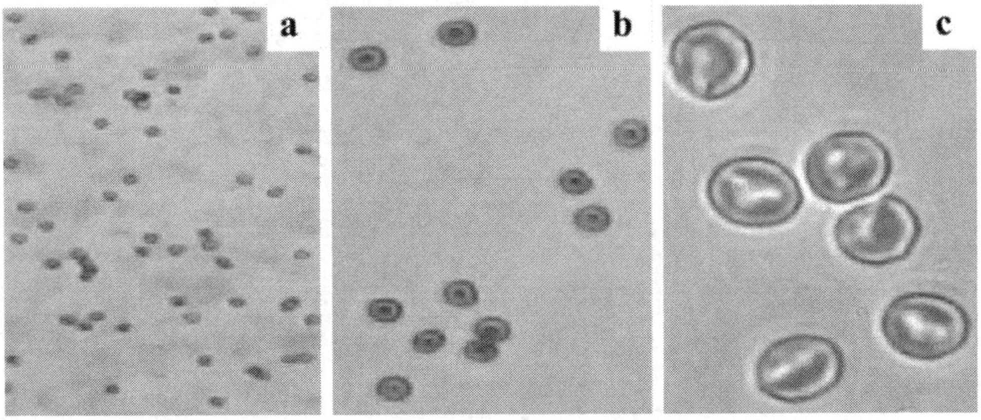

Figure 3. The optical microscope images of the RBCs, (a) 50x of magnification, (b) 100x of magnification, and (c) 400x of magnification.

3.2 Dielectrophoretic force testing of RBCs on the 3D oblique and V-shaped microelectrodes array

Based on the reference, on an 3D oblique microelectrode array, the strongest electric field intensity occurs at the edge of the electrode, and the weakest electric field intensity occurs in the tip space of the electrode [15]. While on a V-shaped microelectrode array, the strongest electric field is identified at the inner tip of the V-shaped electrode, and the weakest electric field is identified at the outer part of the V-shaped electrode [16]. These indicate the existence of a spatial gradient of a non-uniform electric field on oblique and V-shaped electrodes as a requirement for generating the dielectrophoretic force phenomena.

Before the application of an electrical signal, the position of RBCs is about evenly distributed along the microchannel (Figures 4a, 4b, and 4c). When an electrical signal is applied, the RBCs are going to move towards the strong electric field area or the weak electric field area in the oblique and V-shaped Cu-ITO microelectrodes as an effect of the DEP force. Theoretically, there are two types of DEP force namely the positive DEP (pDEP) force and the negative DEP (nDEP) force. When pDEP force occurs on a particle, it will be attracted to the strong electric field zone on electrode, and vice versa when nDEP force occurs on a particle, it will be repelled to the weak electric field zone on electrode [17].

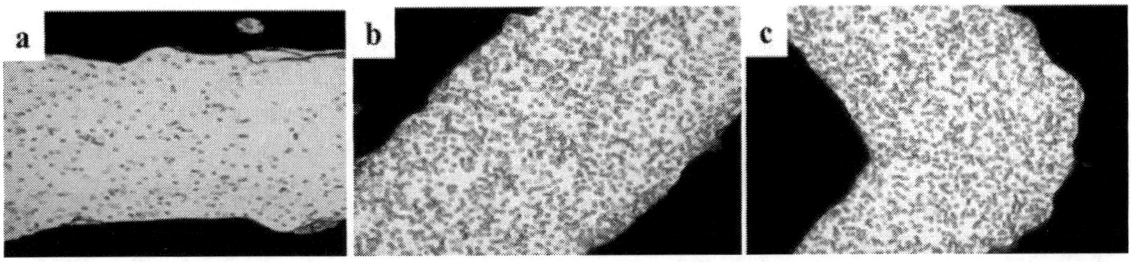

Figure 4. Condition of RBCs on Cu-ITO microelectrodes in microchannel before applying an electrical signal (a) on the space between the oblique electrode tips, (b) on the space between the oblique electrodes, and (c) on the V-shaped electrode

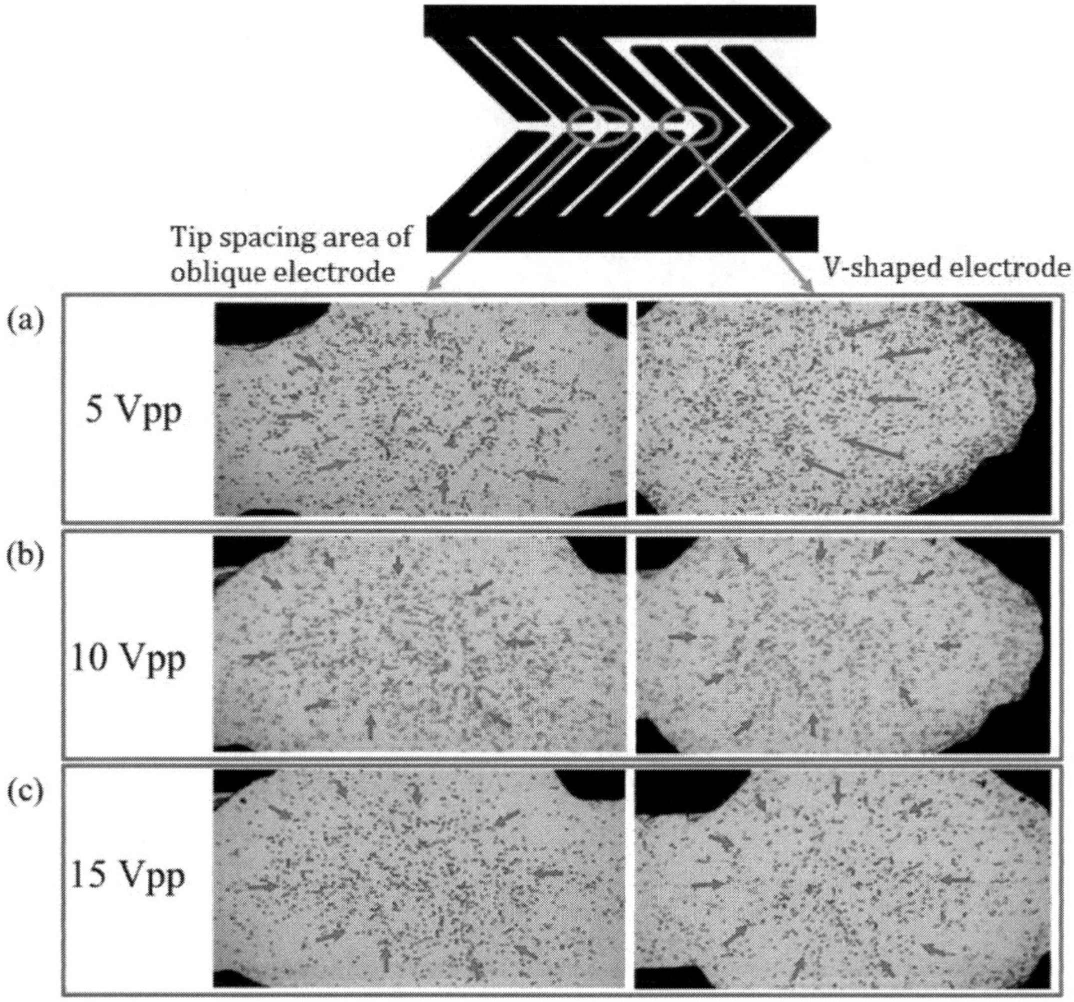

Figure 5. The images of the nDEP phenomena of RBCs on oblique and V-shaped electrode arrays at applying signal of (a) 5 Vpp, (b) 10 Vpp, and (c) 15 Vpp

On the oblique electrode, a positive dielectrophoretic force (pDEP) phenomenon on RBCs will occur when the RBCs are attracted to strongest electric field zone which locates on the edge of the electrode. While a negative dielectrophoretic force (nDEP) phenomenon on RBCs will occur when the RBCs are repelled to the weakest electric field intensity which occurs in the tip space of the oblique electrode. In addition, on a V-shaped microelectrode, the strongest electric field is identified at the inner tip of the V-shaped electrode which causes the attraction of RBCs to this zone as the effect of pDEP force, and the weakest electric field is identified at the outer part of the V-shaped electrode which leads the repulsion of RBCs to the outer part of the V-shaped electrode as the effect of nDEP force. The dielectrophoresis phenomena of RBCs are clearly observed at three variations of voltage magnitude, namely 5 Vpp, 10 Vpp, and 15 Vpp. By applying a voltage of 5 Vpp to the oblique and V-shaped microelectrodes, it was observed that RBCs began to be pushed away from the edge of the electrode to the space between the electrodes tip at a frequency of 5-7 MHz which is stated as a negative dielectrophoretic force (nDEP) phenomenon. However, the positive dielectrophoretic force (pDEP) phenomenon was observed at a frequency of 8–14 MHz which showed that RBCs moved towards the edge of the electrode. In the results of the

IOP Conf. Series: Materials Science and Engineering **1330** (2025) 012001 doi:10.1088/1757-899X/1330/1/012001

further test by applying a voltage of 10 Vpp, the nDEP force phenomenon was obtained at a frequency of 3.5–5 MHz and the pDEP force was observed at a frequency of 6–13 MHz. While applying a voltage of 15 Vpp, the nDEP force phenomenon was observed at a frequency of 2-4 MHz and pDEP at a frequency of 5–11 MHz. The images of nDEP force phenomena with the application of voltages of 5 Vpp, 10 Vpp and 15 Vpp are shown respectively in Figures 5a, 5b, and 5c, while the images of pDEP force phenomena are shown respectively in Figures 6a, 6b, and 6c. The movement of RBCs due to pDEPand nDEP forces follow the arrows and the number of RBCs is higher in the zones shown by the arrows.

Based on the results of dielectrophoresis tests, especially on V-shaped electrodes, it shows that the pDEP phenomenon causes RBCsto be attracted to the strong electric field area, and the nDEP phenomenon causes RBCs to be pushed towards the weak electric field area, namely the inner tip

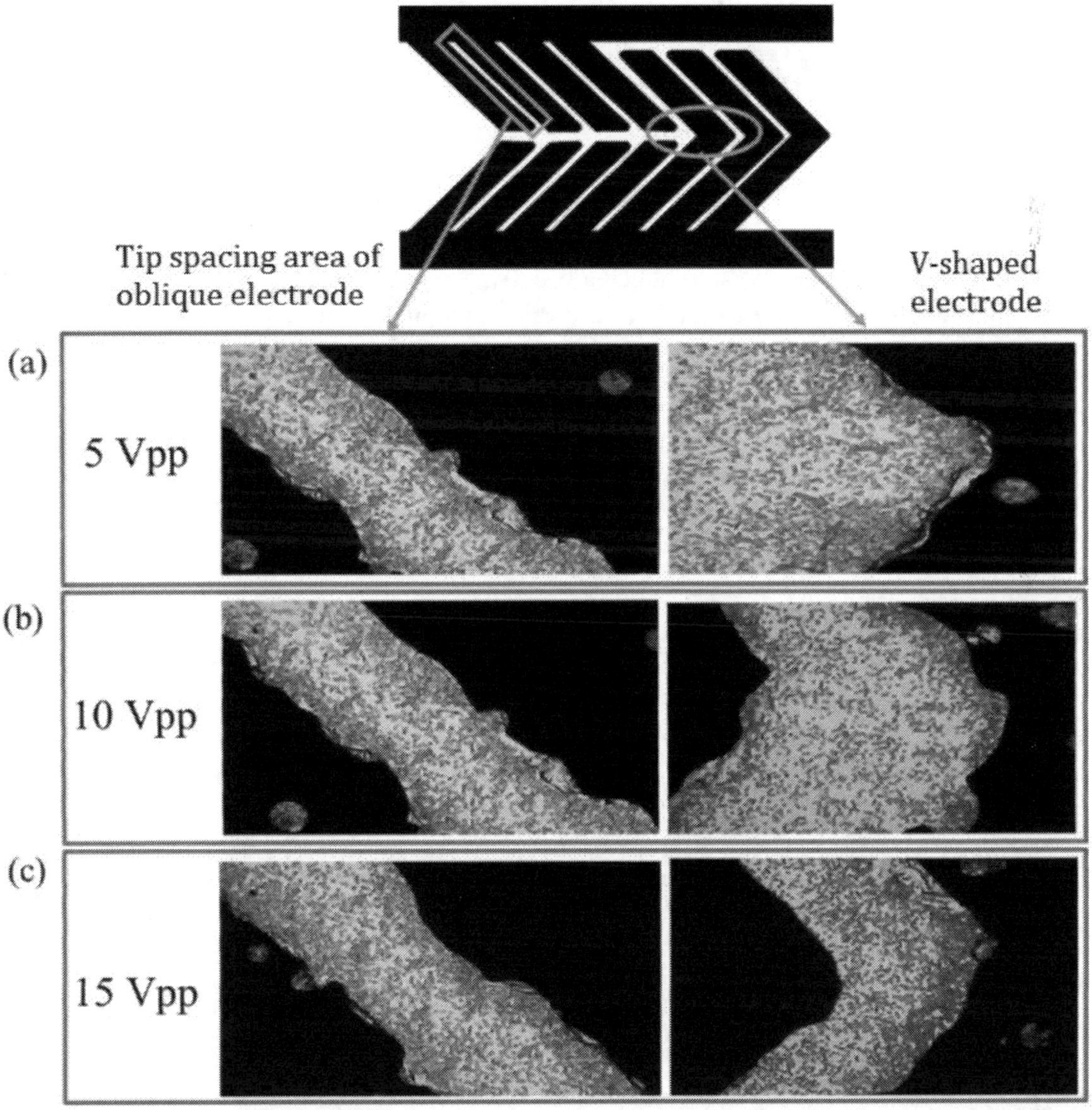

Figure 6. The images of the pDEP phenomena of RBCs on oblique and V-shaped electrode arrays at applying signal of (a) 5 Vpp, (b) 10 Vpp, and (c) 15 Vpp

of the V-shaped electrode. These phenomena are in accordance with the results of previous studies, which explain that the strongest DEP force occurred in the triangle meeting point area, where a high local field could cause RBCs to be attracted to the inner tip area of the V-shaped electrode (pDEP) or pushed from the inner end area of the V-shaped electrode (nDEP) [16].

The combination of 3D oblique and V-shaped Cu-ITO microelectrodes from this study is known to have the potential to be used in RBCs manipulation by the dielectrophoresis method. In the microfluidic channel that flows the biosample, the oblique microelectrode plays a role in directing RBCs when applying a low-frequency electrical signal, which causes nDEP force on RBCs. The flow of RBCs will then travel to the V-shaped electrode. Then the strong pushing effect of the V-shaped electrode due to the nDEP force will cause RBCs to concentrate on the curved part of the V-shaped electrode as a function of the biosample capture. This work was conducted in vitro under controlled environment in laboratory at room temperature (~25-27°C) and humidity of around 75-80%, therefore the free space circumstance effects were not considered. This designed platform also can be employed for other biosamples by setting the operating parameters based on characteristic of utilized biosamples and medium.

4. Conclusions

From this research, a prototype of a lab-on-chip device has been produced that combines an oblique and V-shaped Cu-ITO microelectrode array in a microchannel. By applying the dielectrophoresis method, this device can be used to manipulate, especially to direct and capture RBCs. The results of DEP force testing on RBCs conducted at three voltage applications, namely 5 Vpp, 10 Vpp, and 15 Vpp, obtained nDEP phenomena observed respectively at frequencies of 5-7 MHz, 3.5-5 MHz, and 2-4 MHz. While the pDEP phenomenon was observed respectively at frequencies of 8–14 MHz, 6–13 MHz, and 5–11 MHz. The results of this study show that the combination of oblique and V-shaped microelectrodes has the potential to particularly manipulate RBCs and generally bioparticles using dielectrophoretic method in biosample analysis.

Acknowledgements
The authors would like to thank The Syiah Kuala University, The Ministry of Education, Culture, Research, and Technology Indonesia for funding this research in accordance with the agreement letter for carrying out Penelitian Tesis Magister for Fiscal Year 2024 No: 402/UN11.2.1/PG.01.03/SPK/PTNBH/2024.

References

[1] Emmerich M E P, Sinnigen A S, Neubauer P and Birkholz M 2022 *Biomed Microdevices* 24 (30) 1-19
[2] Li C M, Dong H, Zhou Q, Goh K H 2008 *Biochips fundamentals and applications. in electrochemical sensors, biosensors and their biomedical applications* 307-383
[3] Liao S H, Chang C Y and Chang H C 2013 *Biomicrofluidics* 7 (2), 024110
[4] Farley A, Hendry C and McLafferty E 2012 *Nurs Stand* 27(13), 35–42
[5] Lianidou E S, Markou A, and Strati A 2014 *Circulating tumor cells as a real-time liquid biopsy: isolation and detection systems, molecular characterization, and clinical applications* 4113-4130.
[6] Gross A, Schoendube J, Zimmermann S, Steeb M, Zengerle R and Koltay P 2015 *Int J Mol Sci* 16 16897-16919
[7] Shamloo A, Yazdani A, Saghafifar F 2020 *Eng Life Sci* 1-9
[8] Paie P, Zandrini T, Vazquez R M, Osellame R and Bragheri F 2018 *Micromachines* 9 (5), 200.
[9] Zhao K, Hu M, Baalen C V, Alvarez and Isa L. 2023 J *Colloid Interface Sci* 634 921–929
[10] Afsaneh H and Mohammadi R 2022 *Talanta Open* 5 100092
[11] Tajik P, Saidi, M S, Kashaninejad N and Nguyen N T 2019 *Ind Eng Chem Res 59*(9) pp 3772-3783

IOP Conf. Series: Materials Science and Engineering **1330** (2025) 012001 doi:10.1088/1757-899X/1330/1/012001

[12] Khazanna, Fitriyani S, Safitri R, Lahna K, Yusibani E, Irwansyah and Iswardy E 2024 *J Phys Conf* Vol. 2734, No. 1, p. 012004
[13] Iswardy E, Tsai T C, Cheng I F, Ho T C, Perng G C and Chang H C 2017 *Biosens Bioelectron* 95, 174-180.
[14] Cole L and Kramer P R 2016 *Human physiology, biochemistry and basic medicine* 85–91.
[15] Alatas Y C., Tefek U, Sari B and Hanay M S 2023 *IEEE J Microw* 1-8
[16] Cheng I F, Han H W and Chang H C 2012 *Biosens Bioelectron* 33(1), 36-43
[17] Pohl H A and Crane J S 1972 *J Theor Biol* 37(1), 1-13

IOP Conf. Series: Materials Science and Engineering **1330** (2025) 012002 doi:10.1088/1757-899X/1330/1/012002

Nuclear Energy: A Strategic Path Towards Indonesia's Sustainable Energy Future

Nining Yuningsih[1*], Indah Rosidah Maemunah[2,3], Fisca Dian Utami[4]

[1]SMAN 5 Sukabumi, Indonesia
[2]National Research and Innovation Agency, Indonesia
[3]Graduate School of Engineering, Osaka University, 1-1 Yamadaoka, Osaka, Japan
[4]Politeknik Transportasi Sungai, Danau, dan Penyeberangan Palembang, Palembang, Indonesia

*corresponding author: nining.1924@gmail.com

Abstract

Indonesia's evolving energy landscape highlights nuclear energy as a viable solution for future energy demands. Nuclear power offers a reliable and sustainable alternative, crucial for balancing increasing urban energy consumption and variable rural needs. This study underscores nuclear energy's potential role in meeting Indonesia's future energy requirements, stressing the necessity of strategic energy policy planning to align with these projections.

Comparing energy sources like oil, gas, and nuclear power plants reveals nuclear energy's robustness as an alternative. A 1000 MWth nuclear power plant can fulfil approximately 46-53% of the energy demands in Nusa Tenggara, Maluku, and Papua. Moreover, nuclear power boasts minimal carbon emissions, approximately 0.029 kg/kWh, positioning it favourably in achieving Indonesia's 2060 energy targets and fostering a sustainable energy future.

Keywords: Energy, Emission, Nuclear, Sustainable

Introduction

Based on energy type, Indonesia's final energy consumption in 2022 was dominated by coal (26%), and biodiesel (19%). Additionally, the total primary energy supply in 2022 reached 246 million tonnes of oil equivalent (TOE), with coal utilization still dominating at around 42%. Then, it followed by oil (31%), gas (14%), and new renewable energy (12.3%) of the total primary energy supply. Power generation was still primarily dominated by coal, which accounted for half of the total national capacity. Meanwhile, power generation based on a renewable energy reached only 15%, with approximately 6 GW over the last 10 years. The total CO2 emissions in 2022 were around 696.7 million tons, reflecting an average increase of 4.1% per year. On the energy demand, an energy consumption reached 6,29% that was observed in 2023.

Predictably, the total electricity demand is projected to increase from 313 TWh in 2022 to 479 TWh in 2033. Total electricity production is expected to rise from 351 TWh in 2022 to approximately 538 TWh in 2033. Over the next decade, electricity production will still be dominated by fossil fuels, particularly coal. In 2022, the share of electricity production from fossil power plants by coal was around 71% of total production. However, this share is projected to decrease to 63%. Electricity production from a renewable energy source is expected to grow from 49 TWh in 2022 to 122 TWh in 2033. Currently, the electricity generation from renewable energy sources in 2033 is projected to be dominated by hydropower (38%), geothermal power (33%), and bioenergy power plants, including waste-to-energy: biomass, and biogas plants (21%). The most significant growth in renewable energy production will come from solar power, which is projected to increase by 61%, raising production from 0.4 TWh in 2022 to 68 TWh in 2033. Throughout the projection period, power plants will continue to be the largest source of emissions due to the persistent dominance of fossil fuel usage, particularly coal, which is expected to account for approximately 47% of emissions by 2033.

In 2032, nuclear power plants (NPP) are expected to create the electricity production by 701 GWh. Although, there is currently no national commitment to the construction of NPPs, but demanding of clean energy must be met with Indonesia targets to achieve a net-zero emission by 2060.

The presence of nuclear fuel supply in Indonesia is spread across 26 locations in Sumatra, Kalimantan, Sulawesi, Maluku, and Papua, that is conducted at 9 deposit sites for uranium and thorium. In 2021, the total uranium resource was 89,483 tons of U_3O_8 (equivalent to the thermal generation capacity of 1.85 trillion tons of coal), meanwhile the thorium resource amounted to 143,234 tons. However, the total thorium resource in 2014 was 130,974 tons, including 4,729 tons from Bangka Belitung.

After Fukushima accident in 2011, many countries engaged in strategic manoeuvrers regarding nuclear development, not only due to safety factors but also to enhance their bargaining position in the international market. There are 50 developing countries recently, including 30 countries that still consider involving a nuclear power plant. Indonesia is one of the countries grappling with issues and challenges in implementing its nuclear program [4]. In progress, the Indonesian government plans to build its first High-Temperature Gas-Cooled Reactor (HTGR), PeLUIt-40 [5]. PeLUIt-40 is a nuclear reactor designed in Indonesia to generate heat for electricity production, with a thermal power capacity of 40 MW [6].

Population is one of the most important factors influencing CO_2 emissions and other greenhouse gas emissions. Population growth is expected to impact the increasing of energy consumption and aggregate CO_2 emissions [7]. The key to reducing global CO_2 emissions is the widespread use of renewable energy. Technical innovation is crucial for environmental and economic advancement. The primary goals of alternative energy are to eliminate emissions from fossil fuels, improve air quality, enhance energy security, and promote sustainable economic growth. Progress in innovation plays a vital role in effectively designing policies to reduce carbon emissions [8]. To address climate change, clean energy is a necessity in today's world. The demand for energy is growing faster each decade [9].

Wind, nuclear, hydropower, solar, and biomass are the main low-carbon energy sources estimated throughout their life cycle[10]. By providing consistent and reliable electricity, nuclear power can reduce a country's dependence on non-renewable fuel sources [11]. Currently, nuclear energy is a key economic source of low-carbon electricity and heat, helping to avoid more than one billion tons of CO_2 emissions each year. Meanwhile, it is also supporting the reliability of energy systems, long-term security, and climate resilience [12]. In addition, nuclear energy can facilitate high integration of renewable energy and support both long-term energy security and climate resilience. Specifically, the global nuclear has generating clean energy, and it is also a proven robust in facing extreme climate changed, such as: heatwaves, storms, and droughts. Besides that, nuclear energy has the potential to produce heat and hydrogen that can be used for decarbonizing processes and activities less suitable for electrification [13].

However, despite many countries showing interest in nuclear power throughout history, many of them did not continue their efforts, opting instead to postpone or abandon nuclear power plants for peaceful purposes. Some of reason indicated of major nuclear accidents and political issue to energy policy [14].

The results estimate of electricity production increasing from carbon-intensive fuels, leading to greater CO_2 emissions when the complete cessation of nuclear use is realized. The scenario of the UK nuclear power by 2030 leads to add the emission of 35.2 million tons of CO_2 and increase of electricity costs. The scenario for Belgium nuclear power by 2025 reflects to similar outcomes, requiring a thermal capacity of 5.7 GW, which leads to increased dependence on energy imports and CO_2 emissions [15]. In Sweden, the energy efficiency and renewable energy enable to prevent electricity shortages and begin increasing power surplus, although the surplus amount declines from 2020 to 2040 [16]. Nuclear energy has made a significant contribution to slowing the rise of global CO_2 emissions since the 1970s, globally reducing approximately 66 gigatons (Gt) of CO_2 between 1971 and 2020. Without the contribution of nuclear energy, the total emissions from electricity generation would be nearly 20% higher, and total energy emissions would be 6% higher during that period.

Method

This study employs a literature review method to collect data on Indonesia's energy demand until 2033 [1]. Meanwhile, for the power output generated by nuclear power plants, the DEEP application is utilized [17].

Table 1. Power and desalination plants formulated in DEEP [17].

Energy sources		Power	Heat
NSC	Nuclear Steam Turbine (PWR, PWHR, and SPWR)	v	v
NBC	Nuclear Gas Turbine (GTMHR)	v	v
NH	Nuclear Heat (HR)		v
COAL	Stela Cycle – Coal (SSB)	v	v
OIL	Steam Cycle – Oil	v	v
GT	Gas Turbine/HRSG	v	v
CC	Combined Cycle (Steam Turbine – Gas Turbine)	v	v
FH	Fossil Heat (Boiler)		v
RH	Renewable Heat		v

The structure and models for categories are identical and rely on straightforward thermodynamic cycle calculations. A unique configuration is established by combining various technical and economic parameters. All parameters are adjustable by the user and can be fine-tuned to meet specific requirements [17]. The thermodynamic performance of dual-purpose steam-cycle-based plants is modelled mentioned in DEEP manuals, as follows:

1. **Input Parameters:** The required power output (Pen) and thermal base plant (Qtp) are specified as inputs.
2. **Reject Heat Calculation:** The reject heat load of the single-purpose plant (Qcr) is calculated based on the site-specific average condensing temperature (Tc).
3. **Estimation for Dual-Purpose Plant:** The following figures are estimated for the dual-purpose plant: a. The heat required for operating the desalination plant, which has a capacity of Wdrc and a maximum brine temperature. b. The available reject heat at a modified temperature level (Tmb) necessary for desalination, assuming a hypothetical Carnot or Rankine cycle operating between Tcm and Tc.
4. **Lost Work Estimation:** The lost work due to heat extraction at a higher temperature is estimated as follows: a. In the case of a back-pressure turbine, all heat must be extracted at the new temperature level. b. For an extraction/condensing turbine, only the heat required for the desalination plant is extracted at the new temperature level.
5. **Lost Electricity Calculation:** The lost electricity is equivalent to the amount of electricity that could have been generated if the heat extracted for the desalination plant had been taken from the lowest available temperature (Tc).

Gas-cycle-based power plants (GT, NBC) utilize calculations similar to those of steam cycles, moreover, using the Brayton cycle for Carnot or Rankine cycles. The following equations are used for performance calculations of single-purpose plants. For dual-purpose plants, the calculations are performed in the same manner as those for steam plants, with the key distinction being that lost electricity is equal to zero. Gas turbine-based plants provide what can be considered virtual free heat, as the recovered heat cannot be utilized for further extraction or electricity production.

Results and discussion

Indonesia's Energy Demand: 2023-2033

Indonesia's electricity demand is projected to experience steady growth over the next decade, driven by economic development, population growth, and urbanization. The demand varies significantly across the country's diverse regions, with some areas seeing much more rapid increases than others.

The national electricity demand is expected to rise from around 3.0 x 10^5 GWh in 2023 to over 4.5 x 10^5 GWh by 2033. This growth reflects the expanding needs of industries, households, and infrastructure, especially in more densely populated and economically advanced regions.

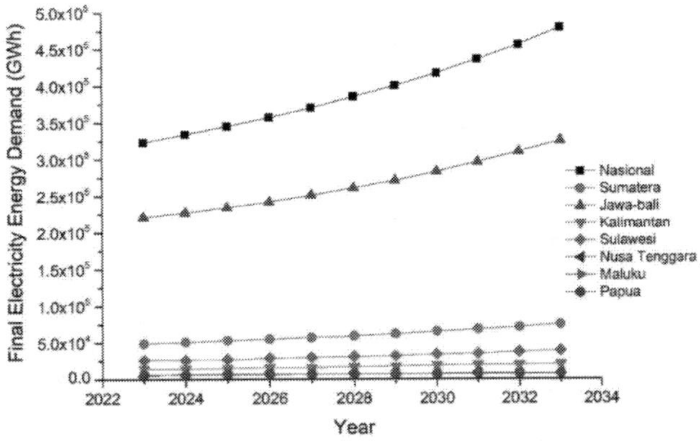

Figure 1. Final electricity energy demand in Indonesia from 2023 to 2033.

As the most industrialized and populous region, Jawa-Bali is the largest contributor to Indonesia's total electricity demand. Starting at around 2.5 x 10^5 GWh in 2023, the region's demand is expected to reach 3.5 x 10^5 GWh by 2033. This sharp increase highlights the concentration of economic activities in Java and Bali, which house the nation's capital and major cities. Secondly, Sumatera is projected to see moderate demand growth, from 5.0 x 10^4 GWh in 2023 to around 7.0 x 10^4 GWh in 2033. This reflects ongoing industrial and urban development in the region, though at a slower rate compared to Jawa-Bali. Thirdly, Kalimantan's energy demand will increase steadily, from 2.5 x 10^4 GWh to 3.5 x 10^4 GWh by 2033. This suggests ongoing infrastructure and industrial expansion, particularly in areas related to mining and natural resources.

Similarly, Sulawesi's demand grows from 2.0 x 10^4 GWh in 2023 to 3.0 x 10^4 GWh in 2033, indicating gradual economic growth, especially in agriculture, manufacturing, and processing industries. These regions represent the smallest shares of national electricity demand. Though there is modest growth in Nusa Tenggara (from 0.5 x 10^4 GWh to 1.0 x 10^4 GWh), the demand in Maluku and Papua remains largely flat. This indicates less industrial activity and slower economic growth in these more remote and less densely populated areas.

The data indicates that Jawa-Bali will continue to dominate Indonesia's electricity consumption, accounting for the largest share of the nation's growing energy needs. Other regions such as Sumatera, Kalimantan, and Sulawesi will also see consistent, albeit slower, growth in energy demand. On the other hand, more remote regions like Maluku and Papua will experience minimal changes in their electricity consumption, reflecting slower economic development in these areas.

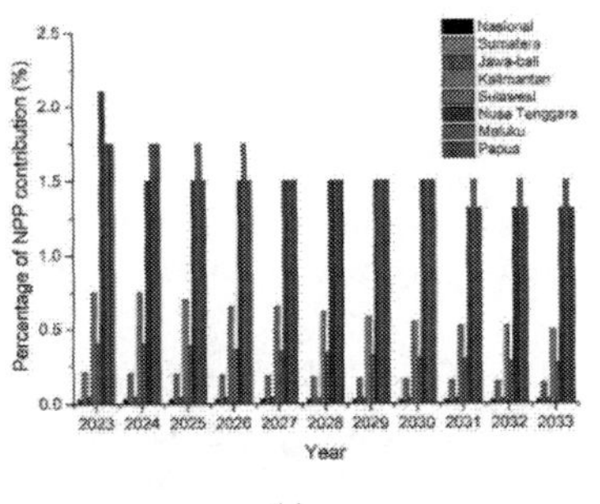

(a)

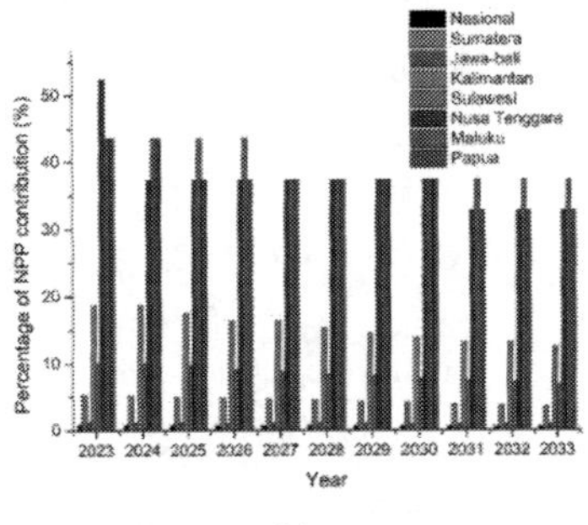

(b)

Figure 2. Projected contribution of Nuclear Power Plants (NPPs) in the gas cycle for energy production in Indonesia with capacities of (a) 40 MWth, and (b) 1000 MWth.

Figure 2 shows the projected percentage contribution of nuclear power plants (NPPs) in the gas cycle to the electricity demand in Indonesia across various regions from 2023 to 2033. The first chart corresponds to a 40 MWth NPP (seen in Fig. 2(a)), while the second chart represents a 1000 MWth NPP (seen in Fig. 2(b)).

The nuclear contribution at a 40 MWth capacity remains quite low in all regions throughout the period, with most regions having percentages below 1%. In Jawa-Bali (Blue Bars) and Sumatera have slightly higher contributions (close to or slightly over 1%). Other regions, particularly Maluku (Purple Bars) and Papua (Magenta Bars), dominate the percentage share but still remain under 2%. Moreover, smaller regions like Nusa Tenggara, Kalimantan, and Sulawesi show very small contributions, close to or below 0.5%. The contribution of a 40 MWth NPP is minimal, even on a national level, staying below 2%. Although it provides some energy to areas like Jawa-Bali and Sumatera, it is insufficient to make a significant impact on energy demand.

In addition, the 1000 MWth NPPs has a high national contribution above 40% in 2023 and fluctuates slightly around that range throughout the period. In Jawa-Bali, Sumatera, and Kalimantan have notable contributions, all starting at around 10-15% in 2023 and remaining steady over the years. Meanwhile, Papua and Maluku show very high contributions, reaching or exceeding 40% of the total demand for most of the years. This suggests that a large portion of their electricity demand can be met by nuclear energy if a 1000 MWth reactor is implemented. And smaller regions like Nusa Tenggara and Sulawesi also see significant contributions in the range of 10-20%. The impact of a 1000 MWth NPP is substantial. Nationally, nuclear power could cover more than 40% of Indonesia's energy demand. In smaller regions such as Papua and Maluku, nuclear energy would be transformative, supplying nearly half of their electricity needs. Even in major areas like Jawa-Bali and Sumatera, nuclear energy could account for 10-15% of demand, which is a considerable contribution.

The 40 MWth plant has only a minimal impact on national and regional electricity demand, contributing less than 2% in any region. On the other hand, the 1000 MWth shows significant potential, particularly at the national level and in smaller, less populated regions, where it could cover up to 40-50% of the electricity demand. Even in more populous regions like Jawa-Bali, its contribution would reach around 10-15%.

Thus, while a 40 MWth NPP is not sufficient to meet substantial energy demand, a 1000 MWth NPP could play a crucial role in the future energy mix, especially in supporting Indonesia's energy transition and diversifying its energy sources.

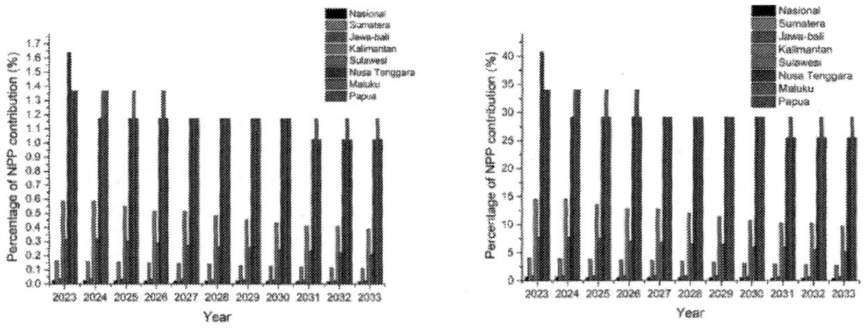

(a) (b)

Figure 3. Projected contribution of Nuclear Power Plants (NPPs) in the steam cycle for energy production in Indonesia with capacities of (a) 40 MWth, and (b) 1000 MWth.

Figure 3 shows the projected contributions of Nuclear Power Plants (NPPs) in the steam cycle for energy production in Indonesia, comparing two different reactor capacities: (a) 40 MWth and (b) 1000 MWth, for the years 2023 to 2033. The results are broken down regionally and nationally, showing contributions in various parts of Indonesia including Sumatera, Jawa-Bali, Kalimantan, Sulawesi, Nusa Tenggara, Maluku, and Papua. For both 40 MWth and 1000 MWth, the national contribution (black bars) increases over time, although it starts to plateau after a few years. In the 1000 MWth scenario (b), the national contribution is significantly larger, starting around 40% in

2023 and dropping gradually to around 35% by 2033. In contrast, in the 40 MWth scenario (a), the national contribution starts at around 1.7% in 2023, decreasing slightly by 2033.

Across both capacities, Papua (purple bars) shows the highest regional contribution to NPP energy production, dominating the charts. Other regions like Jawa-Bali (blue) and Sumatera (red) show relatively smaller contributions in both cases, but they remain constant over time. The contributions from Maluku (pink), Kalimantan (green), Sulawesi (cyan), and Nusa Tenggara (magenta) are minimal but consistent throughout the period.

The higher capacity of 1000 MWth results in a much more significant energy contribution overall, with Papua making a significant impact, especially in the earlier years. In the 40 MWth scenario, the contribution from all regions remains small compared to the larger reactor, showing the increased efficiency and output of the larger NPP. Finally, the 1000 MWth reactor has a far greater impact on Indonesia's energy production, especially in Papua, compared to the 40 MWth reactor, which shows much smaller contributions from all regions.

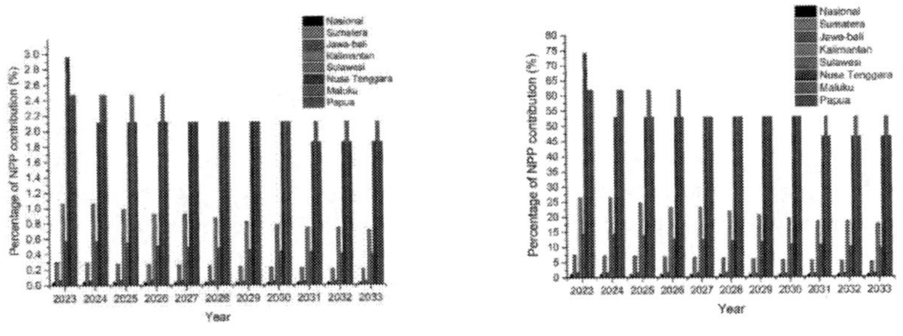

(a) (b)

Figure 4. Projected contribution of Nuclear Power Plants (NPPs) in the gas-steam cycle for energy production in Indonesia with capacities of (a) 40 MWth, and (b) 1000 MWth.

Based on Fig. 4, in both capacities for 40 MWth (Fig. 4(a)) and 1000 MWth (Fig. 4(b)), the national contribution steadily declines over the years. In the 40 MWth capacity, the contribution begins at around 2.8% in 2023, gradually decreasing to approximately 2.3% by 2033.Meanwhile, in the 1000 MWth capacity, the overall national contribution would be larger compared to the 40 MWth, though this particular graph appears to depict only the smaller capacity. Papua (purple bars) continues to dominate regional contributions in both cases, reaching around 2.4% initially and slightly decreasing over time. This shows that Papua plays a crucial role in the gas-steam cycle contribution, similar to its role in the steam cycle contribution. Moreover, Sumatera, Jawa-Bali, and Kalimantan also contribute but remain much smaller compared to Papua, with contributions below 0.3% in both 40 MWth and 1000 MWth capacities. Contributions from Sulawesi, Nusa Tenggara, and Maluku remain minimal but consistent over the projection period.

On other case, the 40 MWth reactor shows a smaller overall contribution than the 1000 MWth, similar to what was observed in the steam cycle results. While Fig. 4 presents data primarily for the 40 MWth capacity, a comparison with a 1000 MWth scenario would result in much larger contributions, as seen in the steam cycle results.

In summary, the gas-steam cycle NPP projections follow a similar trend to the steam cycle projections, with Papua contributing the most regionally. The overall national contribution decreases slightly over the years, while regional contributions remain steady, with Papua leading the pack. The 40 MWth NPP shows lower energy contributions compared to the larger capacity reactors.

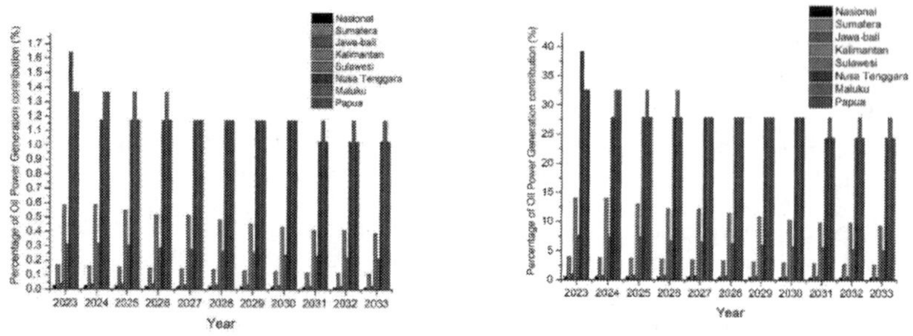

(a) (b)

Figure 5. Projected contribution of Oil power generation contribution percentage for energy production in Indonesia with capacities of (a) 40 MWth, and (b) 1000 MWth.

Beside of that, the oil power generation used in most of countries in the world, including in Indonesia. This aims to comparing energy fulfil from fossil fuel with capacities of 40 MWth (Fig. 5 (a)), and 1000 MWth (Fig. 5 (b)). Both projections show a steady decrease in oil power generation contribution over time across all regions.

In the 40 MWth (first image), the contribution percentages are relatively low, with the national average starting at just above 1.5% in 2023 and decreasing to below 1.0% by 2033. In the 1000 MWth (second image), the contribution percentages are much higher, with the national average starting at around 35% in 2023 and decreasing to below 25% by 2033. Across both figures, regions like Jawa-bali and Papua show higher oil power generation contributions compared to others. The relative contributions by regions remain consistent between the two capacities, but the absolute percentages are significantly larger in the 1000 MWth case. Meanwhile, the rate of decline in oil power contribution is similar in both graphs, but the absolute values are much higher for 1000 MWth. Both projections show a downward trajectory, indicating a shift towards reducing oil-based power generation.

The comparison shows that while the trend of decreasing oil power contribution is consistent across both capacities, the magnitude of oil's contribution to energy production is much higher for the larger capacity (1000 MWth). The national and regional percentages for 1000 MWth are significantly greater but follow a similar pattern of reduction over the same time period.

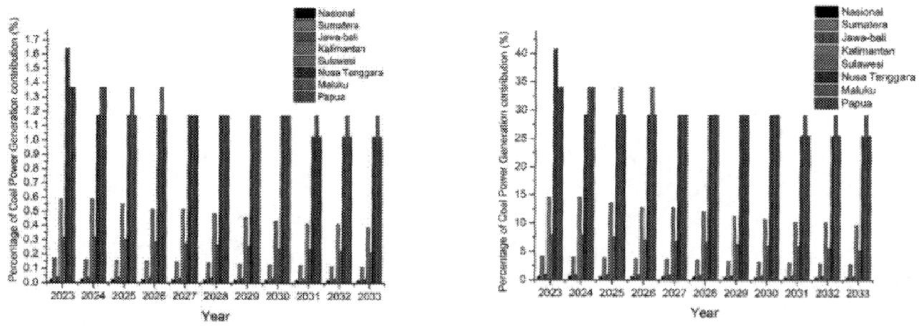

(a) (b)

Figure 6. Projected contribution of Coil power generation contribution percentage for energy production in Indonesia with capacities of (a) 40 MWth, and (b) 1000 MWth.

Other case by using coal power generation, it follows a similar pattern to oil power generation, with a steady decrease from 2023 to 2033. However, coal's contribution starts at a lower level compared to oil for the same capacity 40 MWth (Fig. 6 (a)), and 1000 MWth (Fig. 6 (b)). In the case of coal power generation, regions like Jawa-bali, Papua, and Sumatera consistently show the highest contributions. These regions align with higher energy demands or greater dependency on coal generation. Meanwhile, smaller regions such as Maluku and Sulawesi exhibit lower contributions, similar to the trends seen for oil power generation.

Both coal and oil see a decline over time in all regions, with Jawa-bali and Papua having the largest contributions for both power types, indicating that these regions rely more heavily on fossil fuels for energy production. The downward trend reflects Indonesia's gradual shift away from coal and oil-based energy generation towards other sources.

Conclusions

Nuclear power, as projected in various scenarios, could play a significant role in Indonesia's future energy mix. However, the contribution characterizing by power capacity are 40 MWth and 1000 MWth. For the 40 MWth NPPs, its contribution remains minimal across all regions, typically under 2% nationally and regionally. This capacity will have a limited impact on addressing the growing energy demand. Meanwhile, the 1000 MWth NPPs, these larger reactors show considerable potential, with national contributions starting at around 40% in 2023. Regions like Papua and Maluku could see nuclear power fulfilling up to 40-50% of their electricity needs. Major regions such as Jawa-Bali and Sumatera could see nuclear energy contributing 10-15% of their total demand. This indicates that nuclear power from 1000 MWth plants could play a transformative role in meeting Indonesia's energy requirements, especially in less populous regions.

In other case, both oil and coal power generation contributions are projected to decline over the period, reflecting Indonesia's transition away from fossil fuels. In oil power generation, the contribution of oil power is projected to decrease steadily, with a 40 MWth capacity contributing around 1.5% in 2023 and dropping below 1.0% by 2033. The 1000 MWth capacity shows a higher starting contribution of around 35% in 2023, which also gradually declines to below 25% by 2033. So then, coal power generation follows a similar trend as oil, with contributions decreasing steadily across regions. While coal power starts at a lower level than oil, regions such as Jawa-Bali and Papua show the highest dependence on both fossil fuels for electricity generation. The downward trend for both fuels align with Indonesia's shift towards more sustainable energy sources.

Indonesia's electricity demand will continue to grow, with a strong concentration in economically advanced regions like Jawa-Bali. Large-scale nuclear power plants (1000 MWth) offer significant potential to meet the country's future energy needs, especially in smaller, less populated regions. In summary, Indonesia's energy landscape over the next decade will be shaped by rising demand, particularly in industrial and urban centres, while the country transitions towards reducing its reliance on oil and coal, with nuclear power emerging as a key contributor in the future energy.

Bibliography

[1] S. J. DEN, "Outlook Energi Indonesia 2023," *Jakarta*, pp. 1–85, 2023, [Online]. Available: https://www.esdm.go.id/assets/media/content/content-outlook-energi-indonesia-2019-bahasa-indonesia.pdf.

[2] Kementrian ESDM, A. Wahyu Kencono, M. Dwinugroho, E. Satra Baruna, and N. Ajiwihanto, "Handbook Of Energy & Economic Statistics Of Indonesia 2023," p. 118, 2024.

[3] Y. S. B. Susilo, Suparman, W. L. Widodo, and M. Nasrullah, *INDONESIA NUCLEAR ENERGY OUTLOOK 2015*. Jakarta: Pusat Kajian Sistem Energi Nuklir (PKSEN), 2015.

[4] K. Huda, "Study on Nuclear Power Program in Indonesia: Status Updates of Problems and Challenges," *New Approaches Eng. Res. Vol. 4*, no. July, pp. 102–113, 2021, doi: 10.9734/bpi/naer/v4/2773f.

[5] R. F. S. Budi *et al.*, "Scaling law method for cost estimation of Indonesia's first HTGR (PeLUIt-40) and its implementation strategy," *Nucl. Eng. Des.*, vol. 423, no. April, p. 113195, 2024, doi: 10.1016/j.nucengdes.2024.113195.

[6] D. Salma Salsabila, I. Djoko Irianto, S. Dibyo, and A. Sri Redjeki, "Comparative analysis of coolant mass flow rate for Peluit-40 reactor in energy conversion system: a study of conceptual design with and without a splitter," *Indones. J. Nucl. Sci. Technol.*, vol. 24, no. 2, pp. 66–74, 2023.

[7] P. Petrović and M. M. Lobanov, "The impact of R&D expenditures on CO2 emissions: Evidence from sixteen OECD countries," *J. Clean. Prod.*, vol. 248, no. xxxx, 2020, doi: 10.1016/j.jclepro.2019.119187.

[8] K. Z. Majid Khan, "Shifting Gender Roles in Society and the Workplace: Implications for Environmental Sustainability," *Sherwan*, vol. 1, no. 1, pp. 9–17, 2023, doi: 10.5281/zenodo.7634130.

[9] S. Sadekin, S. Zaman, M. Mahfuz, and R. Sarkar, "Nuclear power as foundation of a clean energy future: A review," *Energy Procedia*, vol. 160. pp. 513–518, 2019, doi: 10.1016/j.egypro.2019.02.200.

[10] S. Paraschiv, "Analysis of the variability of low-carbon energy sources, nuclear technology and renewable energy sources, in meeting electricity demand," *Energy Reports*, vol. 9, no. S11, pp. 276–283, 2023, doi: 10.1016/j.egyr.2023.09.008.

[11] A. Liu, M. Imran, A. A. Nassani, R. H. Binsaeed, and K. Zaman, "Reducing carbon emissions with Geoscience solutions: A look at the contributions of nuclear energy, technology, and green finance," *Geosci. Front.*, vol. 15, no. 4, p. 101698, 2024, doi: 10.1016/j.gsf.2023.101698.

[12] I. E. Agency, "Nuclear Power and Secure Energy Transitions: From today's challenges to tomorrow's clean energy systems," *Intenational Energy Agency*, pp. 1–95, 2022, [Online]. Available: www.iea.org/t&c/.

[13] "Nuclear Energy for Net Zero: Accelerating Investment in Clean Energy Transitions," *Nucl. Energy Net Zero Accel. Invest. Clean Energy Transitions*, 2024, doi: 10.61092/iaea.z743-u8x2.

[14] P. Kim, H. Yasmine, M. S. Yim, and S. S. Chirayath, "Challenges in nuclear energy adoption: Why nuclear energy newcomer countries put nuclear power programs on hold?," *Nucl. Eng. Technol.*, vol. 56, no. 4, pp. 1234–1243, 2024, doi: 10.1016/j.net.2023.11.026.

[15] H. Nam, "Impact of nuclear phase-out policy and energy balance in 2029 based on the 8th

Basic Plan for long-term electricity supply and demand in South Korea," *Renew. Sustain. Energy Rev.*, vol. 122, no. December 2019, p. 109723, 2020, doi: 10.1016/j.rser.2020.109723.

[16] R. E. Technologies, "Energy Scenario for Sweden 2050," *Scenario*, p. 100, 2011, [Online]. Available: http://www.wwf.se/source.php/1409709/Energy Scenario for Sweden 2050_bakgrundsrapport IVL_sep 2011.pdf.

[17] IAEA, *DEEP 5 User Manual*. 2013.

Study of the Impact of Nuclear Power Plants (NPP) and NuScale Reactors on Economic and Environmental Aspects

Kevin Wijaya[1]

[1] Nuclear Physics and Biophysics Research Division, Physics Department, Faculty of Mathematics and Natural Science, Bandung Institute of Technology, Jalan Ganesha 10 Bandung 40132, Indonesia.

E-mail: kevinwijaya3001@gmail.com

Keywords: Desalination, economy, environment, NuScale, nuclear power plant

Abstract

Nuclear Power Plants (NPPs) are one type of power plant that can produce large amounts of electricity and do not produce carbon. Although when operating, NPPs can compete with power plants from other energy sources, they require quite large capital costs. In addition, there have been cases of nuclear accidents that have an impact on the environment. Therefore, this article discusses the economic and environmental analysis of the existence of NPPs. In addition, the NuScale Reactor is also analyzed from the design and development, seawater desalination, and the impact of NuScale on the economy and the environment. From the results of the study, it was found that NPPs require large capital costs but when operating they tend to be smaller, NPPs can also open up jobs from various sectors, have a significant impact on environmental objects such as water quality, soil, air, ecosystems, and habitats. Also obtained are design features that are only owned by NuScale and can answer the Fukushima disaster problem. Then, information was obtained that there are three desalination and reverse osmosis (RO) processes that are more efficient in terms of economy and produce clean water, while if water quality is desired, the multi-effect distillation (MED) process can be used.

1. Introduction

Nuclear Power Plants (NPP) are a type of power plant that can operate for a long period of time. In addition, NPPs can produce large amounts of power and do not produce carbon. However, NPPs require a large investment and considering several nuclear accidents that have occurred such as Chernobyl (1984), Three Miles Island (1979), and Fukushima (2011) which have had an impact on the environment, in-depth studies are needed, especially on the economic and environmental aspects.

One of the reactors currently under development is the NuScale Reactor. This reactor was developed by the NuScale Power Inc. organization. The NuScale reactor has several design features that are not owned by similar reactors. This Small Modular Reactor type reactor is also used for seawater desalination. The desalination process at NuScale uses three different types of processes, including*multi-effect distillation*(MED), multi-stage flash (MSF), and reverse osmosis (RO). Ingersoll et al., 2014 shows comparative data on costs, electrical power generated and the amount of clean water produced from the three types of processes. This article will discuss the economic and environmental analysis related to the existence of nuclear power plants, the analysis of the NuScale Reactor, and the analysis of the desalination process and the impact of seawater desalination on the economy and the environment.

2. Economic Aspect Analysis of Nuclear Power Plants

This section will discuss the types of costs used to support the long-term operation (LTO) program. There are three types of cost drivers to support the LTO program, including technical cost drivers, management cost drivers, and external cost drivers (IAEA, 2018). Table 1 contains information on the sources of each type of cost driver.

Table 1. Sources of each type of cost driver (IAEA, 2018).

Technical Cost Drivers	Management Cost Drivers	External Cost Drivers
Safety enhancements to meet new regulatory requirements	Managing programs	Security of energy supply
Major replacement/repair of safety and non-safety SSCs	Optimizing blackout	Contribution to the decommissioning fund
Operation and maintenance activities	Financing the licensing process	Social acceptance initiative
Management of radioactive waste and spent fuel	Contractor interface management	Environmental and radiological impact assessment
Aging management program (AMP)	Promoting a culture of safety and quality	Carbon policy
	Train and maintain skills	

Full details of the cost components can be seen in the IAEA (2018) reference. In general, some cost aspects that are economic considerations of nuclear power plants are:

- *Capital cost*

*Capital cost*includes the cost of site preparation, construction, manufacturing, commissioning and financing of nuclear power plants. There are several types of capital costs, including overnight costs, construction/investment costs, and financing costs. Capital costs are usually a key component of the overall capital costs of a nuclear power plant project. During the long construction period, when the project is not generating revenue, interest on borrowed funds can increase rapidly. Capital costs are usually calculated at various discount rates in the business plan to determine whether capital expenditures can be recovered. If capital costs are high, capital expenditures will increase disproportionately and can threaten the viability of the project. Figure 1 is a comparison graph of the percentage of Levelized Cost of Electricity (LCOE) against variations in discount rates for several types of energy.

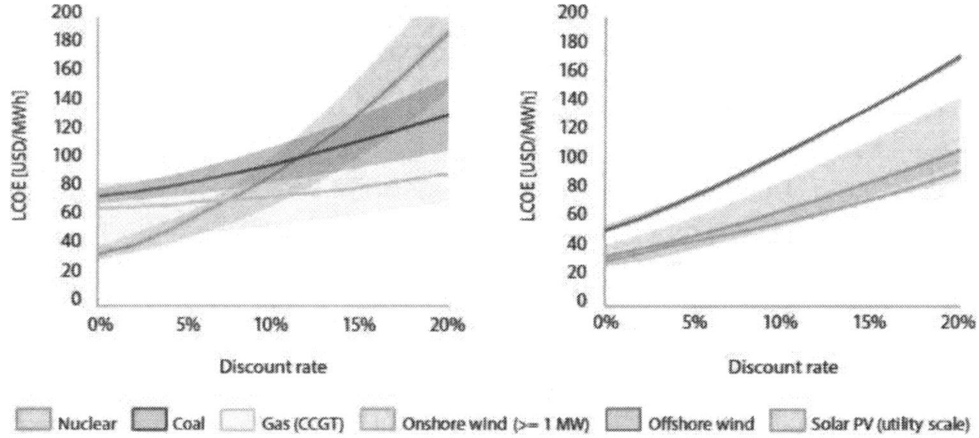

Figure 1. Comparison graph of LCOE against discount rate variations of non-renewable energy (left) and renewable energy (right) (NEA, 2020).

It can be seen in Figure 1 that the discount rate plays an important role in the LCOE value because with the increase in the discount rate, the LCOE also increases, especially nuclear energy which increases quite significantly. This will also affect the capital cost because with the high discount rate, the initial cost or capital required will also be high.

- Power plant operating costs

Operating costs include fuel costs and operating and maintenance costs. Fuel costs consist of spent fuel management and final waste disposal (World Nuclear Association, 2023). Figures 2 and 3 show a comparison graph of fuel prices against uranium prices and a graph of LCOE against fuel price variations, respectively.

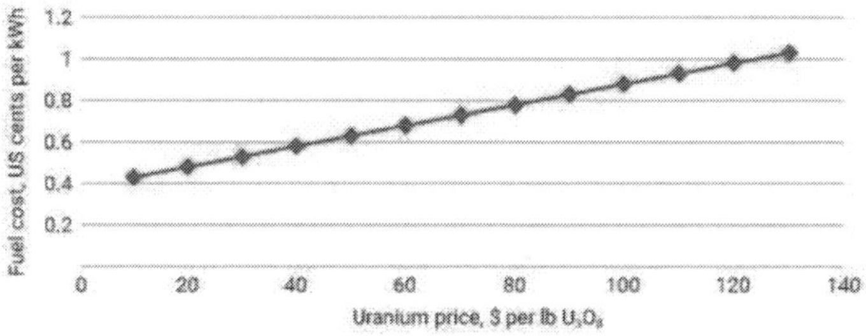

Figure 2. Comparison chart of fuel prices against uranium prices (World Nuclear Association, 2023).

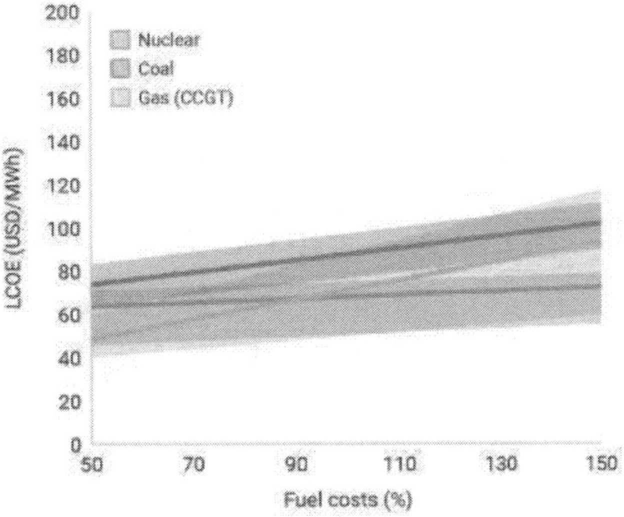

Note: Lines indicate median values, areas the 50% central region.

Figure 3. LCOE graph as a function of fuel price (NEA, 2020).

As seen in Figures 2 and 3, the price of fuel sources will certainly affect the price of fuel. However, changes in fuel prices do not have a significant impact on LCOE. Overall, operating and maintenance costs account for around 66% of total operating costs (World Nuclear Association, 2023).

- External costs

External cost drivers are conditions outside the project that cause costs to arise such as security of energy supply, ensuring social acceptance, environmental impact assessment (EIA), carbon policy, electricity markets, risk assessment and mitigation measures, and decommissioning (IAEA, 2018). Table 2 contains external cost information from the French nuclear fuel cycle.

Table 2. External costs of the French nuclear fuel cycle in m€/kWh (Schieber and Schneider, 2002).

Fuel Cycle Stages	***Discount Rate***		
	0%	3%	10%
Mining and milling	6.45E-02	1.84E-02	6.26E-03
Conversion	9.74E-04	4.78E-04	2.26E-04
Enrichment	1.19E-03	7.90E-04	4.13E-04
Fuel fabrication	1.89E-03	7.35E-04	3.10E-04
Power plant: construction	3.94E-02	3.94E-02	3.94E-02
Operation	4.41E-01	1.68E-02	4.12E-03
Deactivation	1.93E-02	6.91E-03	9.26E-04
Reprocessing	1.92E+00	1.45E-02	1.90E-03
LLW Disposal	4.80E-03	8.52E-06	4.13E-07
HLW Disposal	2.54E-02	6.41E-09	1.12E-10
Transportation	6.54E-04	2.66E-04	1.21E-04
Amount	**2.52**	**0.10**	**0.054**

It can be seen that the amount of external costs is affected by the discount rate. The greater the discount rate, the smaller the external costs required. Although nuclear energy has an LCOE that can compete with other energy, the investment costs required to build a nuclear power plant are very large. This can be seen in Figure 4, for a capacity factor of 80%, the LCOE of a newly built nuclear power plant is higher than that of a nuclear power plant after 10 and 20 years of operation. The amount of nuclear power plant construction costs can be seen in Table 3, the case taken is the estimated cost of building the NuScale Small Modular Reactor (SMR) in Idaho.

Table 3. Estimated NuScale Costs and Expenditures Incurred in Idaho (Black and Peterson, 2018).

General Description	NuScale SMR Costs	Expenditures Sourced or Originated in Idaho
Direct Costs Capitalized	$1,805,616,142	$744,613,813
Structure and Improvement	$612,136,797	$422,374,390
Reactor Plant Equipment	$869,360,876	$234,727,437
Turbine Plant Equipment	$196,121,808	$52,952,888
Power Generation Equipment	$34,982,052	$9,445,154
Heat Rejection System	$62,934,255	$16,992,249
Miscellaneous Factory Equipment	$30,080,354	$8,121,696
Capitalized Indirect Costs	**$663,710,610**	**$663,710,610**
Design Services in Home Office	$130,978,572	$0
Field Construction Management	$60,906,859	$60,906,859
Field Construction Supervision	$246,930,385	$246,930,385
Indirect Field Costs	$224,894,794	$224,894,794
Total Cost of Manufacturing and Construction	**$2,469,326,752**	**$1,408,324,423**

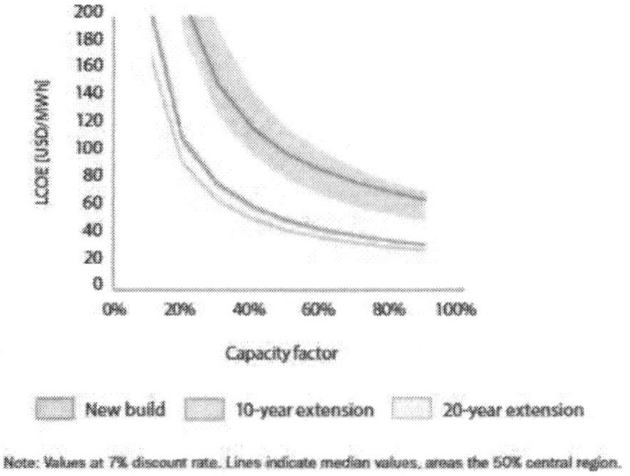

Figure 4. Graph of LCOE against capacity factor with variations in operating time of NPP (NEA, 2020).

On the other hand, the existence of nuclear power plants can create job opportunities so that jobs increase and can help the economy. There are three categories of jobs in the nuclear energy sector, including direct, indirect, and induced. Direct is a job located in a power plant facility such as a nuclear power plant operator, site staff, business personnel, and construction workers. Indirect is a job that supports power plant facilities such as uranium miners and fuel assemblers. Induced is a job created due to direct and indirect jobs (NICE Future, 2022). Here are some data on the number of jobs and the economic impacts resulting from the nuclear program, especially nuclear power plants (see Tables 4 and 5, and Figure 5).

Table 4. Number of jobs from nuclear power plant programs from several countries (IAEA, 2021; NEA and IAEA, 2018; and NEI, 2012)

Country	**Electrical Capacity (GWe)**	***Direct***	***Indirect***	***Induced***	**Estimated Number of Jobs**
France	63	125000	114000	171000	410000
South Korea	18	29400	36700	27400	93500
United States of America	98	70000	-	430000	500000

Table 5. Annual economic impact for NuScale SMR Development (Black and Peterson, 2018).

Job Type	Employee	Income	Value Added	Amount
Direct	8000	$445,595,958	$804,729,176	$1,408,324,423
Indirect	2086	$89,742,130	$139,881,476	$285,697,466
Induced	3377	$108,844,596	$189,203,116	$371,600,867
Amount	13422	$644,181,821	$1,133,813,767	$2,065,622,755

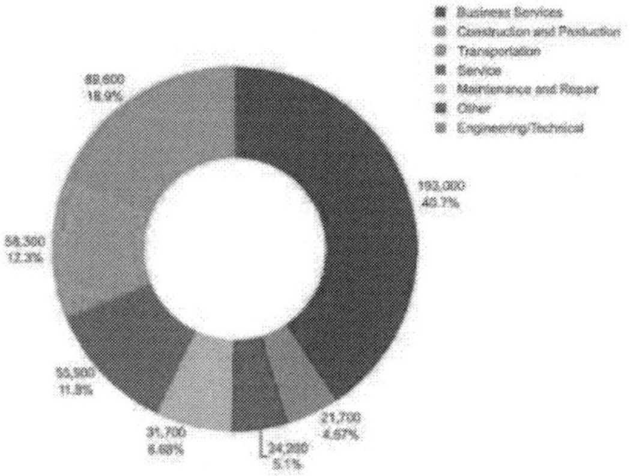

Figure 5. Jobs created, by type, in the United States from nuclear energy (Berkman, Mark, and Dean Murphy, 2015).

3. Environmental Aspect Analysis of Nuclear Power Plants

Nuclear power plants will never be free from environmental aspects. The existence of nuclear power plants will have an impact on the environment. Faishal and Samputra, 2024 conducted a study on the impact of nuclear energy on the environment. The types of problems studied include nuclear disasters such as Chernobyl and Fukushima, fuel management, carbon emissions, comparison of environmental impacts with other energy sources as many as 20 literatures. Of the 20 literatures, 14 literatures show the negative impact of the existence of nuclear power plants on the environment, 1 literature shows the positive impact of the existence of nuclear power plants, and 5 literatures do not show positive or negative impacts of nuclear power plants. Environmental objects used for research include animals, plants, humans, oceans, air, and CO_2 emissions. From the results of the study, almost all environmental objects get negative impacts from the radiation produced except for CO_2 emissions.

Furthermore, there are several environmental aspects that are considered to be impacted by nuclear power plants, including (Fennovoima, 2014):

- Land use and built environment

The construction of a nuclear power plant will change the land use at the actual nuclear power plant site and its surroundings. The legally applicable land use plan indicates the area required for the nuclear power plant. The power plant will affect the city infrastructure because it limits the use of land in the plant protection zone but allows new construction in settlements and villages as well as on highways. The protection zone must not include densely populated areas, hospitals, crowded residences, or important industrial activities that could be affected by an accident at the nuclear power plant. Building holiday homes or recreational activities in the area must ensure that the prerequisites for proper rescue activities will not pose a risk.

- *Landscape and cultural environment*

The impact on the landscape during construction will be caused by heavy traffic required for the transport of large building parts and their requirements, new road connections and the upgrading of existing roads. Tall cranes will be visible in the landscape from a distance

- Soil, bedrock, and groundwater

There is no significant impact on soil or bedrock during normal nuclear power plant operation. Proper engineering, such as drainage systems for overflow and wastewater, will prevent soil contamination. Geological value will be reduced if bedrock is excavated. Groundwater levels and pressures may decrease due to dewatering of structures during construction and operation. Because of the use of explosives and injection into bedrock, the project may affect groundwater quality, especially during construction.

- Flora, fauna and conservation areas

Parts of the forest and coast will be converted into artificial environments, meaning that species in these areas will be lost or altered. Construction will not involve nature conservation areas or coastal grasslands protected by the Nature Conservation Act; nor will construction have any direct impact on them. Construction will cause fragmentation of some of these critically endangered habitat types. During the construction of the power station and the highway, noise may disturb birds in the area. During the operation of the power station, cool warm water is discharged into the sea, indirectly damaging coastal grasslands and making habitats less favorable.

- Water and fisheries systems

For a short period of time, seawater turbidity will occur during the construction of the navigation channel, port area, additional cooling water inlet, cooling water discharge area, and protective wharf. The seabed of the dredged area is mainly composed of sand and gravel, which are coarse-grained materials that settle quickly. During the hydraulic construction work, fishing cannot be carried out in and around the construction area. Construction activities at sea may also block fish from a wider area, disrupting fish migration routes for a long time. Fish may be driven away from a large area by the strong underwater noise generated by the excavation. With very low radioactive emissions, tritium and other gamma and beta emissions will be released into the sea, which will not have any negative effects on humans or the environment.

Warm cooling water, purified process and wash water, and water intake have an effect on the water system. The nutrient loads generated by purified process water, wash water, and sanitary wastewater are very small compared to the loads entering the ocean through local rivers. Seawater temperatures around the discharge site will increase because the cooling water used in the power plant will be discharged into the ocean.

- Emissions to air

To minimize emissions, radioactive gases produced by nuclear power plants will be processed using state-of-the-art technology. To reduce the amount of radioactivity, radioactive gases will be collected, filtered, and contained. Gas containing small amounts of radioactive material will be released into the air in a controlled manner through a ventilation stack. The level of gas emissions will be measured to ensure that it remains below the established limits. Most of the remaining radioactive material released will dissolve in the air.

During the construction of a nuclear power plant, dust is generated by excavation, construction site traffic, and certain operations, such as rock crushing. This impacts air quality, especially at the construction site. Emergency power systems and travel traffic will generate emissions while the nuclear power plant is in operation. Air quality is not expected to be significantly affected by these emissions.

- Waste and waste management

If the facility is well designed and waste management measures are implemented correctly, the handling and final disposal of operational waste will not have a significant environmental impact. Over time, the radioactive substances contained in operational waste will become safe for the environment, and the final disposal facility will be monitored. If the design and implementation of the facility used are appropriate, the handling and storage of spent nuclear fuel will be safe and will not cause a significant environmental impact.

- Traffic and traffic safety

During the construction period, traffic volumes will obviously increase, especially during peak construction years. The nuclear power plant will require a new road to be planned to connect the plant to the main road.

- Noise

During construction and operation of the power plant, noise levels caused by the project will be below standards for residential areas and vacation rentals, according to noise emission modeling.

- Impact of abnormal situations and accidents

If an accident occurs, the impact of the radiation that is spread can harm several aspects, including human health, the possibility of losing a home temporarily or even permanently, prohibition of using agricultural and natural products, loss of habitat, and ecosystem imbalance. Other unusual situations and accidents mainly include chemical and oil leaks that can pollute the soil or groundwater. In addition, radiation-threatening situations can also occur due to fire or human error. Technical measures and personnel training can prevent such situations.

- Power plant decommissioning

If decommissioning is done properly, the impact will be small. The waste generated during the decommissioning phase will be similar to the waste generated during the operation of the plant and can be treated in the same way as the waste generated during the operation process. Not much radioactive waste is generated during the decommissioning of a nuclear power plant.

4. NuScale Reactor Analysis

The NuScale reactor or with the full name NuScale Power Modular and Scalable Reactor is a Small Modular Reactor (SMR) with the Integral Pressurized Water Reactor type. This reactor uses light water as a coolant and moderator respectively and the neutron spectrum used is thermal neutrons. Overall, this reactor is capable of producing 1920 MWth or 540 MW of power. This reactor was developed by the organization or company NuScale Power Inc. In 2003, Oregon State University in collaboration with the Idaho National Engineering Laboratory and Nexant-Bechtel, completed a project to develop an initial reactor design called the "Multi-Application Small Light Water Reactor" (MASLWR). In 2007, NuScale Power Inc. was formed to commercialize the concept, and MASLWR was renamed NuScale Plant to reflect the significant improvements made to the original design. NuScale Power notified the U.S. Nuclear Regulatory Commission in early 2008 that it planned to begin pre-application discussions in an effort to certify the NuScale plant design consisting of twelve modules. In 2011, Fluor Corporation became the majority investor in NuScale Power. They will provide engineering, procurement, and construction services for the deployment plant (IAEA, 2013). The following is a timeline of the NuScale Reactor development tabulated in Table 6 (Langdon, 2019 and NuScale Power, 2020).

Table 6. NuScale reactor development timeline (Langdon, 2019 and NuScale Nonproprietary, 2020).

Year	Information
2003	Precursor concept developed (MASLWR)
2007	NuScale Power, Inc. was created to commercialize the new design.
2008	Pre-application discussion plan in an effort to obtain NuScale power plant design certification
2011	Fluor Corporation becomes lead investor and strategic partner of NuScale
2013	NuScale wins first competitive funding opportunity from US Department of Energy ($226M)
2017	Design certification application (DCA) submitted to the United States NRC
2018	DCA stage 1 review completed
2019	DCA phase 2-4 review completed
2020	Expected completion of DCA review
2023	Commenced fabrication/construction of the first full-scale NuScale power plant (NPP) in the United States
2026/2027	Commercial operation of the first NuScale plant

The NuScale reactor consists of 12 independent modules, each capable of producing 45 MWe of net electrical power. Each module has an Integral Pressurized Water Reactor that operates under natural circulation primary flow conditions. Each reactor is housed in a high-pressure containment vessel submerged underwater in a stainless steel-lined concrete pool. The following table (Table 7) illustrates the key design features of the NuScale Reactor (IAEA, 2013).

Table 7. Technical Data of the NuScale Power Modular and Scalable Reactor (IAEA, 2013).

Multi-Module Generator General Data	
Number of modules	12
Power plant output power	540 MWe
Multi-Module Generator General Data	
Power plant output power	160 MWth or 45 MWe
Gross efficiency	>30%
Operation mode	*Base Load*
Design life of the power plant	60 years
Availability target	>95%
Primary Coolant	Light water
Electrical Circuit Cooling Materials	Light water
Moderator materials (if any)	Light water

IOP Conf. Series: Materials Science and Engineering **1330** (2025) 012003 doi:10.1088/1757-899X/1330/1/012003

Types of cycles	Indirect
Thermodynamic cycle	Rankine
Economic Goals	
Spread mode	Distributed or centralized
Reactor Core	
Active terrace height	2 m
Fuel material	UO2 ceramic pellets
Fuel element type	17×17
Cladding materials	Zircaloy-4 or advanced cladding
Geometry of square array lattice	*Square array*
Enrichment of replenished fuel at core equilibrium	<4.95% weight
Fuel cycle duration	24 months
Average fuel combustion rate	> 30 MWd/kg
Burnable absorber(how to use/ingredients)	Gadolinium
Reactivity control mode	Control rod, boric acid
Reactor mode is turned off	Control rod
Primary Cooling System	
Reactor operating pressure	12.76 MPa
Power Conversion System	
Working media	Steam
Reactor Vessel	
Outer diameter of the cylinder shell	About 2830 mm
Height of vessel	About 20 m
Steam Generator or Heat Exchanger Electric Circuit	
Type	Helical coil
Amount	2
Operation mode	Secondary coolant in the tube
Primary Circulation System	
Circulation type	Natural circulation
Number of pumps	There isn't any
***Pressurizer*, If there are**	
Type	Internal electric heater
Active devices used	Heater and spray
Primary Detention	
Type	Vacuum retainer in
Overall shape	Cylinder
Dimension size (diameter/height)	About 4.6m/24.5m
Design pressure	>5.5 MPa
Residual Heat Dissipation System	
Active/passive system	DHRS/ECCS passive
Generator Configuration and Layout	

Configuration options	Ground based, multi-module, 1-12 power modules
Surface area of factory site	Approximately 160000 m2
Height or underground planting of nuclear island	Underground Containment, Control Room, Spent Fuel Pool
Terrace catcher	Submerged retainer under water in SS lined concrete pool
Protection against aircraft accidents	Underground Containment, Control Room, Spent Fuel Pool; Seismic Reactor Building Category

4.1. NuScale Reactor Description

The NuScale reactor consists of 1-12 modules. Figures 6 and 7 illustrate the design of the single-module NuScale Reactor and the layout of the reactor with 12 modules. The NuScale reactor core consists of 37 fuel assemblies and 16 control rod groups (Ingersoll and Houghton, 2012). The difference between NuScale and other PWRs is that the fuel assembly owned by NuScale is shorter than that of traditional PWRs but the fluid supply is much greater (Reyes, 2011). Then from Figure 7, the NuScale Module is below the surface of the pool water. The function of the pool is to absorb all the decay heat produced for more than 30 days, to be a barrier to additional fission products in the event of a reaction failure, to be a radiation shield outside the container to reduce operational exposure (Ingersoll and Houghton, 2014). The control rod group functions to control the core strength. Water radiates heat in the nuclear core, which produces a low-density fluid that flows upward through the heated ladder. The helical coil of the steam generator is wrapped around the outside of the riser. It acts as a heat sink to cool the water, increasing its density. The buoyancy force, which drives the fluid flow around the loop, is caused by the density difference created by the elevation difference. Because it eliminates pumps, external recirculation piping, and associated valves, natural circulation operation is advantageous in avoiding maintenance and possible breakdowns. In addition, it reduces the amount of electricity used in the internal plant, which is characteristic of conventional LWRs designed to power large recirculation pumps. Elimination of the recirculation system improves overall plant safety and reduces production costs. The helical coil steam generator consists of two separate sets of tubes, each with a feedwater inlet and outlet. Feedwater is pumped into the tubes where it boils, producing superheated steam. The steam generator is replaceable due to its small size. To allow for pressure control, the heating components and spray system are located inside the integrated pressurizer (Reyes, 2011).

IOP Conf. Series: Materials Science and Engineering **1330** (2025) 012003 doi:10.1088/1757-899X/1330/1/012003

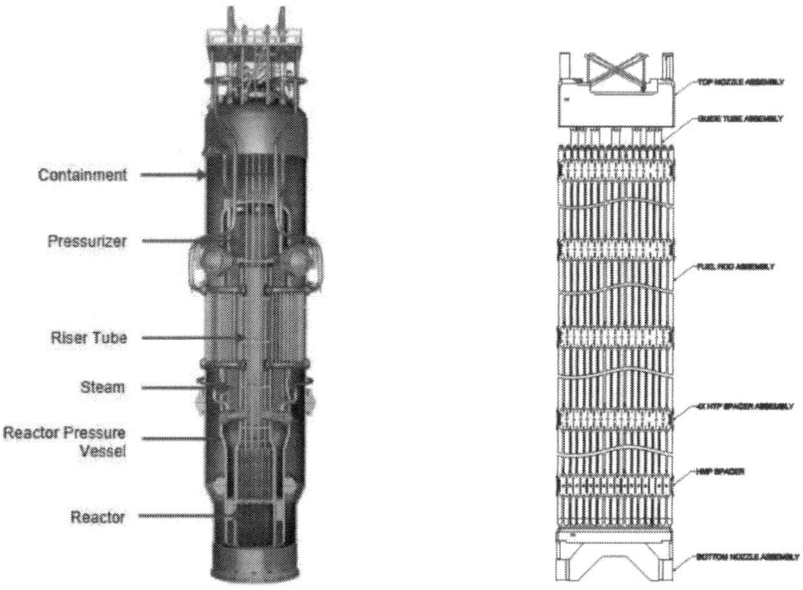

(a) (b)

Figure 6. NuScale Reactor Schematic (a) 1 module and (b) fuel assembly. (NuScale Power, 2020 and IAEA, 2020).

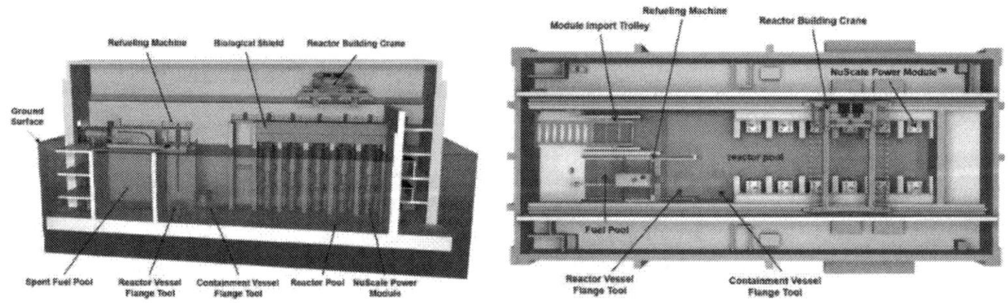

(a) (b)

Figure 7. Schematic of the NuScale Reactor building with 12 modules with (a) cut-away and (b) top views (NuScale Power, 2020).

4.2. NuScale Design Features to Support Nuclear Safety

NuScale reactors have several design features that contribute to improving nuclear safety. Each module uses two independent and redundant passive safety systems. Passive safety systems cool and contain the nuclear core through processes such as natural convection heat transfer, vapor condensation, liquid evaporation, pressure-driven coolant injection, or gravity-driven coolant injection. They do not rely on electricity, signals, mechanical power, or forces such as electric pumps. There are two types of passive safety systems, namely the decay heat removal system (DHRS) and the emergency core cooling system (ECCS) which work in conjunction with the containment heat removal system (CHRS) (Reyes, 2011). Other safety features can be seen in Table 8.

Table 8. NuScale design features that contribute to increased safety (NuScale Power, 2020).

NuScale Design Features	Main Impact	Security Benefits
The reactor coolant system is an integral part of the RPV.	Eliminates large diameter primary coolant pipes	Eliminating the postulated massive LOCA accident spectrum
Natural convection cooled terrace	Eliminates the need for reactor coolant pumps	Eliminate reactor coolant pump failures, such as shaft damage, pump seizure, missile generation, and pump leaks.
Metal containment pressure vessel	The vessels are constructed in accordance with established pressure vessel standards.	Guaranteed containment integrity (liquid core concrete interaction is not possible due to CNV metal). Design and construction as a pressure vessel increases integrity assurance compared to large containment structures.
Modular NSSS and reactor vessel inside CNV	The coolant lost from the RPV remains in the containment and is returned to the RPV by passive means.	No design-based LOCA is postulated to be capable of exposing nuclear fuel. No coolant needs to be added to the NPM for all design-based events.
Evacuated detention	Sub-atmospheric pressure during normal operation	Increased steam condensation rate to remove heat of containment during LOCA events. Any hydrogen released is trapped in the CNV with little oxygen available to produce a flammable mixture.
	There is no insulation on the reactor vessel	Eliminates potential degradation of the core cooling system due to insulation debris and allows cooling outside the container.

Small terrace size	Reduces the need for decay heat removal and fission product inventory.	Increases retention on board; maintains low accident consequences; reduces fission product source life span; thus simplifies emergency planning
Reactor pool	NSSS and CNV are immersed in the reactor pool	Provides long-term passive cooling and improves fission product retention.
Passive safety systems	The safety system cools and depressurizes the CNV in the event of loss of external power.	Electrical power is not required for safety. No operator action is required for safety

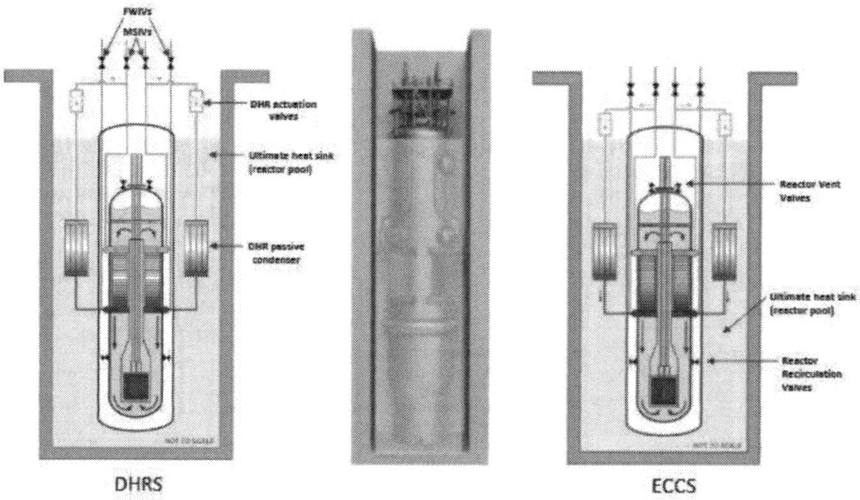

Figure 8. System diagram of decay heat removal system (left) and emergency core cooling system (right) (NuScale Power, 2020)

4.3. The Role of NuScale Reactor Design in Seawater Desalination

There are several desalination processes from seawater to produce clean water that can be consumed. Of all these, three main processes are the most widely and frequently used, including distillation-based technologies, namely multi-effect distillation (MED) and multi-stage flash (MSF), and membrane-based technology, namely reverse osmosis (RO) (IAEA, 2000). RO plants have the easiest coupling because they can operate using only electricity, which is needed to run high-pressure pumps. Therefore, co-locating a desalination plant with a power plant is not important as long as a grid connection is available. However, there are advantages if the power plant and RO desalination plant are located in the same location in terms of shared infrastructure and protection against power grid disruptions.

Both MED and MSF power plants require a thermal heat source such as steam from the secondary side of the nuclear power plant. This steam is typically extracted from the low-pressure turbine stage, which results in a commensurate reduction in the power plant's electrical output and may impact the reliability and flexibility of the power plant and desalination plant operations. In addition, the use of a tertiary heat transfer loop is usually required to ensure that no radionuclides such as tritium are carried over from the reactor's secondary loop to the distillation plant.

For MED itself, a thermo-compressor (TC) is installed and clean steam is needed to run the MED-TC cycle. The steam used is high pressure or known as high pressure (HP), medium or known as medium pressure (MP), and low or known as low pressure (LP). Figure 9 shows a process diagram of the NuScale module combined with the MED-TC distillation cycle (Ingersoll et al., 2014).

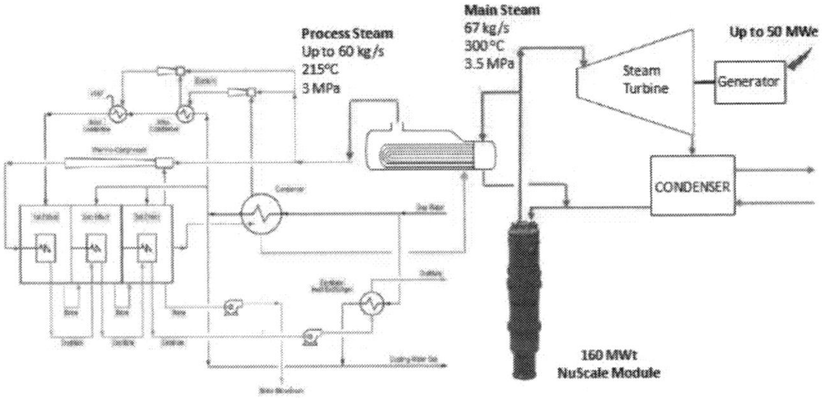

Figure 9. Process diagram of the NuScale module combined with the MED-TC distillation cycle (Ingersoll et al., 2014).

An alternative option is to utilize the extraction steam from the steam turbine. Figure 10 shows a conceptual integration of the NuScale module using controlled extraction into the MSF distillation plant. In this configuration, the extraction steam is extracted from the turbine to supply heat to the reboiler. The steam used in this configuration is high pressure (HP) and low pressure (LP). It should be noted that high pressure (HP) steam is taken before entering the turbine, medium pressure (MP) steam is taken from the controlled extraction of the turbine, and low pressure (LP) steam is taken from the exhaust end of the turbine (Ingersoll et al., 2014).

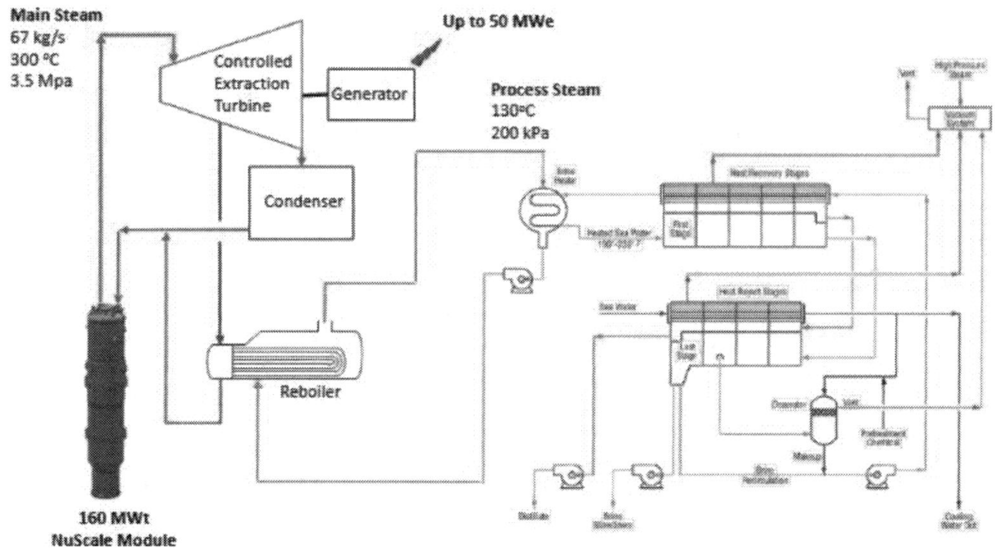

Figure 10. Process diagram for the NuScale module combined with an MSF desalination cycle via a controlled extraction type turbine (Ingersoll et al., 2014).

The final option studied is to combine the NuScale plant with an RO desalination process, as illustrated in Figure 11. In this design, the normal power conversion system of the NuScale plant is left unchanged. The electrical output from the standard turbine-generator system is fed to the RO plant to run the high-pressure pumps and necessary auxiliary equipment. To increase the efficiency of the RO process, the feedwater stream to the RO unit can be preheated with hot water returning from the condenser (Ingersoll, 2014).

IOP Conf. Series: Materials Science and Engineering **1330** (2025) 012003 doi:10.1088/1757-899X/1330/1/012003

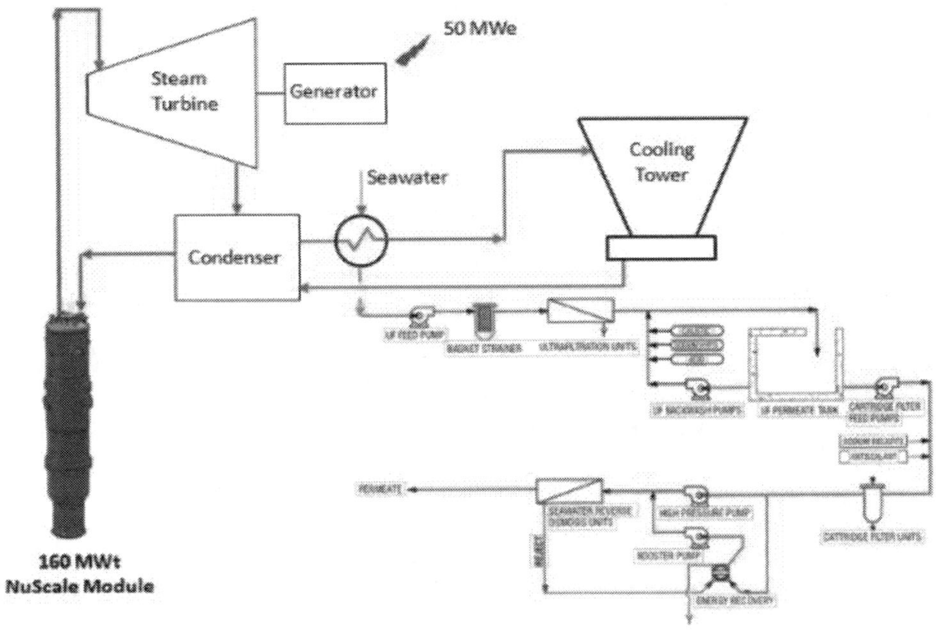

Figure 11. NuScale plant combined with RO desalination cycle (Ingersoll, 2014).

4.4. The Impact of NuScale Reactors on the Economy

The existence of the NuScale Reactor certainly has an economic impact. Table 9 contains financial information owned by NuScale in 2023 and 2024.

Table 9. NuScale financial information (NuScale Power, 2024)

2023	2024
The balance sheet includes cash and equivalents of \$125.4 million (\$5.1 million of which is restricted), and no debt.	In January 2024, NuScale announced cost efficiency measures projected to result in annual savings of approximately \$50-60 million.
Revenue was \$4.6 million and net loss was \$56.4 million for the three-month period ended December 31, 2023, compared to revenue of \$3.4 million and net loss of \$47.2 million, respectively, for the same period in 2022.	NuScale improved its cash position, ending the first quarter with cash and equivalents of \$137.1 million (\$5.1 million of which was restricted), and no debt.
Revenue of \$22.8 million and net loss of \$180.1 million for the twelve months ended December 31, 2023, compared to revenue of \$11.8 million and net loss of \$141.6 million for the same period in 2022.	NuScale reported revenue of \$1.4 million and a net loss of \$48.1 million for the three months ended March 31, 2024, compared with revenue of \$5.5 million and a net loss of \$35.6 million for the same period in 2023.

IOP Conf. Series: Materials Science and Engineering **1330** (2025) 012003 doi:10.1088/1757-899X/1330/1/012003

Based on the information in Table 9, it can be seen that NuScale experienced an increase in cash in the first quarter of 2023 and 2024. However, in general, although NuScale experienced an increase in revenue, the losses experienced also increased so that cost efficiency was carried out which was projected to result in annual savings. In addition, economic analysis data related to seawater desalination was obtained which can be seen in Table 10.

Table 10. Summary of Economic Analysis for Coupled NuScale-Desalination Plant (Ingersoll et al., 2014)

Desalination Technology	MP-MSF	MP-MED	LP-MED	RO
Increased Power Plant Production Level				
Water produced (m3/d)	190000	190000	190000	190000
Net electricity output of the plant (MWe)1	227	293	334	348
***Capital Cost*(in millions)**				
NuScale Power Plant	$1800	$1800	$1800	$1800
Desalination plant	$379	$311	$311	$256
***Operation & Maintenance Cost*[2] (in millions)**				
NuScale Power Plant	$185	$185	$185	$185
Desalination plant	$15.1	$13.3	$13.3	$14.2
Annual Revenue (in millions)				
Annual revenue from water sales (at wholesale price of $1.67/m3)	$101	$101	$101	$101
Annual revenue from electricity sales (at wholesale price of $75/MWh)	$142	$183	$209	$217
Net annual income of coupled plant	$43	$86	$111	$119
Simple payback of coupled plant (years)2	51	25	19	17

1. The net electricity output available to the grid after accounting for reduced generation due to steam extraction and electricity consumed by the desalination process.
2. Does not include financing costs

As can be seen in Table 10, the RO option is the best option if financial efficiency is desired. This is because although the cost required by RO is the lowest compared to other options, the amount of electricity and annual income generated by RO is the largest so that a shorter payback period is required compared to other options. On the other hand, the MP-MSF option is the worst option if financial efficiency is desired. From the NuScale technology itself, there are important features that NuScale has that support the economy of desalination, including natural circulation of coolant and nuclear modularity. Natural circulation eliminates the need for pumps, pipes, and valves so that lower costs are required, while the modularity of 12 can increase the economics of the generator and the flexibility of application to be paired with desalination plants of various sizes.

Next, Table 11 contains comparative information on employment, LCOE, and environmental impacts of coal-fired, gas combined cycle, and NuScale power plants.

Table 11. Comparison of employment, LCOE, and environmental impacts of various power plants (NuScale Power, 2021).

Parameter	Coal	Combined Gas Cycle	NuScale
Employment (per 924 MWe)	143	29	270
LCOE ($/MWh)	73	37	64
SOx (mg/kWh)	6700	300	11
NOx (mg/kWh)	3350	550	9
PM2.5 (mg/kWh)	9210	100	~0
Greenhouse Gas Emissions (CO2 equivalent (g/kWh))	1025	492	15

From an economic perspective, NuScale creates the most jobs compared to the other two power plants, which can support the economy from more workers. Then, it can also be seen that NuScale's LCOE has a value of $ 64 / MWh. This value shows that although it is not the smallest LCOE, NuScale's LCOE can compete with coal and gas power plants. Tightening and increasing carbon capture costs can increase NuScale's attractiveness.

4.5. Impact of NuScale Reactors on the Environment

NuScale reactors also have an impact on the environment. Table 8 shows the safety features of NuScale, these features allow to minimize reactor accidents so that the impact of reactor accidents on the environment such as radiation leaks can be minimized or even prevented. Table 12 explains the advantages of NuScale's cooling features that can overcome the cooling issue during the Fukushima incident.

Table 12. NuScale Design Features Relevant to Fukushima Cooling Issues (Reyes, 2011).

Feature	Fukushima	NuScale
Reactors and containment	Emergency diesel generator required	Emergency diesel generator is not required
	External water supply is required	The container is soaked in water supply for 30 days.
	Coolant supply pump required	Coolant supply pump is not required

	Forced water flow is required for LTC	LTC (more than 30 days) through natural convection into the air
Spent fuel pond	Water cooling from spent fuel	Expanded cooling capacity: four times the amount of water of conventional spent fuel pools per MW of power
	Increased SFP	Embedded SFP
	Limited access to backup water supplies	Accessible water reserves

Then, other environmental impacts are generated from the seawater desalination process. With seawater desalination, the need for clean water can also be resolved. However, considering that this process requires seawater that can affect the marine ecosystem and desalination technology is connected to nuclear power plants, the process of taking seawater and the safety of the technology need to be considered. Figure X shows a comparison of power production to clean water production from several desalination process options.

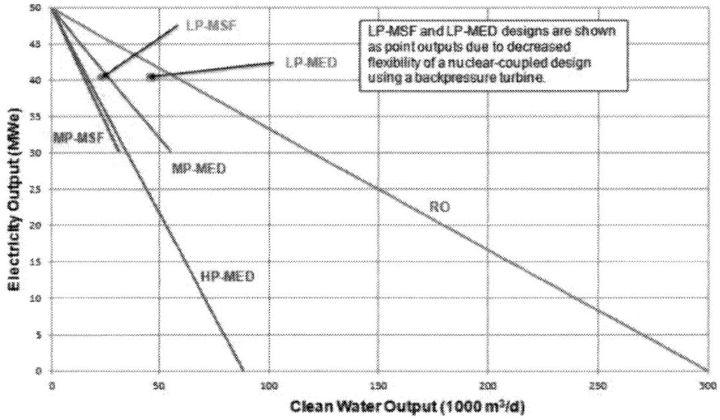

Figure 12. Graph of the relationship between electricity and water output from a single NuScale module combined with various desalination processes (Ingersoll et al., 2014).

The graph shows the advantages that RO has over other options. For the same electrical power, RO produces more clean water than HP-MED. This is because RO has a high conversion efficiency. However, because of this, the quality of water produced by RO is lower than HP-MED. This is because the thermal distillation process can produce water with a higher purity than RO. Therefore, installations with very low quality feedstock or where large quantities of high purity water are required may be better suited to the MED distillation process.

IOP Conf. Series: Materials Science and Engineering **1330** (2025) 012003 doi:10.1088/1757-899X/1330/1/012003

5. Conclusion

Based on the analysis conducted, it can be concluded that nuclear energy has advantages and disadvantages in economic and environmental aspects. From an economic perspective, low operating costs and LCOE values allow nuclear energy to compete with fossil fuels and other renewable energy. However, nuclear power plants require large capital and tend to have a long construction time, which is a weakness of nuclear power plants. In terms of employment, the existence of nuclear power plants has the potential to open up many jobs from various sectors and boost the economy. Meanwhile, from an environmental perspective, the existence of nuclear power plants can affect the quality of water, air, soil, ecosystems, habitats, and noise levels, especially if a nuclear accident occurs. Therefore, an in-depth study is needed if nuclear power plants are to be built.

Then, from the analysis of the NuScale Reactor, it can be concluded that the reactor has several features that the previous PWR did not have to support reactor safety. NuScale in seawater desalination can use 3 process options and in terms of quantity and economy, RO is the main option in the desalination process because it produces more pure water and requires cheaper costs. However, if water quality is targeted, HP-MED is the solution. Furthermore, NuScale's LCOE is able to compete with coal and gas. The NuScale Reactor also has several features that can answer the Fukushima disaster problem.

References

Berkman M and Murphy D 2015 The Nuclear Industry's Contribution to the US Economy *The Brattle Group.* https://www.brattle.com/wp-content/uploads/2017/10/7629_the_nuclear_industrys_contribution_to_the_u.s._economy-3.pdf

Bickel P and Friedrich R (eds) 2005 Externalities of Energy Methodology Rep. EUR-21951 *European Commission, Brussels*

Black G and Peterson S 2018 Economic Impact Report: Construction and Operation of a Small Modular Reactor Electric Power Generation Facility at the Idaho National Laboratory Site *Idaho Policy Institute*

European Commission 2003 External Costs: Research Results on Socio-Environmental Damages Due to Electricity and Transport Rep. EUR-20198 *European Commission, Brussels*

Faishal F and Samputra P L 2024 Impact of Nuclear Energy on the Environment *J. Strateg. Glob. Stud.* https://doi.org/10.7454/jsgs.v7i1.1126

Fennovoima 2014 Environmental Impact Assessment Report for a Nuclear Power Plant *Fennovoima.* https://tem.fi/documents/1410877/2818159/EIA+summary.pdf

International Atomic Energy Agency (IAEA) 2013 NuScale Power Modular and Scalable Reactor *IAEA.* https://aris.iaea.org/PDF/NuScale.pdf

International Atomic Energy Agency (IAEA) 2018 Economic Assessment of the Long Term Operation of Nuclear Power Plants: Approaches and Experience *IAEA, Vienna*

International Atomic Energy Agency (IAEA) 2021 Assessing National Economic Effects of Nuclear Programs *IAEA, Vienna*

International Energy Agency (IEA) and Nuclear Energy Agency (NEA) 2020 Projected Costs of Generating Electricity 2020 Edition *IEA/NEA, France.* https://www.oecd-nea.org/upload/docs/application/pdf/2020-12/egc-2020_2020-12-09_18-26-46_781.pdf

Ingersoll D T, Houghton Z J, Bromm R and Desportes C 2014 NuScale small modular reactor for co-generation of electricity and water *Desalination.* https://doi.org/10.1016/j.desal.2014.02.023

Langdon K 2019 NuScale Small Modular Reactor (SMR) Overview: INPRO Dialogue Forum on Opportunities and Challenges in Small Modular Reactors *IAEA.* https://nucleus.iaea.org/sites/INPRO/df17/IV.5-KenLangdon-NuScale.pdf

Nuclear Energy Agency (NEA) and International Atomic Energy Agency (IAEA) 2018 Measuring Employment Generated by the Nuclear Power Sector *OECD Publishing.* https://doi.org/10.1787/9789264305960-en

Nuclear Energy Institute (NEI) 2020 Nuclear Energy in a Low-Carbon Energy Future *NEI.* https://www.nei.org/resources/reports-briefs/nuclear-energy-in-a-low-carbon-energy-future

Nuclear Innovation: Clean Energy Future 2022 Nuclear Energy—Providing Power, Building Economies *Clean Energy Ministerial*

NuScale Nonproprietary 2020 Status Report – NuScale SMR (NuScale Power, LLC) *IAEA.* https://aris.iaea.org/PDF/NuScale-NPM200_2020.pdf

NuScale Power 2021 NuScale SMR Technology: An Ideal Solution for Repurposing US Coal Plant Infrastructure and Revitalizing Communities *NuScale Power.* https://www.nuscalepower.com/-/media/nuscale/pdf/publications/nuscale-smr-technology-an-ideal-solution-for-coal-plant-replacement.pdf

NuScale Power 2024 NuScale Power Reports Fourth Quarter and Full Year 2023 Results *NuScale Power.* https://www.nuscalepower.com/en/news/press-releases/2024/nuscale-power-reports-fourth-quarter-and-full-year-2023-results

NuScale Power 2024 NuScale Power Reports First Quarter 2024 Results *NuScale Power.* https://www.nuscalepower.com/en/news/press-releases/2024/nuscale-power-reports-first-quarter-2024-results

Reyes J N Jr 2011 NuScale Plant Safety in Response to Extreme Events *NuScale Power Inc.*, Oregon Vol. 178

Schieber C and Schneider T 2002 The External Cost of the Nuclear Fuel Cycle in *Externalities and Energy Policy: The Life Cycle Approach, Workshop Proceedings OECD, Paris*

World Nuclear Association 2023 Economics of Nuclear Power *World Nuclear Association.* https://world-nuclear.org/information-library/economic-aspects/economics-of-nuclear-power#capital-costs

Periodic safety review of RSG-GAS: strategy and evaluation to manage research reactor aging

Endiah Puji Hastuti[1,2*], Topan Setiadipura[2], Dwi Irwanto[1] and Abdul Waris[1]

[1] Faculty of Mathematics and Natural Sciences Dept., Bandung Institute of Technology, Bandung-Indonesia-401323,
[2] Research Centre for Nuclear Reactor Technology, Research Organization for Nuclear Energy, National Research and Innovation Agency (BRIN), Building No. 80 BJ Habibie Science and Technology Area, Setu, South Tangerang, Indonesia-15314

*E-mail: endi001@brin.go.id

Abstract. According to the IAEA Research Reactor Database, approximately 50% of the world's 225 research reactors—including those in Indonesia—have been in operation for over 50 years, with around 20% exceeding 60 years. As research reactors age, their structures, systems, and components (SSCs) are subject to degradation, potentially impacting safety and performance. Operators also face challenges related to outdated instrumentation and support systems that may no longer conform to original specifications. In Indonesia, the national regulatory body (BAPETEN) mandates a Periodic Safety Review (PSR) every ten years to evaluate and ensure continued safe reactor operation. The PSR of the 30 MW GA Siwabessy multi-purpose research reactor (RSG-GAS) reviewed operational, maintenance, and aging management records from 2005 to 2015. Key findings included recurring issues in the power-range neutron detectors and obsolescence in the instrumentation and control systems. A thorough inspection of the aluminum reactor tank liner in 2023, conducted with IAEA support, verified its structural integrity. All recorded incidents were rated at or below Level 1 on the INES scale, with no radiological consequences. The review informed several safety enhancements and aging management strategies, confirming the reactor's compliance with current safety standards and supporting its continued licensing and operation.

Key words: Periodic safety review, Aging management, RSG-GAS, SSC

1. Introduction

The Multi-Purpose Reactor GA Siwabessy (RSG-GAS) is a pool-type research reactor with a nominal thermal power of 30 MW and a peak neutron flux of approximately 2.0×10^{14} n/cm^2·s. Developed to support a broad range of nuclear science and engineering activities, RSG-GAS plays a critical role in advancing nuclear fuel technology, reactor physics, and the production of radioisotopes for both medical and industrial applications. It enables the irradiation of targets for the synthesis of essential isotopes such as Molybdenum-99 (Mo-99), Iodine-131 (I-131), and Iridium-192 (Ir-192), which are widely used in nuclear medicine and non-destructive testing.

The reactor core comprises 40 standard fuel elements and 8 control fuel elements, all fabricated with uranium enriched to approximately 19.75% U-235 [1]. The configuration of the reactor core is presented in Figure 1. The fuel manufacturing process, carried out by the National Nuclear Energy Agency (BATAN), adheres to the guidelines of the Reduced Enrichment for Research and Test Reactors (RERTR) program. This global initiative promotes the conversion of research reactors from high-enriched uranium (HEU) to low-enriched uranium (LEU) to enhance nuclear proliferation resistance. However, such conversion requires an increase in uranium density in the fuel to offset the lower fissile content. Whereas HEU fuels such as UAl_x/Al typically achieve uranium densities around 2.3 gU/cm^3, LEU fuel designs demand densities of up to 8.0 gU/cm^3. Consequently, advanced fuel types such as uranium silicide (U_3Si_2) have been developed and deployed to meet performance requirements under LEU constraints [2].

RSG-GAS achieved first criticality in July 1987 and reached its nominal full-power operation by March 1992. The reactor has since served as Indonesia's leading research facility for nuclear science and technology. By 2015, it had operated for 28 years, prompting a Periodic Safety Review (PSR) covering the operational period between 2005 and 2015. PSRs are structured ten-year evaluations conducted to identify any emerging safety concerns, verify the effectiveness of ageing management programs, and ensure continued regulatory compliance. Although originally designed for power reactors, PSRs have become an integral component of international safety standards for research reactors, endorsed by the International Atomic Energy Agency (IAEA) [3].

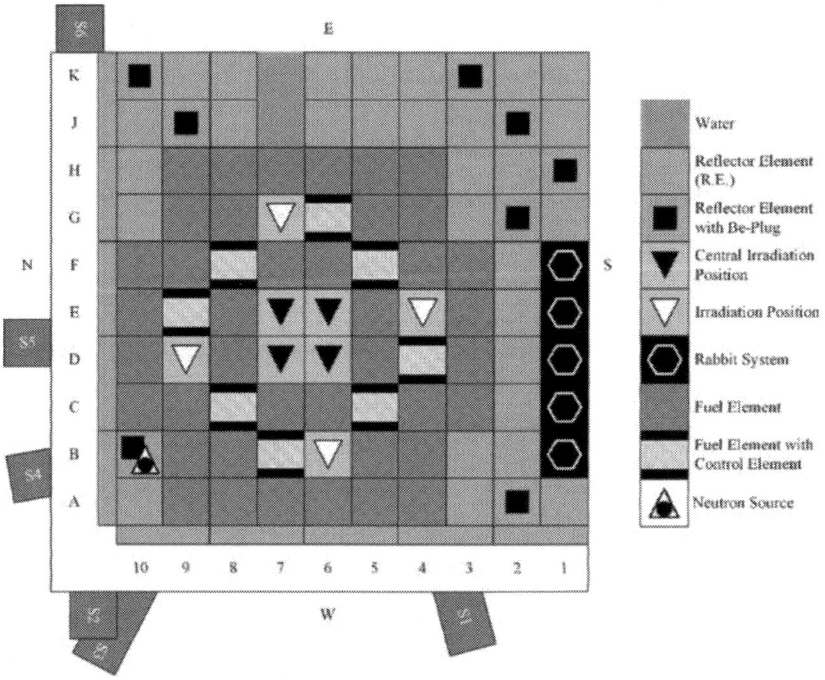

Figure 1. RSG-GAS Core configuration

One of the central concerns in reactor sustainability is ageing, defined as the gradual degradation of structures, systems, and components (SSCs) due to prolonged exposure to stressors such as neutron irradiation, thermal cycling, corrosion, and mechanical fatigue. Over time, these factors may reduce component reliability, increase maintenance needs, and pose risks to safe reactor operation [4]. Accordingly, ageing management programs are essential to detect, monitor, and mitigate degradation before it compromises safety functions [5].

The scientific objective of this paper is to evaluate the long-term operational safety and sustainability of the RSG-GAS reactor through a focused analysis of its ageing management practices and outcomes from the 2015 Periodic Safety Review. This study also seeks to assess the adequacy of fuel conversion strategies, the integrity of core components, and the effectiveness of monitoring systems in mitigating life-limiting conditions. By integrating design evaluation, operational experience, and regulatory frameworks, this paper aims to contribute to the global discourse on safe long-term operation of ageing research reactors.

According to the IAEA's Research Reactor Database, as of June 2023, there were 223 research reactors in operation, 11 under construction, 16 in design stages, and 58 permanently shut down, with hundreds more in various stages of decommissioning [6]. Many of these facilities, including RSG-GAS, were originally designed for a lifespan of 30 to 60 years. The status of research reactors worldwide is illustrated in Figure 2. Their continued operation depends on robust safety assessments and timely upgrades. PSRs play a pivotal role in this context by evaluating facility modifications, operational events, corrective measures, and updates to licensing and safety documentation. These assessments ensure that safety margins remain intact and that the reactor meets modern regulatory expectations throughout its operational lifecycle [7].

This study of RSG-GAS thus provides a case-specific insight into the challenges and solutions associated with the ageing of research reactors and supports the broader adoption of best practices in maintaining safety and reliability across the international research reactor fleet.

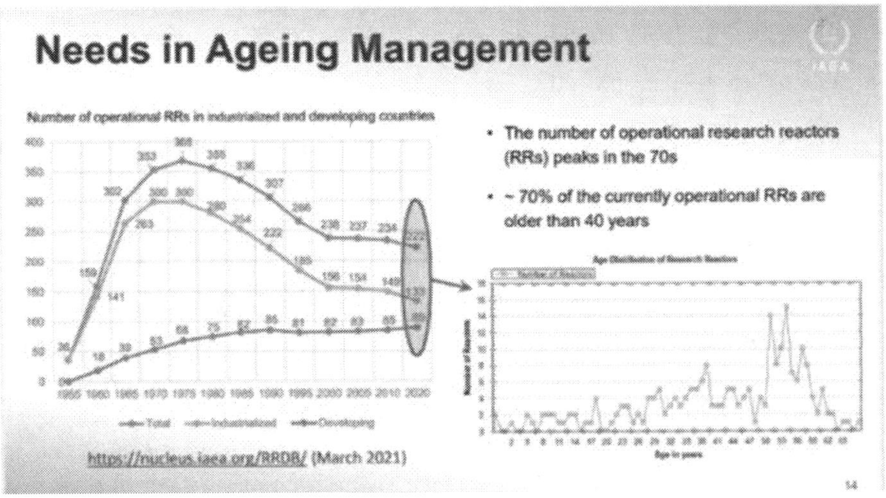

Figure 2. Status of world research reactors in 2021 (source IAEA 2023) [8]

Indonesia operates three research reactors under the authority of the National Research and Innovation Agency (formerly the National Nuclear Energy Agency). These include the 30 MWth RSG-GAS reactor, located in the Puspiptek area, South Tangerang, which achieved first criticality in 1987; the TRIGA 2000 reactor in Bandung, which began operation in 1965 at 250 kW, was uprated to 1 MW in 1971, and to 2 MW in 2000; and the 100 kW Kartini reactor in Yogyakarta, which reached first criticality in 1979 [8]. All three reactors are over 30 years old, making them subject to ageing management and the requirement to conduct Periodic Safety Reviews (PSRs). The operational status of these reactors is summarized in Table 1 and visually represented in Figure 3.

Table 1. Overview of Research Reactors in Indonesia

Reactor Name	Location	Thermal Power	First Criticality	Current Status
RSG-GAS	Puspiptek, South Tangerang	30 MW	July 1987	Operational
TRIGA 2000	Bandung	2 MW (upgraded from 250 kW)	1965 (original), upgraded in 1971 & 2000	Operational
Kartini	Yogyakarta	100 kW	1979	Operational

Triga Bandung Reactor Kartini Reactor Yogyakarta GA Siwabessy Reactor (RSG-GAS)

Figure 3. Operational research reactors in Indonesia under the National Research and Innovation Agency

2. Bapeten chairman regulation No. 2 Year 2015

The Nuclear Energy Regulatory Agency of Indonesia (BAPETEN) issued Chairman Regulation No. 2 of 2015, mandating that all non-power reactor licensees conduct and submit a Periodic Safety Review (PSR). This ten-year review cycle began with the evaluation period from 2005 to 2015. A licensee refers to an individual or entity authorized by BAPETEN to utilize nuclear energy. Regulation No. 2/2015 governs the verification and safety review of non-power reactors, aligning with the IAEA Safety Standard No. SSR-3 (formerly No. 99) on periodic safety reviews for research reactors [9]. The regulation aims to ensure safety oversight during the reactor's construction, commissioning, and operational phases.

Safety evaluations must be conducted on all safety-related structures, systems, and components (SSCs) that have been in service for extended periods. These SSCs are considered critical if their failure or malfunction could result in radiation exposure to personnel or the public. The PSR encompasses reviews of: (a) reactor design; (b) the current condition of SSCs; (c) equipment qualification; (d) ageing; (e) safety performance and operational experience feedback; (f) nuclear safety management and emergency preparedness; and (g) radiological impact on the environment. Figure 4 illustrates the structural components of the PSR content. Licensees are required to implement a management system to ensure that the safety review is conducted by qualified and competent personnel through a structured and systematic approach.

Figure 4. Structure of the Periodic Safety Review (PSR) as required by BAPETEN Chairman Regulation No. 2/2015

3. Analysis and Strategy for Establishing the First PSR of RSG-GAS

The objectives of the RSG-GAS Periodic Safety Review (PSR) are threefold: (a) to fulfil regulatory requirements for the preparation and submission of the PSR report; (b) to provide systematic feedback and evaluation on the effectiveness of reactor operations, system maintenance, and safety controls; and (c) to support the renewal process of the reactor's operating license issued by BAPETEN.

The Periodic Safety Review (PSR) of RSG-GAS was the first of its kind conducted for a research reactor globally, prompting the IAEA to organize a technical meeting for member states to share experiences and monitor implementation. The review evaluated ten years of reactor operation and safety-related structure, system, and component (SSC) maintenance data, spanning from 2005 to 2015. In accordance with BAPETEN regulations, the licensee was given three years to complete the PSR. The preparation involved several key steps: initiating discussions between the regulatory body and the licensee to clarify PSR requirements; forming a dedicated PSR team and collecting operational cycle data; compiling reports on ageing management, maintenance activities, and component refurbishments; assigning writing responsibilities based on team members' technical expertise; maintaining ongoing coordination with BAPETEN to ensure compliance; submitting the draft PSR for regulatory review; revising the report based on feedback; and conducting final consultations to support the issuance of a renewed operation license valid until 2030. These coordinated actions ensured that the PSR was prepared systematically and aligned with national and international safety standards. The analysis and strategy for preparing the first RSG-GAS PSR are shown in Figure 5.

Ageing in research reactor components affects reactor vessels, concrete structures, nuclear fuel and cladding, cooling systems, reflectors, instrumentation and control systems, and other metallurgical parts. Reactor vessels may degrade due to radiation, pressure, and heat, assessed via visual inspection, wall thickness measurements, non-destructive testing (e.g., ultrasonic, radiographic), and chemical analysis. Concrete structures face cracking and weathering from radiation and environment, evaluated using crack mapping, scanning, and strength testing. Fuel and cladding degrade from radiation and coolant interactions, requiring post-irradiation analysis and mechanical-chemical testing. Fuel follows a cycle from reactor core to interim storage and then to long-term storage.

Cooling systems—comprising pipes, pumps, and heat exchangers—are susceptible to corrosion, erosion, and deposit buildup, and are evaluated through visual inspection, pressure testing, ultrasonic methods, and thermal analysis. Reflector materials such as beryllium degrade from neutron irradiation and coolant interactions, assessed via non-destructive testing and neutronic calculations. Instrumentation and control (I&C) systems face ageing challenges including obsolescence, electronic failure, and sensor drift, mitigated through routine calibration, performance simulations, and system upgrades [10], [11]. Metallurgical components experience fatigue, creep, and thermal stress, and are assessed using tensile and hardness tests, as well as microstructural analysis through optical or electron microscopy. These evaluations ensure ageing-related degradation is identified and addressed to uphold reactor safety and performance. The operation of RSG-GAS remains aligned with its design principle of minimizing radiation exposure to the public and operators during both normal and accident conditions.

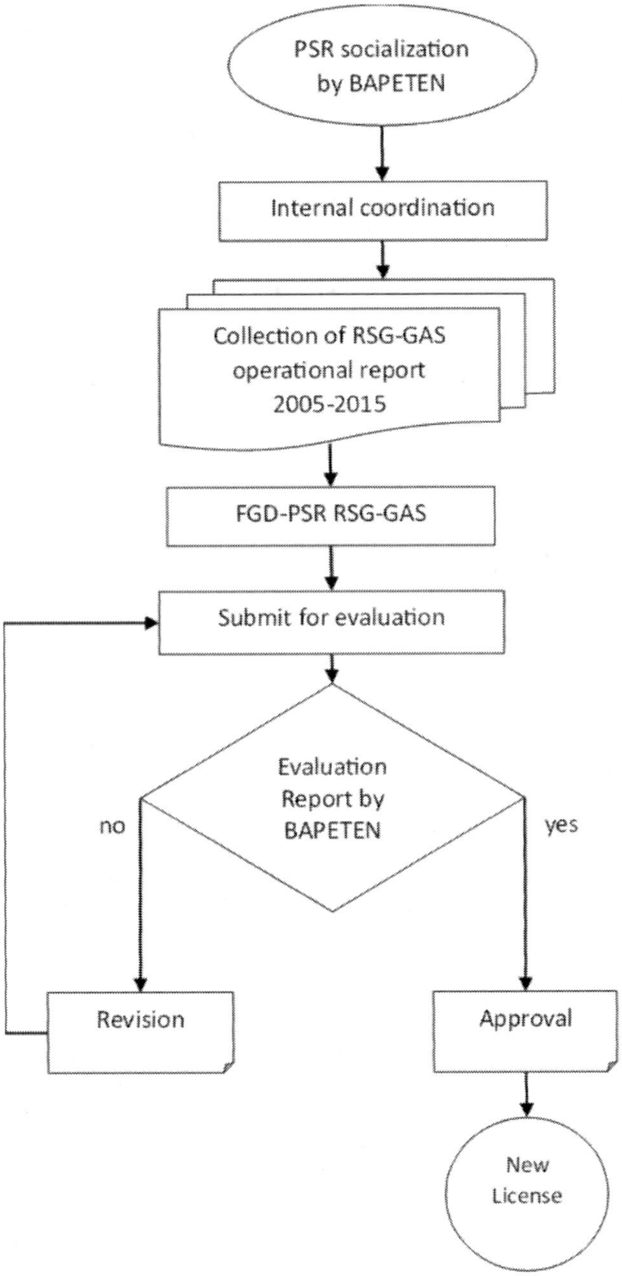

Figure 5. Analysis and strategy for preparing the first RSG-GAS PSR

4. Evaluation Results

4.1 RSG-GAS design evaluation

RSG-GAS was designed in accordance with requirements of design standards set by the IAEA for research reactor facility. Quality standard selected for structure, system, and component (SSC) [12]. The ten-year RSG-GAS operational report includes data from Core 55 to Core 89, covering nuclear, thermohydraulic, radiation parameters, and records of reactor utilization, incidents, and operational disturbances. Each core represents a fuel cycle aligned with planned fuel replacement patterns. In abnormal conditions, the reactor protection system functioned reliably, automatically shutting down the reactor core. A total of 27 incidents occurred during this period [13], classified under the International Nuclear and Radiological Event Scale (INES) as Scale 1 (anomalies) or Scale 0 (deviations), with no radiological impact or safety system failure. These events, while minor, were effectively managed and used to improve reactor safety. The INES scale, illustrated in Figure 6, ranges from minor anomalies to major accidents, such as the Fukushima Daiichi incident in 2011.

The review confirms that RSG-GAS operations complied with all safety requirements, and the reactor design meets national and international standards. Operational parameters remained within acceptable limits, with minor deviations still tolerable. All analyses and procedures followed BAPETEN and IAEA guidelines [15]. No degradation was found in the performance of safety-related SSCs, confirming that the reactor was operated and utilized safely during the 2005–2015 period.

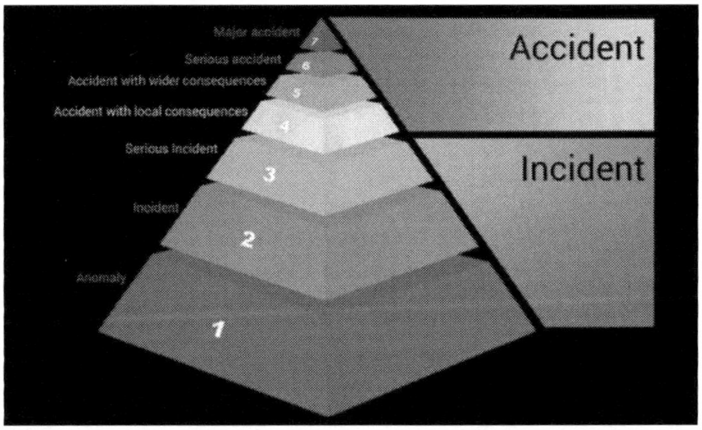

Figure 6. The IAEA International Nuclear and Radiological Event Scale (INES) for classifying nuclear and radiological incidents based on safety significance.

4.2 Maintenance and Incident Data

In line with the ageing management programme, the performance of all safety-related structures, systems, and components (SSCs) at the RSG-GAS reactor is routinely monitored, maintained, and evaluated in accordance with the RSG-GAS Maintenance and Repair Manual. Both periodic and non-periodic maintenance activities are carried out to ensure continued compliance with operational standards. Maintenance implementation follows a structured plan, with particular emphasis on instrumentation and control systems, including signal processing modules, output amplifiers, and actuator interfaces. Tasks beyond internal technical capabilities are outsourced to certified third-party contractors under formal agreements.

Between 2005 and 2015, a total of 422 maintenance activities were executed under standard operating procedures, primarily consisting of semi-annual and annual tasks. Preventive maintenance constituted 313 tasks (74%), while 109 tasks (26%) were corrective, demonstrating a proactive approach to preserving system reliability and functional integrity. During this period,

27 reactor scrams were recorded; however, all events were classified as Level 0 or 1 on the IAEA International Nuclear and Radiological Event Scale (INES), indicating no significant safety or radiological consequences. Notably, no radiation-related incidents occurred. These results affirm the effectiveness of the maintenance programme in maintaining operational safety, meeting safety requirements, and ensuring the reactor system remains within its design and performance limits. A summary of maintenance activities and operational events is provided in Table 2, and the distribution of preventive and corrective tasks is illustrated in Figure 7.

Table 2: Summary of Maintenance and Incidents (2005–2015)

Parameter	Value
Total Maintenance Tasks	422
Preventive Maintenance	313 (74%)
Corrective Maintenance	109 (26%)
Reactor Scrams	27
Radiation-related Incidents	0

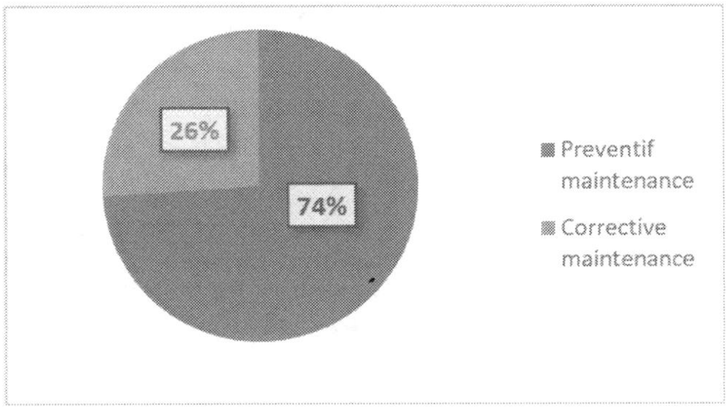

Figure 7. Distribution of Maintenance Activities at RSG-GAS (2005–2015)

4.3 Comparative Benchmarking

Compared to the 62-year-old TRIGA reactor in Belgium and the 18-year-old OPAL reactor in Australia, RSG-GAS exhibits a higher frequency of reactor scram events. The average annual scram rate of RSG-GAS (2.7 per year) is higher than that of BR2 (~2) and significantly higher than OPAL (<1), indicating a relatively greater frequency of automatic reactor shutdowns, as shown in Table 3. This elevated scram rate may be attributed to the aging of critical systems, particularly obsolescence in the instrumentation and control (I&C) modules, which can lead to spurious trips or reliability issues in reactor protection systems. While a higher scram frequency can suggest operational vulnerability, the safety impact remains limited if each event is properly managed. Importantly, all RSG-GAS scrams between 2005 and 2015 were categorized at INES Level 0 or 1—signifying either no safety significance or minor anomalies without radiological impact. According to IAEA guidelines (e.g., NS-G-4.2 and SSG-10), scram frequency is not in itself a regulatory breach but is a key performance indicator monitored during Periodic Safety Reviews (PSRs) and Aging Management Programs (AMPs). The consistent compliance of RSG-GAS with IAEA PSR standards affirms that, despite a comparatively higher scram rate, the reactor maintains an acceptable safety margin and effective incident response protocols. Nonetheless, the data underscore the importance of modernizing obsolete SSCs to enhance reliability and reduce unplanned reactor trips.

Table 3: Comparative Safety Indicators with Peer Research Reactors

Reactor	Years Operated	Avg. Annual Scrams	Major SSC Obsolescence	IAEA PSR Compliance
RSG-GAS	37	2.7	I&C Modules	Yes
BR2 (Belgium)	62	~2	Control Rod Drives	Yes
OPAL (Australia)	18	<1	None	Yes

4.3 Equipment qualification and classification

RSG-GAS equipment is categorized into safety, quality, and seismic classes [17]. Maintenance records confirm that periodic and non-periodic activities have been performed to sustain qualification, along with appropriate environmental controls. No critical components have required replacement. Evaluation of surveillance data shows that equipment and SSCs remain qualified from installation through operation.

4.4 Aging mechanisms and mitigation in RSG-GAS

Ageing in RSG-GAS is managed by closely monitoring primary cooling water chemistry to control corrosion, erosion, fatigue, and material degradation [18], [19]. Pitting corrosion is common in aluminum pool structures, and erosion occurs in high-flow piping bends. Neutron-induced brittleness is minimal due to low fluence, and fatigue is limited by low mechanical loads. Preventive measures include regular water quality checks, resin filtration, and repositioning beryllium reflector blocks to balance neutron exposure.

An ageing-related incident in Core 54 involved corrosion at the JKT03 unbalance load detector cable, leading to an unplanned shutdown. Corrective actions included dry air purging, design modifications, and integration into the neutron detector management program.

Obsolescence in instrumentation and control (I&C) systems presents replacement challenges due to discontinued components. To ensure continued safety, outdated systems must be upgraded with modern technology, and a detailed refurbishment plan, particularly for the reactor protection system, is essential.

4.5 Safety Performance and Operating Experience in RSG-GAS

A significant modification in RSG-GAS was the transition from U_3O_8-Al to U_3Si_2-Al fuel elements, maintaining the same fuel density (2.96 gU/cm^3) to increase uranium content and extend fuel life [19]. This conversion introduced changes in core reactivity and heat transfer characteristics. The transition was managed through a mixed-core approach, allowing the use of both oxide and silicide elements, which optimized fuel usage and maintained core equilibrium. Although the cooling system remained unchanged, thermohydraulic calculations were conducted to confirm that safety margins remained within limits defined in the Safety Analysis Report (SAR) [20].

Over the last decade, several reactor scrams have occurred due to the reactor protection system detecting parameter deviations. Each event was managed through a structured response: incident handling and reporting, root cause investigation, safety review, corrective actions, and procedural revisions. These scrams had no radiological impact and were treated as learning opportunities to strengthen reactor safety. Mitigation steps are illustrated in Figure 8.

4.6 Safety Policy and Emergency Preparedness in RSG-GAS

RSG-GAS upholds a safety policy that prioritizes the protection of individuals and facilities, aiming for zero accidents. This commitment aligns with regulatory requirements, mandating licensees to ensure work safety, radiation protection, and reactor operational safety [21]. Safety performance is monitored through operational parameters, with systems maintained within established safety margins. Effective operation relies on the responsibility and commitment of all stakeholders, supported by a robust safety culture.

The nuclear emergency preparedness program encompasses infrastructure and mitigation functions, regularly reviewed for adequacy. Coordination with relevant institutions is essential for implementing preparedness measures and mitigating potential radiological impacts. Evaluations of emergency response drills from 2006 to 2015 indicate that organizational structures, responsibilities, coordination mechanisms, facilities, equipment, and procedures are adequately established. Continuous improvements are pursued in areas identified as needing enhancement.

Safety analyses have considered scenarios involving increased fuel element melting, from one to five elements. Over the 2005–2015 period, safety management and emergency preparedness programs were systematically evaluated through management reviews and audits to ensure effective implementation.

4.7 Environmental Monitoring and Emergency Preparedness at RSG-GAS

To safeguard the public and environment from potential radiological impacts, BAPETEN issued Chairman Regulation No. 7/2013, establishing environmental radioactivity thresholds and permissible release limits . RSG-GAS conducts continuous environmental monitoring to detect any significant increases in surrounding radioactivity levels. Should such increases occur, predefined mitigation procedures are implemented promptly.

The RSG-GAS nuclear emergency preparedness program is designed to address emergencies at the reactor's alert level, coordinating an integrated response involving the entire nuclear emergency response team. As a 30 MW research reactor with radiological hazards classified under Category II, RSG-GAS's emergency protocols align with the Provincial Preparedness Program developed by BPBD and the National Preparedness Program by BNPB, as stipulated in Government Regulation No. 54/2012. Annual emergency simulations involve radiation protection officers, trained medical personnel, regional and national disaster management agencies, and BAPETEN.

In the event of a nuclear accident, such as the melting of five fuel elements—a beyond design basis accident leading to radiation release—the RSG-GAS Director declares an emergency, activating nuclear emergency procedures to mitigate risks and protect the public. Data from

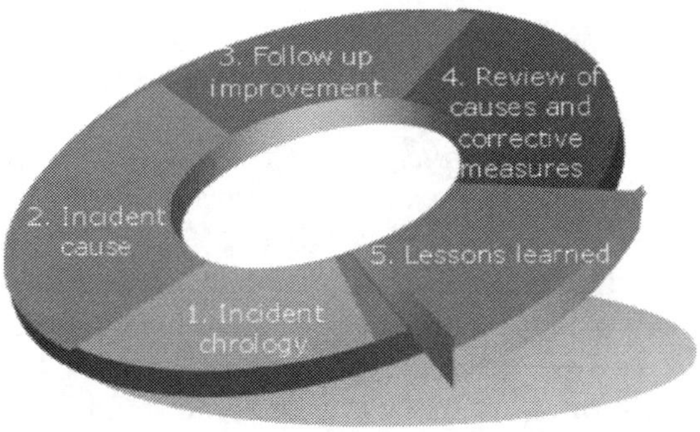

Figure 8. Mitigating Procedures of Incident

environmental radioactivity monitoring serve as critical inputs for safety reviews, reinforcing protective measures for both the public and the environment.

5. Conclusion

The reactor operation report for each cycle provides comprehensive historical data on operations, utilization, nuclear fuel, SSC maintenance, radiation protection, waste management, and nuclear preparedness. This ensures that the reactor operates strictly within prescribed safety limits, with any anomalies addressed and resolved before resuming operations. Such systematic documentation supports the safe and efficient operation of the reactor while maintaining regulatory compliance. The maintenance and repair history highlights the need for attention to the power channel detector and the obsolescence of the I&C system, while inspections of the reactor liner, conducted with IAEA technical cooperation and expertise in 2023, confirm its integrity remains intact.

Periodic safety reviews of the RSG-GAS reactor operation have been conducted to assess the acceptability of its continued operation. The review results confirm that all safety requirements are met, with the reactor core operation and system design adhering to applicable standards. Strengthening defence-in-depth principles at all organizational levels is emphasized to address potential internal and external threats effectively. Over the last decade, operational parameters have generally complied with safety standards, with a few deviations remaining within tolerable margins. The use of computer programs, experimental protocols, and review procedures aligns with BAPETEN and IAEA regulations, ensuring consistent adherence to safety and operational guidelines.

References

[1] E. P. Hastuti *et al.*, "Map of Radioisotope Production and Batan Research Reactor Utilization," *J. Teknol. Reakt. Nukl. Tri Dasa Mega*, vol. 23, no. 3, p. 105, 2021, doi: 10.17146/tdm.2021.23.3.6288.

[2] Supardjo, T. Surbakti, E. P. Hastuti, and Et.al., *Teknik Pembuatan dan Pengujian Bahan Bakar Dispersi U3Si2/Al Densitas Uranium Tinggi*, 1st ed. Jakarta: Batan Press - Badan Tenaga Nuklir Nasional (BATAN), 2021.

[3] IAEA, "Project Experiences in Research Reactor Ageing Management, Modernization and Refurbishment," 2014.

[4] G. Bignan and J.-Y. Blanc, "Supporting infrastructures and research reactors: status, needs and international cooperation, IAEA ICERR (International CEntres based on Research Reactors) and IGORR (International Group on Research Reactors), FP7 and H2020 JHR access rights," *EPJ Nucl. Sci. Technol.*, vol. 6, p. 26, 2020, doi: 10.1051/epjn/2019022.

[5] IAEA, "Ageing Management for Research Reactors," *Specif. Saf. Guid. No. SSG-10*, 2010.

[6] S. Martorell, I. Martón, A. Sánchez, and S. Carlos, "Harmonisation of surveillance requirements and maintenance in a context of ageing and obsolescence based on reliability, availability and risk information," *Reliab. Eng. Syst. Saf.*, vol. 202, no. March, p. 106992, 2020, doi: 10.1016/j.ress.2020.106992.

[7] IAEA., "Research Reactors : Summary of an International Conference Held in Vienna, 16-20 November 2015.," no. November 2015, p. 66, 2017.

[8] IAEA, "Periodic Safety Review for Research Reactors," 2020, [Online]. Available: https://www.iaea.org/resources/safety-standards

[9] Badan Pengawas Tenaga Nuklir, "PERATURAN KEPALA BADAN PENGAWAS TENAGA

NUKLIR No. 2 Tahun 2015 tentang Verifikasi dan Penilaian Keselamatan Reaktor Nondaya," *Bapeten*, pp. 1–17, 2015, [Online]. Available: https://jdih.bapeten.go.id/id/dokumen/peraturan/peraturan-kepala-badan-no-2-tahun-2015-tahun-2015-tentang-verifikasi-dan-penilaian-keselamatan-reaktor-nondaya

[10] E. Kibrit and A. Aquino, "Sustainability Management System Model for Operating Organizations of Research Reactors," *Int. Nucl. Saf. J.*, vol. 4, no. 4, pp. 23–37, 2015.

[11] H.-P. Berg, C. Toth, and M. Röwekamp, "Periodic Safety Review: A Tool to Maintain and Enhance Safety," *Int. Conf. Nucl. Eng.*, no. ICONE20-POWER2012-54441, pp. 111–120, 2013, doi: https://doi.org/10.1115/ICONE20-POWER2012-54441.

[12] E. P. Hastuti, "Evaluation of Classification of Structure, Systems, and Components on Aging Management of RSG-GAS," *J. Teknol. Reakt. Nukl. Tri Dasa Mega* vol. 23, pp. 43–57, 2011.

[13] I. Kuntoro, "Evaluation on the Performance of Safety Systems of the RSG-GAS Reactor During 25 Years Operation.," 2014.

[14] IAEA, "Applications of Research Reactors," *Encycl. Nucl. Energy*, pp. 8–25, 2021, doi: 10.1016/b978-0-12-819725-7.00110-0.

[15] A. S. Catur, S. Sutrisno, and S. Susana, "The design of RSG-GAS reactor maintenance information system application based on WEB," *Reakt. - Bul. Pengelolaan Reakt. Nukl.*, vol. 16, no. 2, pp. 32–41, 2019, doi: 10.17146/bprn.2019.16.2.5651.

[16] A. S. Catur and D. S. F. Djunaidi, "Maintenance of structures, system and component of the RSG-GAS reactor as an implementation of BAPETEN Regulation NO.5/2011," *Pros. Semin. Nas. Teknol. dan Apl. Reakt. Nukl. PRSG Tahun 2012 ISBN 978-979-17109-7-8*, no. 5, pp. 66–78, 2012.

[17] E. P. Hastuti, S. Bakhry, E. Sihombing, and Sudarmono, "Aging management experience to support reliability of RSG-GAS operation and future NPP in Indonesia," *J. Phys. Conf. Ser.*, vol. 2193, no. 1, 2022, doi: 10.1088/1742-6596/2193/1/012008.

[18] BAPETEN, "Peraturan Kepala Badan Pengawas Tenaga Nuklir R Nomor 8 Tahun 2008 Tentang Ketentuan Keselamatan Manajemen Penuaan Reaktor Non Daya," 2013.

[19] S. Pinem, T. M. Sembiring, and P. H. Liem, "Neutronic and thermal-hydraulic safety analysis for the optimization of the uranium foil target in the RSG-GAS reactor," *Atom Indonesia*, vol. 42, no. 3, pp. 123–128, 2016, doi: 10.17146/aij.2016.532.

[20] I. Kuntoro, S. Pinem, T. M. Sembiring, D. Haryanto, and S. Purwanto, "Evaluation of Equilibrium Core Operation of the Rsg-Gas Reactor," *J. Teknol. Reakt. Nukl. Tri Dasa Mega*, vol. 23, no. 1, p. 15, 2021, doi: 10.17146/tdm.2021.23.1.6150.

[21] T. S. A. S. Catur, "Study on application of aging program management and RSG-GAS pool component as an implementation of Bapeten regulatory 8/2008," *Pros. Semin. Nas. Teknol. dan Apl. Reakt. Nukl. PRSG Tahun 2012 ISBN 978-979-17109-7-8*, no. 8, pp. 86–92, 2013.

[22] BAPETEN, "jdih.bapeten.go.id," BAPETEN. [Online]. Available: file:///C:/Users/Win10/Downloads/Bapeten_Perka.7.2013.pdf

[23] BAPETEN, "jdih.bapeten.go.id." [Online]. Available: https://jdih.bapeten.go.id/id/dokumen/peraturan/peraturan-pemerintah-nomor-54-tahun-2012-tentang-keselamatan-dan-keamanan-instalasi-nuklir

[25] SCK CEN, "BR2 - Belgian Reactor 2," [Online]. Available: https://www.sckcen.be/en/infrastructure/br2-belgian-reactor-2

[26] ANSTO, "OPAL multi-purpose reactor," [Online]. Available: https://www.ansto.gov.au/facilities/opal-multi-purpose-reactor

Analysis Of Burnable Poison Doping on Fuel Molten Mixture for Molten Salt Reactor (MSR) Fuji-12

Hilarius Adiwarna(b), Cici Wulandari(a,b), Syeilendra Pramuditya(a,b), Sidik Permana(a,b*), Syaiful Bakhri (c)

a) Nuclear Physics & Biophysics Research Division, Department of Physics, Faculty of Mathematics and Natural Sciences, Institut Teknologi, Bandung, Indonesia
b) Department of Nuclear Science and Engineering, Faculty of Mathematics and Natural Sciences, Institut Teknologi, Bandung, Indonesia
c) Research Center for Nuclear Material and Radioactive Waste Technology (BRIN), South Tangerang 15314, Indonesia
*psidik@itb.ac.id

Nuclear Power Plants (NPPs) are a type of power generator that supports Sustainable Development Goals (SDGs) by producing clean, affordable, and climate-friendly energy. This study simulated the MSR FUJI-12 to determine suitable SRAC 2006 parameters with a PIJ module for cell calculation and CITATION module for core calculation. The simulation results showed an initial keff of 1.013046 for a $^{233}UF_4$ concentration of 0.22%, with a deviation of 0.0354% from the target of 1.0134. However, the final keff remained below 1, at 0.9455, prompting an increase in the $^{233}UF_4$ concentration to 0.28%. One of the challenges identified was the high Power-Peaking-Factor (PPF) of 3.49, that may damage the inside coating of reactor chamber. Using Protactinium as burnable poison and actinide producer to stabilize the keff, it was concluded that the doping on molten mixture is 0.003% to avoid un-criticality on MSR FUJI-12. However, the impact of doping is unsignificant to the PPF. It is then proposed using other types of burnable poison, such as Americium (Am) and Gadolinium (Gd). However, as it was known that only Protactinium that produces burnable actinide to fuel the core, the impact is negligible except for the Gd, as its doping is enough to decrease the initial keff and to increase the conversion ratio (CR) in MSR from 0.835 to 0.933.

Keywords: MFR, MSR FUJI-12, Protactinium, $^{233}UF_4$

1. Introduction

Greenhouse gases (GHG) are a collection of atmospheric gases that capture heat similar to the glass walls of a greenhouse [1]. This collection includes water vapor, carbon dioxide, methane, and nitrous oxide that perform as effective global insulators[2]. Carbon dioxide is one of the biggest contributors to GHG, and electricity production is the biggest producer of CO_2 [3]. Coal and natural gas-powered plants burn the fuel and create smoke that contains a lot of CO_2s.

Nuclear reactor is a type of power plant fueled by fissile fuels that can be burned to boil water and spin a turbine connected to a generator. Nuclear reactor has been developed since 1960 and spans from generation I to generation IV. In this paper, we will focus on a generation IV reactor called the Molten Salt Reactor (MSR). MSR is a generation IV reactor that combines its fissile, fertile, and coolant materials into a molten mixture[4] to enhance the reactor's efficiency and have the potential for breeding [5]. This reactor also provides a passive security system that can dump the mixture if any accidents happen using a freeze valve[6].

Burnable poison (BP) is a material with a large cross section that can absorb neutrons and reduce the fission reaction in the initial criticality [7]. This addition can ensure the stability of nuclear reactor in the beginning of cycle, as high initial criticality is prevalent in most reactor. One type of BP is actinide, a burnable poison that can absorb neutrons and produce fissile fuel to prolong the life span of the life span of a nuclear reactor. These materials are ^{231}Pa, ^{241}Am, ^{237}Np, ^{238}Pu, and ^{240}Pu [8].

This study will focus on thermal MSR for many reasons. First, the coolant used in thermal reactor, like light water (H_2O), heavy water (D_2O), or carbon dioxide (CO_2), is easier to make rather than the coolant for fast reactor, such as sodium or lead-bismuth eutectic[9], [10]. Second, thermal reactors only require low-enriched fuel, typically 3-5% ^{235}U for light water reactor and even no enrichment for heavy water reactors, to stay critical compared to fast reactors that need a highly enriched fuel, around 15-20% ^{235}U, to compensate for the low neutron absorption cross-section the fast neutron energy group can achieve.

Before this study, other types of thermal MSR were done, such as FUJI-U1, FUJI-U3, and FUJI-12. MSR FUJI-U1 is a type of MSR that can use different grades of plutonium on the fuel salt, that usually count as spent fuel in other reactors, such as Light Water Reactor (LWR)[11]. MSR FUJI-U3 is a type of MSR that has a three-division active core system with graphite as moderator[12]. MSR FUJI-12 is one-division active core system with graphite as reflector[13].

In this study, MSR FUJI-12 will be further simulated to improve the effective multiplication factor (k_{eff}) and CR. MSR FUJI-12 has been studied by the previous researcher with the utilization of plutonium and minor actinides in the fuel salt [14]. This study will experiment the effect of burnable poison that can stabilize the keff of the reactor on molten fuel salt.

This study is conducted to simulate MSR FUJI-12 using Standard Reactor Analysis Code (SRAC) 2006 with a PIJ module for cell calculation and a CITATION module for core calculation using Japan Evaluated Nuclear Data Library (JENDL) 4.0 [15]. By changing parameters of reactor design, we will strive to produce results as close as possible to references. Afterwards, by adding burnable poison to the molten mixture, the results will be observed and maximized depending on its effects. By using different types of burnable poison, it is aimed at enhancing MSR neutronic parameter outputs, such as factor multiplication (k_{eff}) and conversion ratio (CR).

2. Reactor specification

In this study, the MSR parameter that is used are adopted from MSR FUJI-12[13]. The geometry of the active core is a cylinder consisting of hexagonal channel that is moderated by Graphite. The fuel used a combination of Uranium-233 (^{233}U), Thorium (Th), and coolant such as, F-Li-Be as shown in Table 1. Graphite is used to slow down neutrons into thermal energy for a higher chance of fission reaction. This reactor consists of three heat exchangers to avoid any radioactive material leaking into the reservoir. As shown in Figure 1, the first loop consists of fuel salt, the second loop consist of molten salt and the third loop consisting of water for vapor creation[6].

Table 1. The parameter of MSR FUJI-12 [13]

Reactor parameter	Specification
Thermal capacity	350 MWth
Net electric generation	150 MWe
Thermal Efficiency	43%
Number of coolant loops	2
Core	
• Radius/ height	2.0/ 4.0 m
• Graphite fraction	70 vol%
Reflector	
• Thickness radius/ axis	0.5/0.6 m
• Graphite fraction	99 vol%
Average power density	7.0 kWth/ liter
Specification of heat exchanger	

• Thermal capacity	350 MW
• Tube length	6.8 m
• Total heat transfer area	760 m^2
• Fuel salt volume	1.19 m^3
• Fuel salt speed in heat exchanger	3.14 m/s
• Fuel salt temperatures (Inlet/ outlet)	980 K/ 840 K
• Coolant volume	7.25 m^3
• Coolant speed	1.22 m/s
• Coolant temperature (Inlet/outlet)	765 K/ 670 K
• Total length of primary loop	7.2 m
• Fuel salt speed in primary loop	1.2 m/s
Fuel salt	
Composition	
• LiF	Balance
• BeF_2	16.0 mol%
• ThF_4	12.0 mol%
• $^{233}UF_4$	0.22 mol%
Volume in reactor	15.7 m^3
Total volume	20.2 m^3
Flow rate	0.55 m^3/s
Temperature Inlet/outlet	840 K/ 980 K

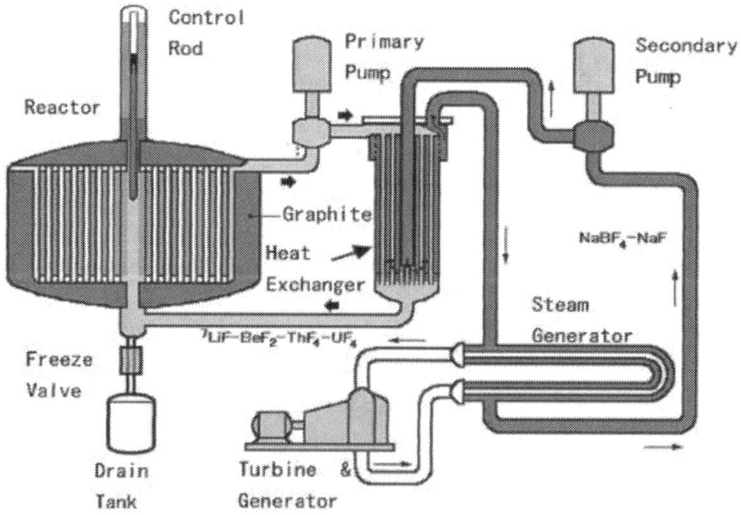

Figure 1. The diagram of MSR FUJI-12[13]

In Figure 2, the core chamber is divided into two regions, reflector, made out of high-vol% of graphite, and an active core, filled with fuel rods, representing the simplified version of the active core reactor. Reflector made of 99%vol graphite to reflect the escaping neutron back into the core and core consisting of different percentage of fuel fraction, where the fission reaction happens. In this study, different types of burnable poisons are implemented to maximize neutronic parameter outputs, such as factor multiplication (k_{eff}) and conversion ratio (CR).

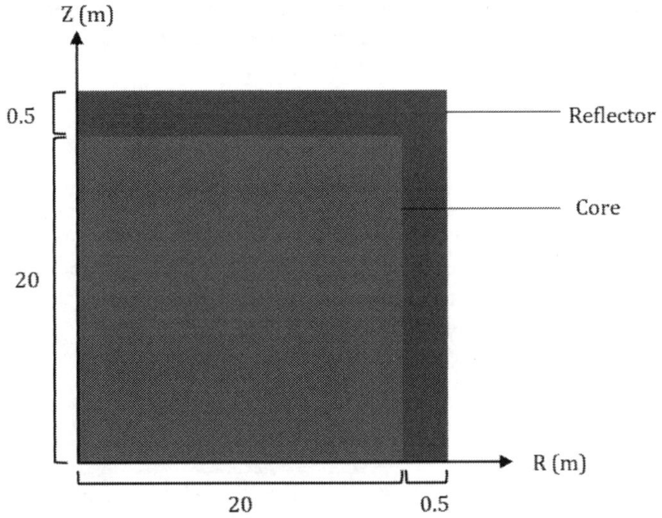

Figure 2. Diagram of reactor chamber of MSR FUJI-12

As seen in the Figure 3, the core geometry is FUJI-12 is ilustrated. This core is simplified for easier neutronic simulation. The core consisted of one region of molten mixture, with reflector on the top and bottom of the core, and surrounded by plenum [13]. This is what SRAC 2006 will simulate in this study.

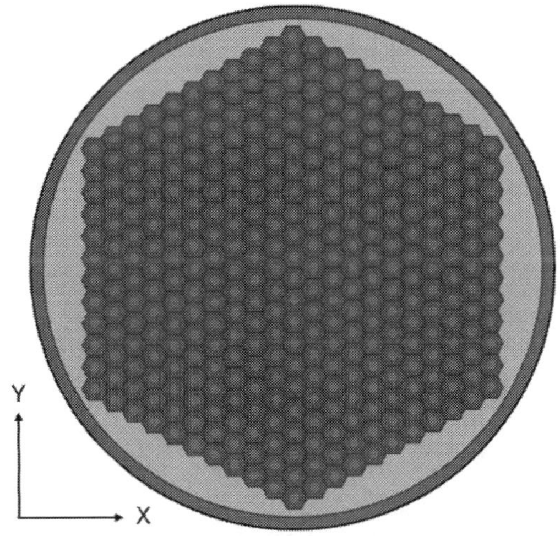

Figure 3. Diagram of a neutronic core configuration of MSR FUJI-12

3. Calculation methods

In this study, parameter validation will be conducted by comparing initial keff result with references. Then, by applying a small percentage of burnable poison, it is aimed to improve MSR output, such as k_{eff} and CR. The percentage of salt for BeF_2 dan LiF were fixed to 16 and 71.78 mol%, while the percentage combination of ThF_4, UF_4, and burnable poison is 12.22 mol%. The percentage of burnable poisons are varied to achieve a better k_{eff} and CR on MSR, especially for reactivity parameters that influence the reactor safety aspects. Making sure the initial k_{eff} is as close to 1 as possible for safety purposes and increasing the CR as neutrons are more used to convert fertile material rather than fission reaction. First, the survey will be conducted using Protactinium and the best percentage will be used for Americium and Gadolinium as presented in Table 2.

SRAC 2006, we will use in this study are PIJ and CITATION modules. PIJ module is used to calculate the collision method probability technique, while CITATION is used to calculate the neutron dynamics inside the core [15]. The neutron energy groups are divided into 107 groups, that is compressed into 30 groups, 24 groups of fast neutron groups and 6 groups of thermal neutron groups. This study is conducted by simulating the reactor for 2700 days with 100 days for each step, with each step's temperature kept at 900 K from the average of the inlet and outlet [13], [17]. The core will be divided into 64 X 32 grid nodes in the axial and radial directions [18].

Table 2. Protactinium percentage surveys for MSR FUJI-12

Material	Fuel volume fraction (%)
Protactinium	0.01
	0.005
	0.003
	0.002
	0.001

4. Results and discussion

4.1. Parameter Validation on SRAC 2006 PIJ Module

Using the parameter given, it can be concluded that the initial keff on PIJ module after simulation is 1.0130. Comparing this value with Yamamoto's study [18], which is 1.0134, the difference is 0.036%. This difference could emerge from the number of nuclides in JENDL 4.0 [16] and 3.2 [19] is different because for each iteration of JENDL, the number of nuclides always increases. However, as shown in Figure 4, the reactor did not stay critical for 2700 days. To help increase the criticality, increasing uranium percentage on fuel salt is needed. Referencing Wulandari's work, the uranium percentage will be increased to 0.28%. As shown in Figure 5, reactor can stay critical for 2700 days[20], as it is the usual operation time for MSR before the component inside the core corroded and became a hazard for security.

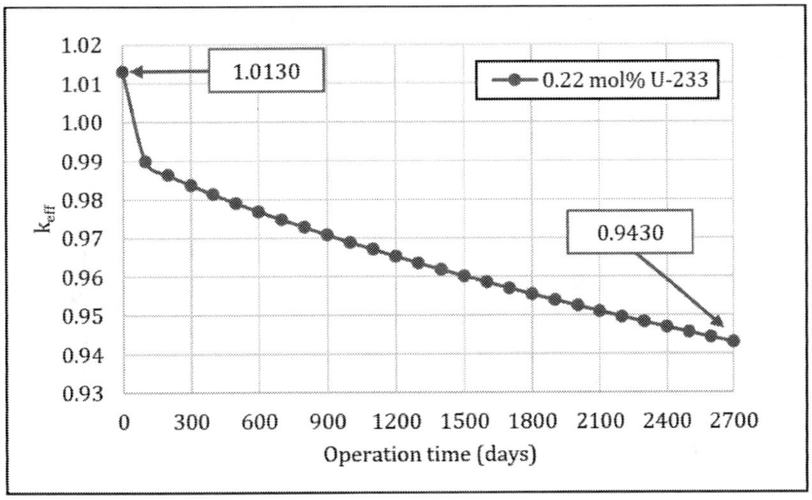

Figure 4. Effective multiplication factor (keff) for PIJ module using 0.22 mol% ^{233}U

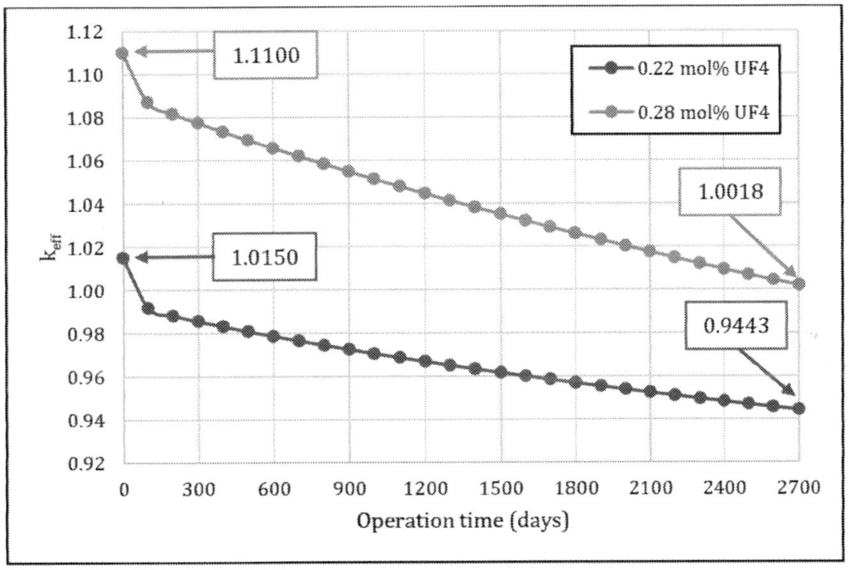

Figure 5. Effective multiplication factor (k_{eff}) for CITATION module using 0.22 mol% (blue) and 0.28 mol% (orange) ^{233}U

4.2. Burnable Poison Doping on Molten Mixture

Burnable absorbers or burnable poison (BPs) are substances with high thermal neutron absorption cross sections arranged inside the core and slowly depleted during reactor operation by absorbing neutrons that will lower excess reactivity[8]. This type of material has been applied in reactors, such as Gd_2O_3 on VVER [21] and Er_2O_3 in RMBK [22]. However, there is another type of BP, which is actinide burnable poison, a type of BP that can absorb neutrons and convert into fissile nuclides [7]. One of these actinide BPs is protactinium (Pa). By applying protactinium (Pa) on fuel salt with a 0.28 mol% ^{233}U, as shown in Figure 6, the excess reactivities are going to decrease alongside the decrease of protactinium (Pa). This happened because the number of BPs that can decrease the amount of fission reaction by absorbing neutrons can affect the number of loose neutrons free to react with fissile fuels.

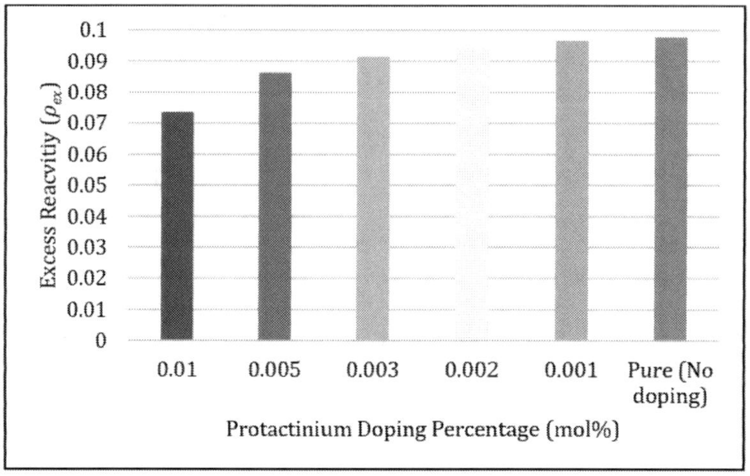

Figure 6. The effect of Protactinium doping on excess reactivity

As shown in Table 3, the maximum dosage for Protactinium doping before the ending keff go under criticality is 0.003 mol%. This percentage only lowered the initial keff by 0.693% and dropped the CR by 0.707% compared to reference. This result didn't show the full effect of burnable poisons on fuel salts.

Table 3. The effect of Protactinium on initial and ending keff for PIJ module

PaF_4 (mol%)	Initial k_{eff}	Ending k_{eff}
0	1.1080	1.0005
0.01	1.0791	0.9936
0.005	1.0941	0.9976
0.003	1.1003	0.9993
0.002	1.1035	1.0001
0.001	1.1067	1.0009

4.2.1. Applying Americium (Am) and Gadolinium (Gd) for Doping

Using the same percentages of Am and Gd in fuel salt, keff and CR are observed and compared with its pure and protactinium counterparts. The effects of Am and Pa are similar because both are from the actinide BPs groups, meaning that the effect of both materials is still rather small to observed. However, Gadolinium output a different effect, by lowering the initial k_{eff} by 26.24% from its high neutron absorbing cross section [23] and at the same time increasing the CR of the reactor by 11.87%, this is in line with the CR formula [24] as shown in Figure 7 and Figure 8.

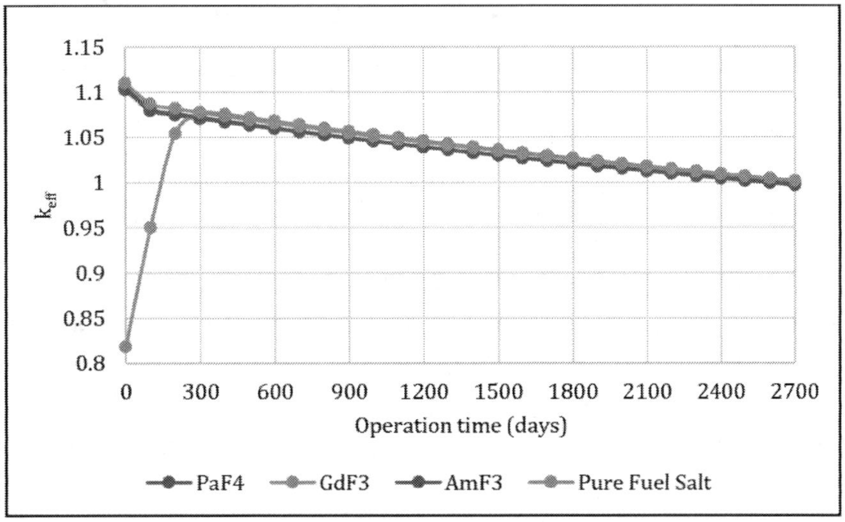

Figure 7. Different effects of burnable poisons on k_{eff} from PIJ module

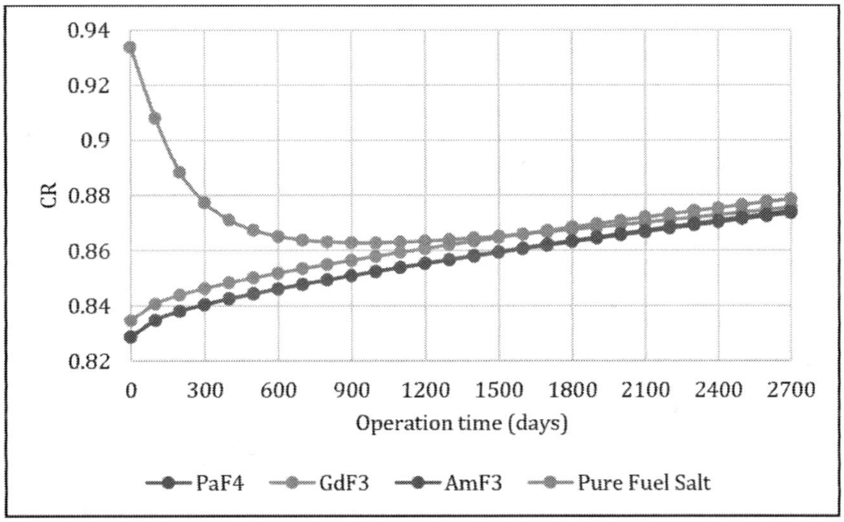

Figure 8. Different effects of burnable poisons on CR

5. Conclusion

A study on burnable poison doping in MSR FUJI-12 using a simulation on SRAC 2006 with PIJ and CITATION module is conducted. This doping will affect some parameters on reactor, such as k_{eff} and CR. Firstly, parameters used in this study are valid because compared with references the difference is only 0.036%, however with 0.22 mol% of ^{233}U on molten mixture, reactor can only stay critical for less than 100 days. By increasing the percentage to 0.28 mol%, the reactor can stay critical for 2700 days.

From the survey, only 0.003 mol% Protactinium doping can be applied on this reactor before making the reactor uncritical, however this doping does not apply a huge difference because of its small percentages. By applying other types of BP such as, Americium and, especially Gadolinium, the reactor behaves differently. Gadolinium has a high neutron absorbing cross section that lowers the initial k_{eff} and increases the initial CR in the beginning of the cycle.

Acknowledgements

The author would like to thank Prof. Sidik Permana for his guidance throughout this research, Dr. Wulandari for her helpful advices and insights, which greatly increase the quality of this paper, and Bandung Institute of Technology for providing resources and facilities.

References

[1] D. Kweku *et al.*, "Greenhouse Effect: Greenhouse Gases and Their Impact on Global Warming," *J Sci Res Rep*, vol. 17, no. 6, pp. 1–9, Feb. 2018, doi: 10.9734/JSRR/2017/39630.

[2] U.K. Met Office, "Warning: A guide to climate change," 2011, *Met Office Hadley Centre*.

[3] IPCC, *Climate Change 2022 - Mitigation of Climate Change*. Cambridge University Press, 2023. doi: 10.1017/9781009157926.

[4] T. Letcher, Ed., *Future Energy : Improved, Sustainable and Clean Options for our Planet*, Third. Laurel House, Stratton on the Fosse: Elsevier, 2020.

[5] M. Carelli and D. Ingersoll, Eds., *Handbook of Small Modular Nuclear Reactors*. Elsevier, 2014. doi: 10.1016/C2013-0-16379-9.

[6] T. Dolan, Ed., *Molten Salt Reactors and Thorium Energy*. Elsevier, 2017.

[7] S. Xu, T. Yu, J. Xie, L. Yao, and Z. Li, "Burnable Poison Selection and Neutronics Analysis of Plate Fuel Assemblies," *Front Energy Res*, vol. 9, Aug. 2021, doi: 10.3389/fenrg.2021.729552.

[8] G. G. Kulikov, E. G. Kulikov, A. N. Shmelev, and V. A. Apse, "Protactinium-231 – New burnable neutron absorber," *Nuclear Energy and Technology*, vol. 3, no. 4, pp. 255–259, Dec. 2017, doi: 10.1016/j.nucet.2017.10.002.

[9] E. Lewis, *Fundamental of Nuclear Reactor Physics*. Academic Press, 2008.

[10] L. Massimo, *Physics of High-Temperature Reactors*. Pergamon Press, 1976.

[11] C. Wulandari, A. Waris, S. Permana, and S. Pramuditya, "Plutonium and Minor Actinides Utilization in FUJI-U1 Molten Salt Reactor," *J Phys Conf Ser*, vol. 1204, p. 012132, Apr. 2019, doi: 10.1088/1742-6596/1204/1/012132.

[12] C. Wulandari, A. Sofyan, A. Waris, and S. Permana, "Natural Uranium Utilization in FUJI-U3 Molten Salt Reactor," *J Phys Conf Ser*, vol. 1772, no. 1, p. 012023, Feb. 2021, doi: 10.1088/1742-6596/1772/1/012023.

[13] N. Suzuki and Y. Shimazu, "Reactivity-initiated-accident analysis without scram of a molten salt reactor," *J Nucl Sci Technol*, vol. 45, no. 6, pp. 575–581, 2008, doi: 10.1080/18811248.2008.9711881.

[14] A. Waris, I. K. Aji, Novitrian, R. Kurniadi, and Z. Su'ud, "Plutonium and minor actinides utilization in thorium molten salt reactor," in *AIP Conference Proceedings*, 2012, pp. 115–118. doi: 10.1063/1.4725445.

[15] K. Okumura, T. Kugo, K. Kaneko, and K. Tsuchihashi, *SRAC2006: A Comprehensive Neutronic Calculation Code System*. 2007.

[16] K. SHIBATA *et al.*, "JENDL-4.0: A New Library for Nuclear Science and Engineering," *J Nucl Sci Technol*, vol. 48, no. 1, pp. 1–30, Jan. 2011, doi: 10.1080/18811248.2011.9711675.

[17] Y. Honma, Y. Shimazu, and T. Narabayashi, "Optimization of flux distribution in a molten-salt reactor with a 2-region core for plutonium burning," *Progress in Nuclear Energy*, vol. 50, no. 2–6, pp. 257–261, Mar. 2008, doi: 10.1016/j.pnucene.2007.11.073.

[18] T. Yamamoto, K. Mitachi, and T. Suzuki, "Steady State Analysis of Small Molten Salt Reactor * (Effect of Fuel Salt Flow on Reactor Characteristics)."

[19] T. NAKAGAWA *et al.*, "Japanese Evaluated Nuclear Data Library Version 3 Revision-2: JENDL-3.2," *J Nucl Sci Technol*, vol. 32, no. 12, pp. 1259–1271, Dec. 1995, doi: 10.1080/18811248.1995.9731849.

[20] C. Wulandari, "Evaluasi Pemanfaatan Plutonium dalam MSR Jenis Fuji-12," Bandung, 2015.

[21] IAEA-TECDOC-844, *Characteristics and Use of Urania-Gadolinia Fuels*. Vienna: IAEA, 1995.

[22] J. A. Renier, "Development of Improved Burnable Poisons for Commercial Nuclear Power Reactors," Oak Ridge, TN (United States), Apr. 2002. doi: 10.2172/814398.

[23] R. C. Lloyd, E. D. Clayton, and L. E. Hansen, "Criticality of Plutonium Nitrate Solution Containing Soluble Gadolinium," *Nuclear Science and Engineering*, vol. 48, no. 3, pp. 300–304, Jul. 1972, doi: 10.13182/NSE72-A22487.

[24] D. Vondy, *On Nuclear Fuel, Mass Balances, Conversion Ratio, Doubling Time and Uncertainty*. Oak Ridge: Oak Ridge National Laboratory, 1976.

IOP Conf. Series: Materials Science and Engineering **1330** (2025) 012006 doi:10.1088/1757-899X/1330/1/012006

Preliminary Numerical Investigation of Cooling Fins Feature in the Frozen Salt Melting Performance

Anni Nuril Hidayati[1*], Asril Pramutadi Andi Mustari[2], Yulia Mifftah Huljanah[3], Nina Widiawati[1]

[1] Research Center for Nuclear Reactor Technology, Research Organization for Nuclear Energy, National Research and Innovation Agency, Building No. 80, BJ Habibie Integrated Science Area, South Tangerang 15310, Indonesia
[2] Nuclear Physics and Biophysics Research Division, Physics Department, Faculty of Mathematics and Natural Science, Institut Teknologi Bandung, Jalan Ganesha 10, Bandung 40132, Indonesia
[3]Bachelor Program of Physics Department, Faculty of Mathematics and Natural Science, Institut Teknologi Bandung, Jalan Ganesha 10, Bandung, 40132, Indonesia

*E-mail: anni030@brin.go.id

Abstract. The opening time of the freeze plug safety feature in a molten salt reactor is a crucial aspect that must be carefully considered. This relates to the reactor core's drainage during a blackout or abnormal temperature rise due to a Loss of Coolant Accident (LOCA) while in operation. This study analyses the impact of varying the geometry of the distance between the freeze plug and cooling fins on the opening time. Calculations were performed using a particle-based Lagrangian numerical simulation called MPS, Moving Particle Semi-implicit. The calculations were conducted using two-dimensional geometry, with the initial design being an inline arrangement. The MPS improvement involves forced convective heat transfer between the liquid fuel, the freeze plug, and the cooling fins. At the same time, conductive calculations were performed among the freeze plug particles during melting. The results show that a larger distance between the fin and plug has an insignificant effect on the opening time. This study is essential to determine the variations that yield optimal values for supporting the safety system of molten salt reactors.

1. Introduction

The need for advanced nuclear reactors, especially Molten Salt Reactors (MSRs), to minimize the environmental impact is in line with the increasing global demand for energy. MSRs are of great interest because they are safe, work at high thermal efficiencies, and can have closed fuel cycles [1]. The management system of the reactor's temperature is a critical feature of MSR technology that would ensure effective, safe operation and efficient heat removal produced by fission reactions. Combining cooling fins and freeze plugs is an innovative solution to enhance thermal management in MSRs. Cooling fins have been used for a long time to improve the surface area of heat transfer, thereby increasing heat removal from the reactor core to the ambient environment. On the other hand, freeze plugs are passive safety features designed to melt at certain temperatures to release molten salt into a safe containment in case of a temperature excursion [2]. These cooling fins enhance the function of the freeze plugs further, providing a dual benefit of improved heat transfer efficiency during normal operations and acting as a reliable safety measure during accidents. In the process, these features integrated with MSRs will significantly

IOP Conf. Series: Materials Science and Engineering **1330** (2025) 012006 doi:10.1088/1757-899X/1330/1/012006

enhance their safety and performance. This paper investigates the design, implementation, and performance of cooling fins incorporated with freeze plugs in a molten salt reactor.

The Moving Particle Semi-Implicit (MPS) method has emerged as a Lagrangian mesh-free computational technique for simulating fluid dynamics and heat transfer problems, particularly in complex geometries and free surface flows[3]. Developed initially for incompressible fluid flow analysis, the MPS method has been extensively adapted and improved for various applications, including thermal management systems in nuclear reactors. However, the integrated design study of cooling fins and freeze plugs using the MPS method has not been conducted yet. Therefore, this study aims to simulate the performance of integrated design cooling fins and freeze plugs when constant high-temperature molten fuel comes from the core. The freeze plugs and cooling fins material were $LIF\text{-}ThF_4$ and Hastelloy N, respectively. The frozen molten salt opening time was analyzed by including a fundamental parameter: delta, the distance between the cooling fins and freeze plug. The results were compared to the reference, which utilized the Finite Element Method FEM [4], and further analyzed.

2. Method

2.1 Integrated Design of Cooling Fins and Freeze Plug

Figure 1 illustrates the integration design of cooling fins and freeze plug. The feature automatically opens when abnormal conditions are detected from the reactor core. When a fluid flows over a surface, a thin layer forms right next to the boundary where the velocity decreases gradually from the fluid's free-stream velocity to zero [5]. Within this boundary layer, heat transfer occurs primarily by conduction due to the reduced fluid velocity. Because convective heat transfer is more efficient than conductive transfer, this boundary layer poses a resistance to heat transfer. Consequently, the thermal conductivity of the fluid and the thermal boundary layer's thickness significantly impact the heat transfer coefficient, which directly measures the resistance

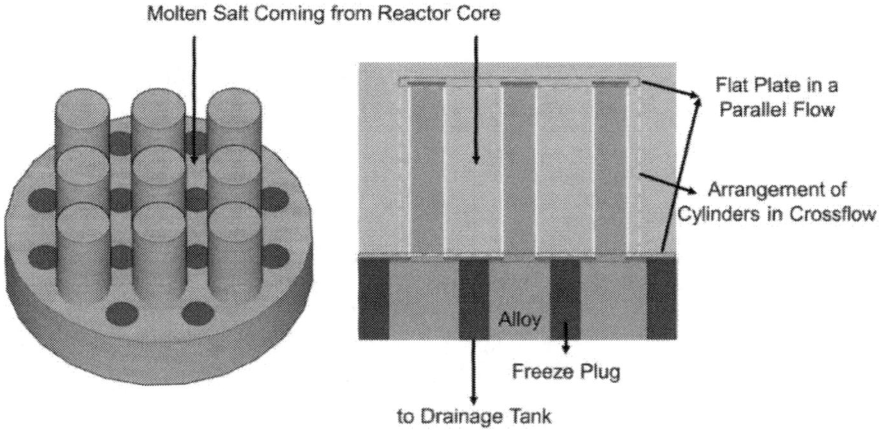

Figure 1. Illustration of cooling fins and freeze plug integrated design in 3D (left) and 2D (right) with the approximation flow happened near the interface

to heat transfer across the surface. Thus, in this case, both conduction and convection must be carefully considered in the calculation.

Forced convection heat transfer is modeled by applying a heat flux in direct contact with the flowing fluid at the boundaries [6]. In this case, the forced convection is divided into two forms: a flat plate in a parallel flow and an arrangement of cylinders in a crossflow [7], [8]. Equations (1) and (2) represent the average Nusselt number of the freeze plug's top surface and the heat transfer coefficient in parallel flow, respectively. This Equation is valid for Re $< 3 \cdot 10^5$. Meanwhile, in this preliminary study, the inline arrangement of the cylinder is used to arrange the cooling fins, and its Nusselt number is shown by equation (3).

$$\langle Nu \rangle_{top} = \frac{1}{L}\int_0^L Nu_x \, dx = 0.221 \left(\frac{\rho v D}{\mu}\right)^{1/2} \left(\frac{v}{\alpha}\right)^{1/3} \quad (1)$$

$$\langle k \rangle_{top} = \langle Nu \rangle_{top} \frac{\lambda}{D} \quad (2)$$

$$\langle Nu \rangle_{fin} = C(Re)^{1/2}(Pr)^{1/3}$$

$$\text{with } C = \left(0.25 + e^{-0.55\,\zeta_L}\right)\zeta_L^{\;0.212}\zeta_T^{\;0.285} \quad (3)$$

2.2 Numerical Simulation Method

The Moving Particle Semi-Implicit (MPS) method is a mesh-free computational approach used primarily for simulating incompressible fluid dynamics and heat transfer problems. It represents the fluid domain using discrete particles, each carrying physical properties such as mass, velocity, and temperature. The interactions between these particles are governed by conservation in fluid

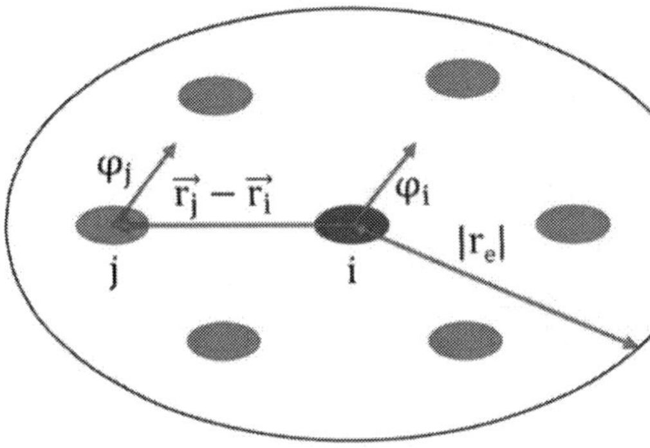

Figure 2. Particle interaction within r_e in MPS method

mechanics and heat transfer principles, as shown in equations (4) and (5), respectively. Meanwhile, the incompressible flow is represented by Equation (6) of constant fluid density.

$$\frac{D\vec{u}}{Dt} = -\frac{1}{\rho}\nabla P + \nu\nabla^2\vec{u} + \vec{F} \quad (4)$$

$$\frac{DT}{Dt} = \frac{1}{\rho C_p}(k\nabla^2 T + Q) \quad (5)$$

$$\frac{\partial\rho}{\partial t} + \nabla.(\rho\vec{u}) = 0 \quad (6)$$

The interaction of particles in the MPS method is delineated in Figure 2. Mathematically, the interaction is defined using a kernel function w(r), Equation (4), which determines the influence of neighboring particles based on their distance. The w(r) is in quadratic form to avoid the singularity at the target particle. All discretization of mathematical operations, such as gradient, divergence and Laplacian are adopted from the previous research [9].

$$w(r) = \begin{cases} \left(1 - \frac{r}{r_e}\right)^2 & 0 < r \leq r_e \\ 0 & r_e \leq r \end{cases} \quad (7)$$

Figure 3 illustrates the modified MPS flowchart. For each simulation time increment, the PND for each particle is calculated, followed by velocity and position corrections, particularly for the moved fluid particles. Subsequently, calculations are performed on each particle involved in the

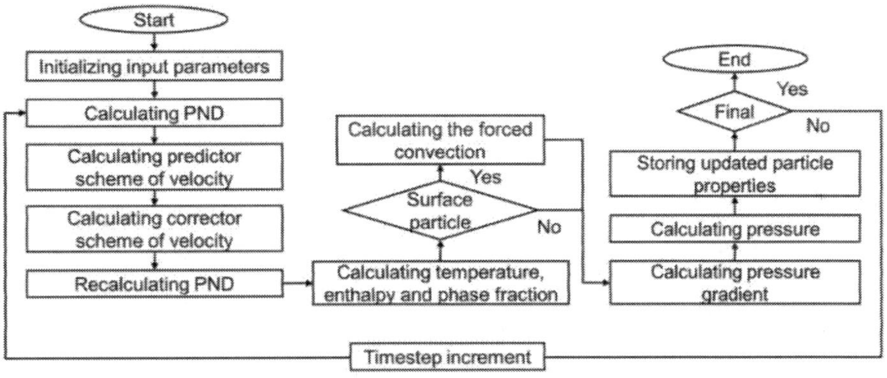

Figure 3. Initial condition of improved MPS simulation

heat transfer process. Equation (5) is divided into two components: the conduction term and the convection term. The conduction term is represented by Laplacian heat transfer. At the same time, the forced convection component involves the heat flux transfer from molten salt to Hastelloy N and the freeze plug at the surface. Finally, the pressure of each particle is calculated. The computation ends once the required time is achieved.

Figure 4 depicts the initial geometry during the MPS simulation. The colors red, blue, grey, and black refer to particles of molten salt, freeze plug, Hastelloy N, and dummy wall, respectively. The dummy wall particles represent a boundary separating the actual geometry from the void region; no pressure and temperature calculations will be performed in this region. This configuration has the adiabatic conditions around the freeze plug. The molten salt temperature is taken at a constant 973 K throughout the simulation. This value is the highest safety limit

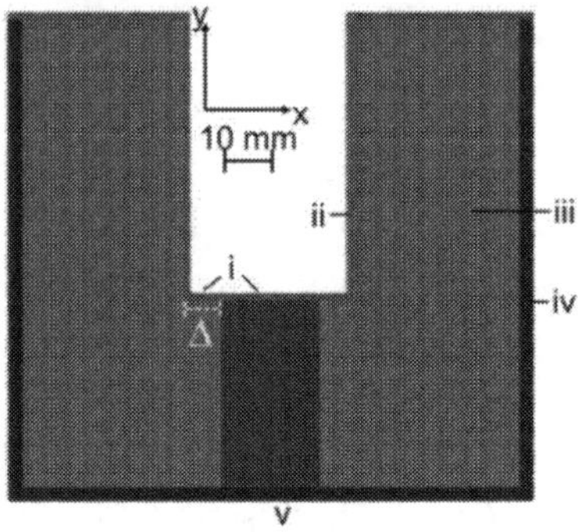

Figure 4. Initial condition of MPS simulation.

Table 1. Parametric survey

Parameter	Value
Cooling fins	Hastelloy N
Freeze Plug	$LiF\text{-}ThF_4$
P (mm)	70
H_{fin}	60
D_{fin}	35
H_{plug}	40
D_{plug}	14-35
PDR	2-5
S_L	70
ζ_L	2
ζ_T	2
Δ	0-10.5

acceptable for the molten salt reactor project. The simulation proceeds up to 300 s. The survey parameter ranges from 0 mm to 10.50 mm as shown in Table 1. The initial temperature of the other particles is set at 77 K according to the reference under comparison. The figure indicates the numbers with the following simulation conditions:

(i) The heat flux in this region is implemented as forced convection using Equation (2) and Newton's law of cooling. $\phi_q'' = k\ (T_\infty - T_s\ )$.

(ii) The heat flux in this region is calculated from the forced convection across a tube bank using Equation (3) to determine the Nusselt number.

(iii) Conduction occurs between the alloy particles and the frozen salt.
(iv) The dummy wall is used to implement adiabatic conditions.
(v) In addition to being thermally insulated, the molten freeze plug is regarded as stationary, thus interactions with the air beneath the plug are neglected.

3. Results and Discussion

Figure 5 shows the simulation result of melting the freeze plug with a delta value of 7 mm. During the first 100 seconds, the major contribution to the phase change is from the molten salt at the upper surface of the freeze plug and the liquid phase concentrated in the middle part of the plug. It should be noted that the frozen salt, molten salt, and mushy particles are represented as blue, turquoise, and red, respectively. This indicates that the phase change at the plug's edges was hastened after the seconds reached 100 due to the heat transfer from the cooling fins. The mingling of the cooling fins with the molten salt increased energy absorption from the molten salt. Since Hastelloy N had good conductivity, it could rapidly transfer heat to the plug and thus melted its edges quite fast. Figure 6 verified the result from Figure 5, which was the temperature

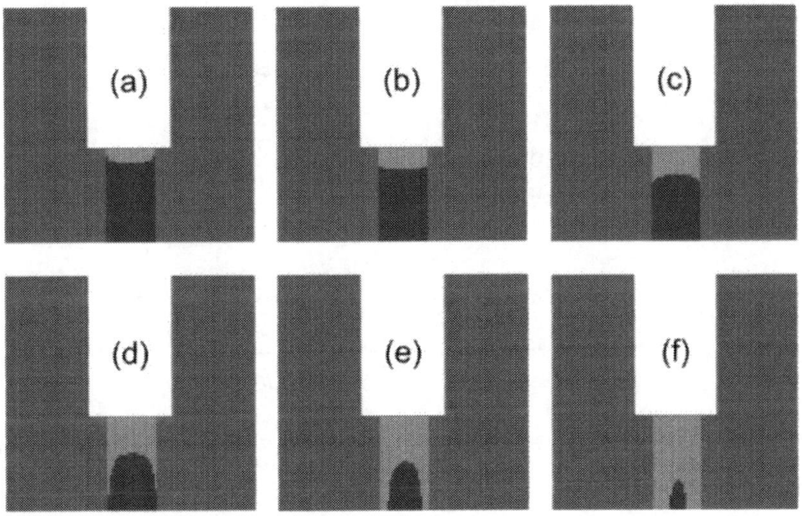

Figure 5. Simulation result of melted freeze plug based on particle changing at (a) 50 s (b)100 s (c) 150 s (d)200 s (e) 250 s and (f) 300 s.

distribution of the plug's melting process. The temperature of the surface of the plug and the cooling fin's surface increased significantly in the initial 150 s. This occurs due to the dominant effect of the forced convection from the molten salt to the surface. In the meantime, the conduction between the plug particles and the Hastelloy N appears with the temperature gradient caused by the forced convection.

Figure 7 shows the comparison of opening times of the freeze plug at different delta values between the MPS method and reference using the FEM method. Both showed that the delta value significantly affected opening time. The higher PDR value increased the speed of opening of the freeze plug. If there were no gap between the cooling fins and the freeze plug, the opening time would increase due to the forced convection reduction at the Hastelloy N surface surrounding the freeze plug. However, it should be noted that increasing the delta value may not lead to a significant acceleration of the opening plug. An increased delta value for constant P-value decreased D_{plug}, then prolonging reactor core draining time. However, this study revealed the

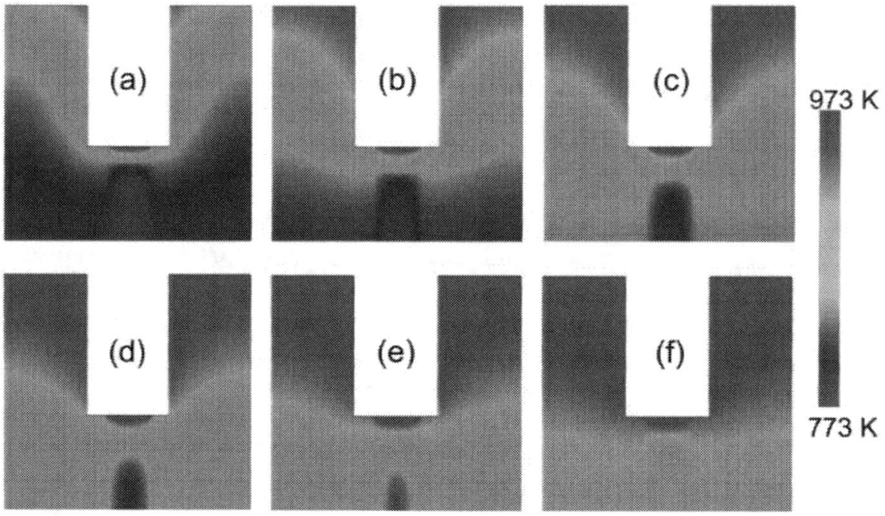

Figure 6. Temperature distribution during simulation at (a) 50 s (b)100 s (c) 150 s (d)200 s (e) 250 s and (f) 300 s.

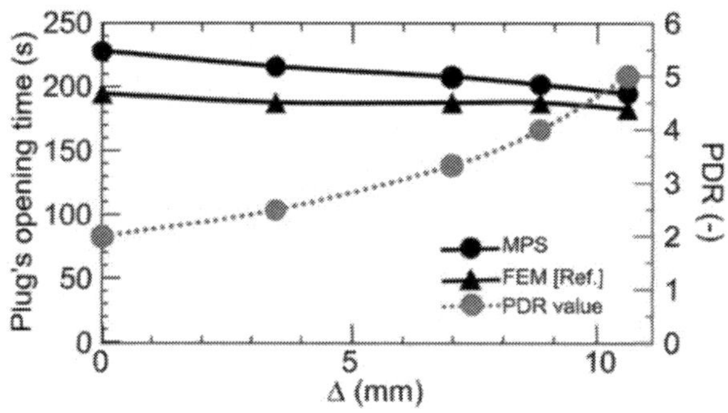

Figure 7. Comparison between current study and reference of freeze plug's opening time.

relative error to the reference around 6.55% - 16.92%. Both methods showed good agreement and produced a freeze plug opening time of less than 4 minutes. The duration was considered as a safe within the limits for fuel mitigation.

Figure 8 (a) and (b) depicts the simulation results for the phase change at 200 s in the center and the edge of the freeze plug. The indicator value of 1 and 0 denoted the liquid and solid phase, respectively. Therefore, the left side and the right side of the line represented the portion of the melted and remaining plug, respectively. The center position's phase change was observed to be slower compared to the edges at this time. At a delta value of 10.50 mm, the edges have melted, meanwhile more than half of the plug's height in the center remained frozen. For delta values of

8.75 mm, 7.00 mm, 3.50 mm, and 0.00 mm, the phase percentage in the center of plug were same. However, at the plug's edge, a smaller delta value resulted a less phase change. This observation

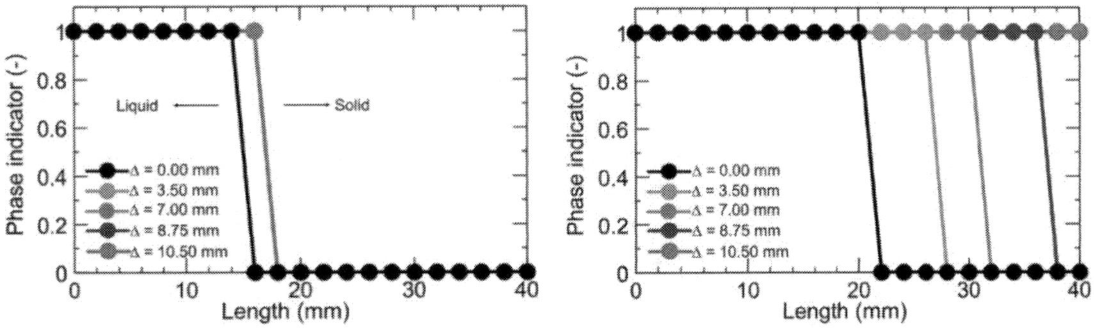

(a) (b)

Figure 8. Phase indicator at t = 200s (a) center of plug (b) edge of plug

was consistent with the plug opening time shown in Figure 7, which indicates that conduction became the dominant process towards the end of the simulation, influencing the phase transition.

4. Coclusions

A 2D simulation of an inline arrangement of cooling fins integrated with a freeze plug opening was conducted, yielding relative errors ranging from 6.55% to 16.92% when compared with reference data. The findings highlight the critical role of the distance (Δ) between cooling fins and the freeze plug in enhancing forced convection heat transfer during the initiation stage. However, the opening time did not decrease significantly. These results underscore the importance of optimizing this parameter to improve system efficiency. Future research should focus on reducing relative errors by varying particle sizes and incorporating non-constant temperature values to better capture residual heat behavior during abnormal or accident conditions. Additionally, integrating the analysis with reactor core draining dynamics to the drainage tank would provide a more comprehensive understanding of freeze plug behavior and its implications for reactor safety. This work contributes valuable insights into the design and operation of molten salt reactors under severe conditions..

References

[1] J. Serp *et al.*, "The molten salt reactor (MSR) in generation IV: Overview and perspectives," *Prog. Nucl. Energy*, vol. 77, pp. 308–319, Nov. 2014, doi: 10.1016/J.PNUCENE.2014.02.014.

[2] B. M. Chisholm, S. L. Krahn, and A. G. Sowder, "A unique molten salt reactor feature – The freeze valve system: Design, operating experience, and reliability," *Nucl. Eng. Des.*, vol. 368, p. 110803, Nov. 2020, doi: 10.1016/J.NUCENGDES.2020.110803.

[3] S. Koshizuka, K. Shibata, M. Kondo, and T. Matsunaga, "Chapter 2 - Fundamentals of Fluid Simulation by the MPS Method," S. Koshizuka, K. Shibata, M. Kondo, and T. B. T.-M. P. S. M. Matsunaga, Eds. Academic Press, 2018, pp. 25–109.

[4] F. R. Deurvorst, "Design of Cooling Fins on the Freeze Plug," 2017, [Online]. Available: https://api.semanticscholar.org/CorpusID:199536467.

[5] S. Aliu, O. M. Amoo, F. I. Alao, and S. O. Ajadi, "Mechanisms of heat transfer and boundary layers," *Appl. Heat, Mass Fluid Bound. Layers*, pp. 23–53, Jan. 2020, doi: 10.1016/B978-0-12-817949-9.00010-4.

[6] B. Zohuri, "Forced Convection Heat Transfer BT - Thermal-Hydraulic Analysis of Nuclear Reactors," B. Zohuri, Ed. Cham: Springer International Publishing, 2017, pp. 323–345.

[7] A. Mostafavi and A. Jain, "Theoretical analysis of unsteady convective heat transfer from a flat plate with time-varying and spatially-varying temperature distribution," *Int. J. Heat Mass Transf.*, vol. 183, p. 122061, Feb. 2022, doi: 10.1016/J.IJHEATMASSTRANSFER.2021.122061.

[8] D. Sumner, "Two circular cylinders in crossflow: A review," *J. Fluids Struct.*, vol. 26, no. 6, pp. 849–899, Aug. 2010, doi: 10.1016/J.JFLUIDSTRUCTS.2010.07.001.

[9] A. N. Hidayati, A. P. A. Mustari, D. Irwanto, and A. Waris, "Experimental and numerical analysis of impingement temperature effect on the sacrificial plate's hole formation during nuclear severe accident," *Prog. Nucl. Energy*, vol. 162, p. 104782, Aug. 2023, doi: 10.1016/J.PNUCENE.2023.104782.

IOP Conf. Series: Materials Science and Engineering **1330** (2025) 012007 doi:10.1088/1757-899X/1330/1/012007

Study of Prospecting NPP Cogeneration as District Heating and Cooling in Indonesia

A Muksin Aji Setiawan[1,2*], Sidik Permana[1] and Topan Setiadipura[2]

[1]Master Program in Nuclear Science and Engineering Department,Faculty of Mathematics and Natural Science, Bandung Institute of Technology, Jalan Ganesha 10, Bandung, 40132, Indonesia
[2] Research Center for Nuclear Reactor Technology, Research Organization of Nuclear Energy, National Research and Innovation Agency, PUSPIPTEK Area Building 80, 15310, Banten, Indonesia

*E-mail: muks006@brin.go.id

Abstract. Metropolitan areas like Jakarta are often associated with dense populations and significant energy consumption challenges. During working hours, the city sees the movement of up to 4 million people, creating a distinct energy demand pattern compared to surrounding satellite areas. Energy usage peaks during the summer months due to the high demand for air conditioning, highlighting the need for a stable electricity supply. Jakarta's current reliance on coal as its primary energy source becomes problematic when supply chains are disrupted or coal becomes economically unsustainable. Moreover, coal usage contributes to long-term environmental degradation. In alignment with the government's target of achieving net-zero carbon emissions, there is a growing urgency to adopt renewable energy alternatives. Nuclear power presents itself as a viable solution for Indonesia, given its high energy output. The cogeneration capability of Nuclear Power Plants (NPPs) can be utilized for district heating and cooling (DHC), decreasing the city's reliance on fossil fuels, particularly for cooling. This shift could significantly cut down on air pollution caused by fossil fuel combustion, thus enhancing air quality in Jakarta. This study explores the feasibility of integrating NPPs in major Indonesian cities, factoring in geographical, demographic, economic, and environmental aspects. Ultimately, the initiative aims to help Jakarta progress toward a cleaner and more sustainable urban environment.

1. Introduction

The accelerating impacts of global warming are becoming increasingly tangible, particularly in equatorial nations such as Indonesia. As global temperatures rise, urban centers face compounded challenges from extreme heat events to declining air quality further stressing the infrastructure and energy systems of densely populated areas[1]. Indonesia, a low-middle income countries with a population nearing 300 million, is witnessing a rapid increase in energy demand driven by economic growth, industrial development, and internal migration[2]. This demand is disproportionately concentrated on Java Island, especially in Jakarta, the country's economic and administrative epicenter[3].

IOP Conf. Series: Materials Science and Engineering **1330** (2025) 012007 doi:10.1088/1757-899X/1330/1/012007

Jakarta as a city experiences intense human mobility, with up to 4 million commuters moving in and out of the city daily during working hours. This large-scale population movement, combined with the effects of urbanization and high ambient temperatures that can reach up to 40°C, creates spikes in electricity demand particularly for cooling systems. The current electricity system, dominated by fossil fuels such as coal, is increasingly strained under these peak conditions. Indonesia's heavy reliance on coal not only poses a risk in terms of supply stability and economic volatility but also contributes significantly to air pollution and carbon emissions, exacerbating climate-related challenges[4][5].

Transportation-related emissions further aggravate the problem, as limited public transportation infrastructure drives increased use of private vehicles and ride-hailing services. The heating and cooling sector, particularly space cooling in tropical urban environments, accounts for a significant portion of energy consumption globally reaching around 50%, with space and water heating/cooling alone comprising nearly 79%[3][6][7]. This pattern highlights the urgent need for alternative, cleaner energy sources capable of supporting large-scale, stable, and efficient power and thermal energy supply.

While renewable energy sources like solar and wind are critical for Indonesia's long-term energy transition, their intermittency and land-use requirements pose challenges for dense cities like Jakarta. In this context, nuclear energy presents a compelling alternative due to its high capacity factor, stable output, and minimal land footprint. Nuclear Power Plants (NPPs) offer not only reliable, carbon-free electricity but also the potential for cogeneration, where thermal energy byproducts are utilized for secondary applications an area of growing interest in sustainable urban energy planning[8][9][10]. This general aplication of NPPs shown in **Figure 1**.

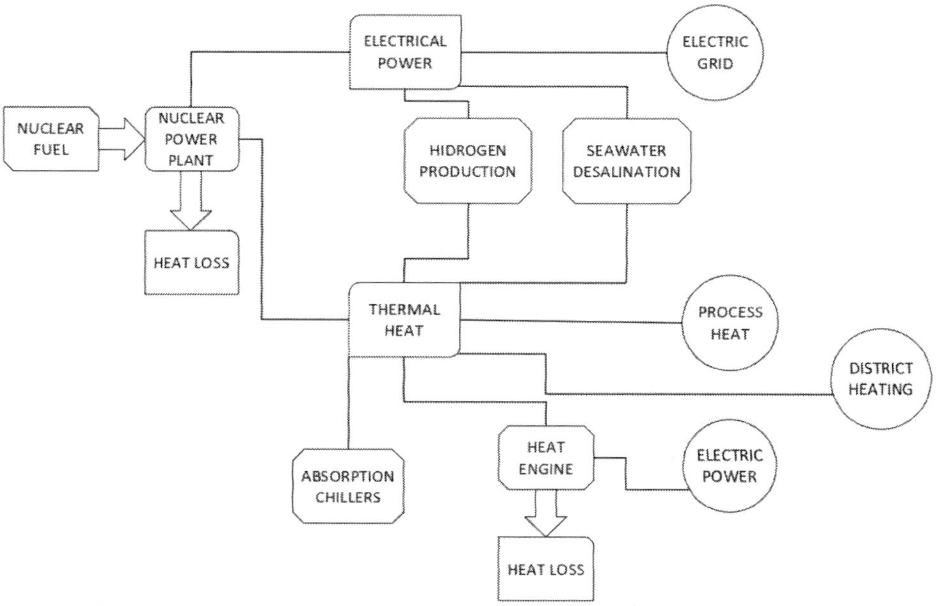

Figure 1. NPP Cogeneration application

One of the most promising applications of cogeneration is District Heating and Cooling (DHC), which distributes thermal energy for heating and cooling multiple buildings from a centralized source[1][11]. Coegeneration aspect base on temperature operation range of the NPPs in various type, and its dominant application are described in **Figure 2**. In Jakarta's tropical climate, the cooling function of DHC systems is especially relevant. When powered by nuclear cogeneration, DHC systems can provide large-scale, energy-efficient cooling services, reducing the demand on conventional air conditioning systems, cutting fossil fuel use, and improving air quality in urban environments. The use of waste heat from NPPs not only increases overall energy efficiency but also supports emissions reduction strategies crucial for Indonesia's net-zero carbon targets [12].

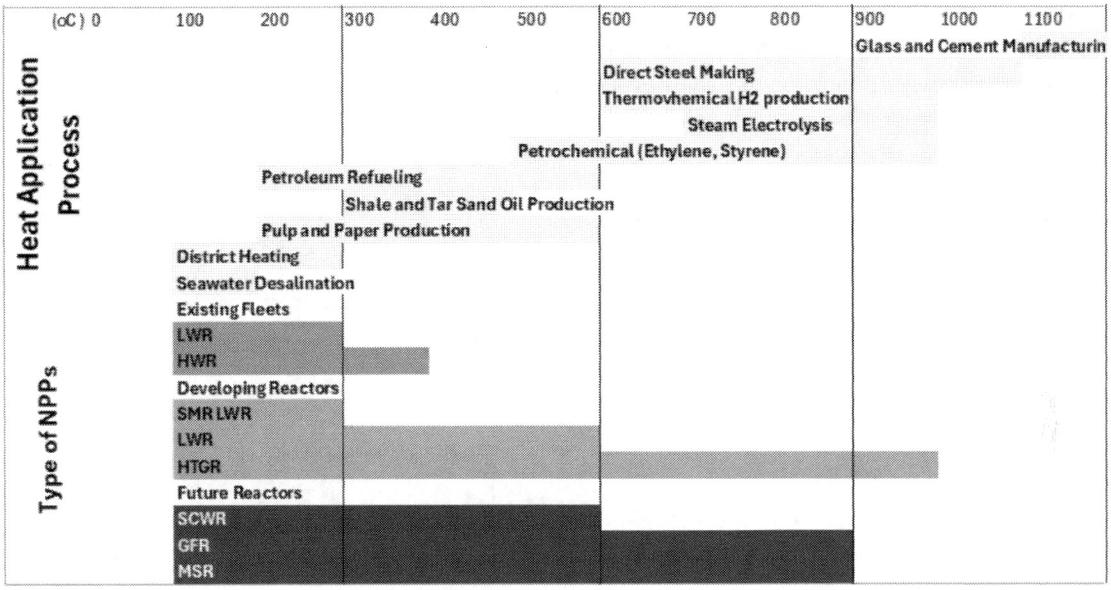

Figure 2. Temperature Range of Different type of NPP and Heat Application Process

This study explores the feasibility of implementing nuclear-powered DHC systems in Jakarta, using the city as a case study for high-density urban environments in tropical developing nations. Through a comprehensive literature review, this research assesses global examples of NPP-based DHC systems, evaluates their potential environmental and economic benefits, and considers their applicability within the Indonesian context. The insights gained aim to inform Jakarta's energy transition strategy and could serve as a foundation for future implementation in the country's new capital city in Kalimantan.

2. Methodology

This research follows a systematic approach, beginning with the collection of relevant data from diverse sources. These sources primarily focus on the utilization of DHC systems in various countries, providing essential reference material for the study. Once the data is gathered, a comprehensive analysis is performed to identify both the benefits and limitations of implementing DHC systems in large urban settings, with a specific emphasis on Indonesia. This analysis offers valuable insights into the operational aspects of DHC systems, highlighting their advantages and challenges in the context of Indonesian cities.

The next step is to gain a thorough understanding of how DHC systems function and the unique obstacles they may face when implemented in large Indonesian cities. This understanding is crucial for developing effective policies and strategies tailored to the specific needs of these urban areas. Despite potential limitations in the available data, researchers proceed to draw informed conclusions and hypothesize about the effectiveness of certain policies or strategies for implementing DHC in Indonesia. These conclusions and hypotheses provide the foundation for further research and subsequent decision-making processes.

The ultimate goal of this research is to utilize the collected information to make well-informed decisions and provide recommendations regarding the implementation of DHC systems in Indonesia. This process takes into account insights from data analysis and the hypotheses formed. It is important to note that limitations in the data should not hinder hypothesis formulation; instead, researchers are encouraged to work creatively within the available data to generate valuable insights and recommendations. This flexible approach ensures that gaps in the data are addressed, allowing for the exploration of the feasibility and effectiveness of DHC systems in large Indonesian cities.

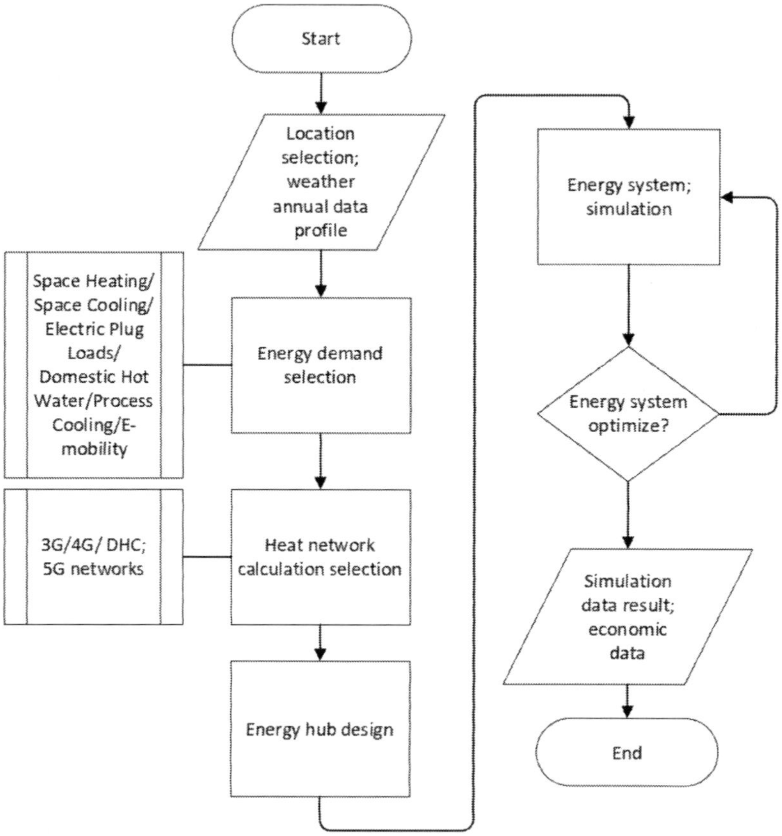

Figure 3. Flow Chart of The Simulation

The research process incorporates a simulation workflow, as depicted in **Figure 3**. A simple simulation is conducted using the nPro Tools calculation program, which is a web-based application designed to estimate optimal scenarios for district heating designs, required investments, and energy systems[13]. It is specifically designed for the development of heating, cooling, and Fifth-generation District Heating and Cooling (5GDHC)/energy networks, ensuring high efficiency in modern energy system design. By incorporating annual weather data, the simulation provides an overview of the conditions in the target area. This data includes variables such as ambient air temperature, wind speed at 10 meters, global horizontal irradiance, direct normal radiation, soil surface temperature, and other climatological components needed to simulate energy requirements for designing a district heating system. Data for the simulations can be obtained from platforms like Ladybug Tools, EnSimS, EnergyPlus.net, and Climate.OneBuilding.org, which offer free downloadable weather data for various locations worldwide.

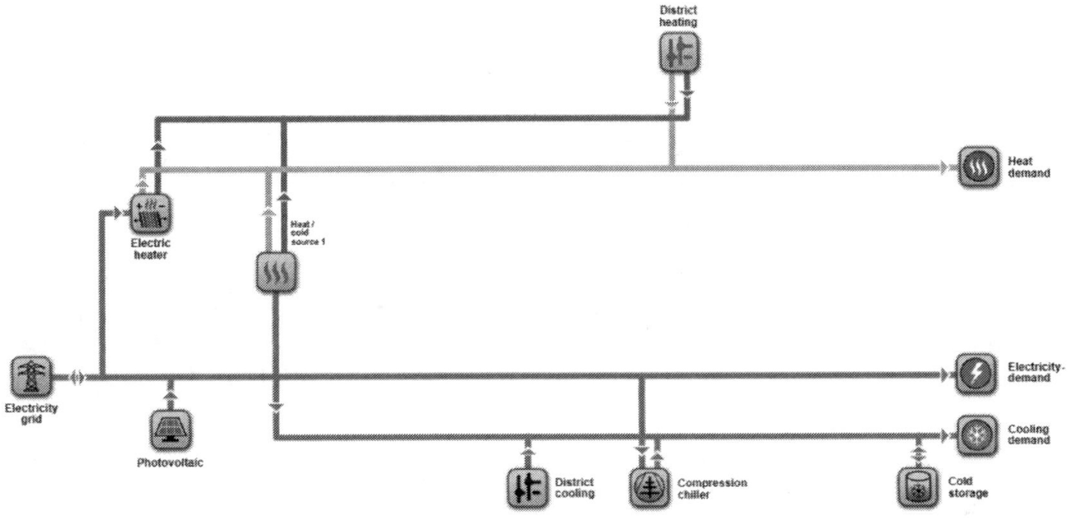

Figure 4. Initial Simulation for Design of Energy System

Figure 4 illustrates the simulation scenario. The simulation compares two different scenarios: Scenario 1, which includes both heat and cooling demands, and Scenario 2, which prioritizes only cooling demands. As the simulation is designed to model the application of NPPs in a real-world scenario, adjustments are made due to the absence of a dedicated NPP component in the software. Instead, the heat/cold source is used to represent the heat produced by NPP cogeneration. The heat generated by the NPP is assumed to be converted into heat in the heat/cold storage system. Additionally, the system is integrated with photovoltaic components to accommodate the combination of renewable energy sources with nuclear-generated heat.

3. Result and Discussion

3.1 temperature rising, air polution and nuclear cogeneration

Air pollution, a critical issue for urban environments, significantly impacts various aspects of human life, both directly and indirectly[14]. The changing climate, reflected in altered temperature, precipitation, air pressure, and cloudiness, exacerbates the problem [15]. In particular, the demand for energy especially for heating and cooling shows a direct correlation with outdoor temperatures. As the climate warms, the need for energy increases, driving a seasonal fluctuation in energy consumption [15]. Fossil fuel reliance is a major contributor to this phenomenon, as its burning to meet energy demand directly increases air pollution[16]. This interdependency between human activity, energy use, and environmental change is illustrated in **Figure 5**.

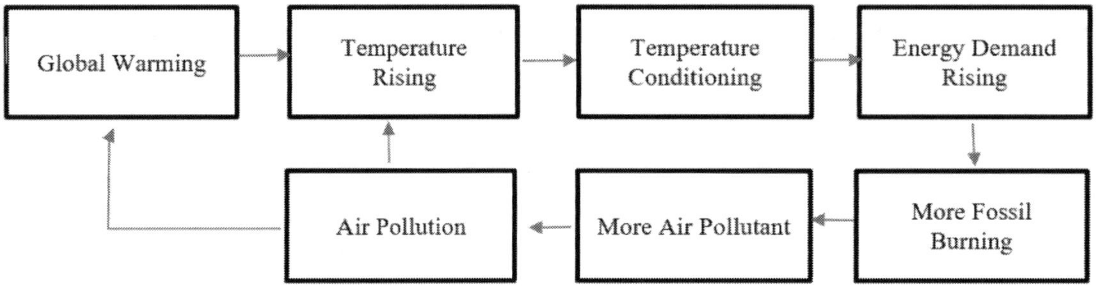

Figure 5. Simple Correlation Global Warming and Air Pollution

Transportation, as a significant contributor to ozone concentration, further interacts with factors such as air temperature, wind speed, and solar radiation to shape air quality [17] [18]. The complexity of these environmental variables necessitates a comprehensive understanding when analyzing the impacts of energy systems on urban air quality[19]. The Covid-19 pandemic lockdown highlighted a temporary shift in energy patterns, with a focus on renewable energy as a means of economic recovery. However, this shift was less pronounced in a low-middle income countries, where renewable energy adoption faces unique challenges [2][20].

Empirical evidence suggests that nuclear energy could play a crucial role in mitigating air pollution and reducing CO_2 emissions. As countries globally work to reduce their reliance on fossil fuels, the transition to cleaner energy sources must accelerate [12][21]. Nuclear energy, with its ability to generate consistent, low-carbon electricity, emerges as a key solution. Notably, the implementation of cogeneration in NPPs has shown promise in enhancing energy efficiency, as NPPs can supply both electricity and heat, reducing reliance on fossil fuels and improving air quality [22–25].

This study explores the potential benefits of implementing DHC systems powered by NPPs in Jakarta. Given the city's high population density, limited land availability, and significant cooling demand, the research examines the technical and economic feasibility of integrating NPP-based cogeneration systems with DHC in urban areas. The use of Small Modular Reactors (SMRs) is identified as a promising option due to their flexibility, low power output, and smaller land requirements. While the study does not focus on the construction of NPPs near Jakarta, it highlights the broader potential of NPP cogeneration in addressing the city's energy challenges, not only for electricity generation but also for meeting heating and cooling needs.

The implementation of DHC systems powered by nuclear cogeneration presents several key benefits. These include a reduction in carbon emissions, contributing to Jakarta's climate goals, and a decreased reliance on fossil fuels, which often depend on volatile international markets. Additionally, DHC systems can improve building energy ratings, enhancing property values and promoting the adoption of more sustainable practices within the city [7]. Furthermore, a continuous, reliable heat supply reduces maintenance needs, contributing to the long-term economic sustainability of the system[26]. With the integration of NPP-based cogeneration, Jakarta can move toward cleaner, more efficient energy systems that align with national and global decarbonization efforts.

However, successful implementation requires addressing several challenges. The selection of appropriate land for the nuclear plant must consider socio-economic factors, transportation infrastructure, disaster resilience, and community acceptance[1][7]. Additionally, regulatory frameworks, safety protocols, and the integration of renewable energy sources must be considered to ensure the sustainability of the system[27]. While NPPs provide a reliable energy source, the long-term planning and financial investments necessary for such projects cannot be overlooked. Careful financial planning, including cost comparisons with existing cooling technologies, is essential to ensure the economic viability of DHC systems [28].

The benefits of cogeneration in NPPs extend beyond electricity generation, with the potential for waste heat recovery, which is particularly relevant in tropical countries like Indonesia. DHC systems, especially when integrated with high-temperature storage solutions, can mitigate urban heat island effects and reduce reliance on traditional cooling methods[8][29]. During extreme heat events, the implementation of DHC can not only reduce energy consumption but also protect public health by mitigating heat-related deaths [1][14]. This highlights the importance of good system design, including considerations for cooling methods such as absorption chillers, mechanical chillers, and cold storage [30].

Finally, the implementation of DHC as part of a broader energy strategy could be pivotal for decarbonization in densely populated urban areas like Jakarta. By transitioning to cleaner, more sustainable energy systems, the city could reduce its carbon footprint, improve air quality, and create economic opportunities[28]. However, the successful deployment of DHC systems will require careful planning, financial investment, and long-term regulatory certainty[31]. The integration of SMRs and cogeneration into Jakarta's energy infrastructure could serve as a model for other rapidly urbanizing cities in Indonesia and beyond [1][31] [32].

3.2 Economic Evaluation and System Efficiency

Reliability remains a critical consideration in thermal energy systems, necessitating some level of backup heat capacity[11]. In District Cooling (DC) systems, chilled water is typically distributed with a design temperature differential of approximately 10°C between the supply and return streams [15]. This temperature range must be maintained to ensure optimal performance and energy efficiency. To assess the economic viability of different thermal energy systems, including DHC powered by nuclear cogeneration, the Levelized Cost of Energy (LCOE) and Levelized Cost of Heat (LCOH) provide comprehensive metrics [33,34]. These indicators account for all cost parameters over the life cycle of a power plant, including initial investment (I_a), annual maintenance (M_a), operational expenses (O_a), revenues (R_a), discount rate (r), and energy output (Q_a), as shown in Equation (1):

$$\frac{LCOH}{LCOE} = \frac{\sum_a^{a_{end}} (I_a + M_a + O_a - R_a)(1+r)^{-a}}{\sum_a^{a_{end}} Q_a (1+r)^{-a}} \quad (1)$$

Net Present Value (NPV), meanwhile, measures the total value of projected cash flows (CF_a) discounted over time, as formulated below:

$$NPV = \sum_a^{a_{end}} CF_a \, (1+r)^{-a} \quad (2)$$

In terms of system performance, the overall efficiency of a cogeneration system can be expressed as the ratio of the sum of net electricity supplied to the grid and thermal energy used for DHC to the total heat generated in the reactor [11]. Energy flows in DHC systems are inherently bidirectional and often involve multiple energy sources, requiring adaptive control strategies. To avoid over-dimensioning the heating and cooling infrastructure, co-simulation techniques integrating district energy models with building-level simulations are beneficial. These simulations help evaluate the feasibility of potential customer connections and their influence on overall thermal efficiency. However, a trade-off must be managed between simulation accuracy and computational efficiency [35].

3.3 Technological Evolution of District Heating and Cooling Systems

The advancement of technology and the application of cogeneration provide a unique selling point for a new generation nuclear power plant model. District Heating and Cooling (DHC) systems have undergone significant evolution over the past decades. The Fifth Generation District Heating and Cooling (5GDHC) is regarded as a breakthrough in enabling low-temperature, highly efficient, and renewable-integrated thermal networks [6] [36]. An overview of the intended application can be seen in **Figure 6**. One key advantage of 5GDHC is its compatibility with decentralized energy production, bidirectional flows, reversible heat pump substations, and real-time metering, contributing to a thermal efficiency index of 1.69. In this context, electricity accounts for approximately 19% and thermal integration for 41% of total energy needs [37–40].

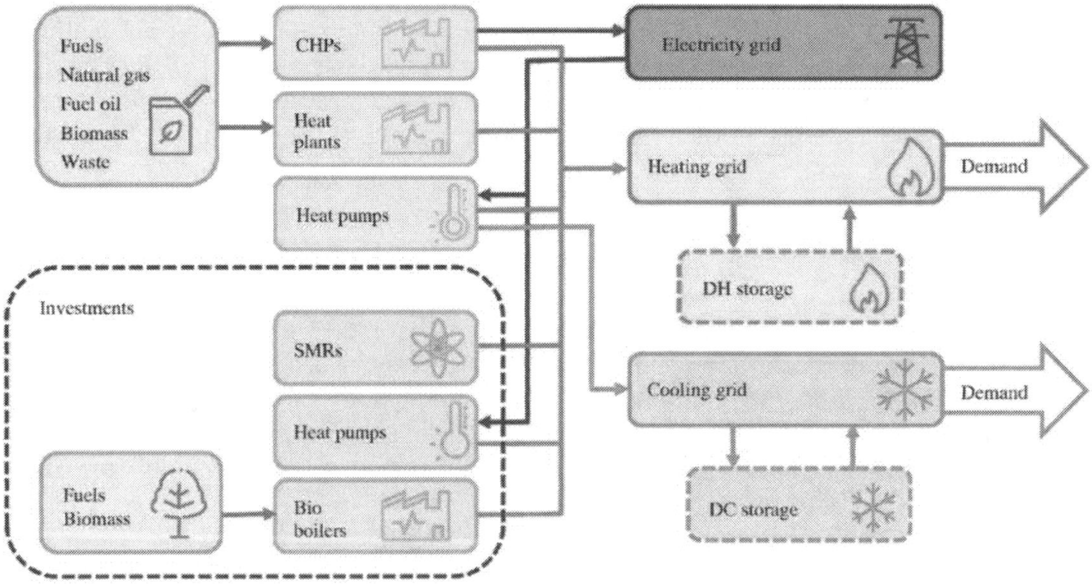

Figure 6. Simplified Structure of DHC Energy System Model [26]

The main characteristics of each generation of district heating are summarized as follows [41]:

a. 1st Generation (1GDH) High-temperature steam (<200°C), typically from coal-fired plants.
b. 2nd Generation (2GDH) Pressurized hot water (>100°C), mainly fossil-fueled.
c. 3rd Generation (3GDH) Hot water (~100°C), integrated with CHP and heat storage.
d. 4th Generation (4GDH) Low-temperature (50–80°C), focused on renewables and waste heat.
e. 5th Generation (5GDH) Ultra-low temperature, decentralized, integrated with heat pumps, and designed for thermal sharing.

While promising, 5GDHC implementation faces challenges. For example, in the Baltic region, existing 3GDH infrastructure covers a large portion of the population, making the transition to 5GDHC economically demanding due to the significant upfront investment required [42,43].

3.4 Regional Case Studies and Potential

A successful case study in Singapore demonstrates the viability of optimized District Cooling Systems (DCS) in tropical climates. The simulation model developed provided a new framework

IOP Conf. Series: Materials Science and Engineering **1330** (2025) 012007 doi:10.1088/1757-899X/1330/1/012007

for design optimization, balancing technical and economic objectives. The optimized DCS scenario reduced life cycle costs by 7% to 21%, yielding savings of 30 to 107 million SGD, and shortened the payback period by over 20%. These improvements were achieved by leveraging central chiller plants and thermal energy storage. Additionally, integrating solar PV technologies further supports carbon reduction and operational efficiency [10][15].

Globally, the outlook for district cooling is increasingly positive. In regions with high cooling demands, such as the Middle East, and in environmentally conscious regions like Europe and North America, DCS is seen as a cornerstone for achieving decarbonization targets[44]. Advances in smart grids, energy storage, and the integration of renewables are further enhancing system flexibility and environmental performance [45].

3.5 District Heating and Cooling Initial Simulation

The simulation study was conducted using Jakarta as the reference location for implementing a District Heating and Cooling (DHC) system. To simplify the initial model and allow for manageable computation, several assumptions were adopted:

a. Uniform Building Load: All buildings are assumed to have the same energy demand per square meter, enabling simulation outcomes to depend solely on total building area.
b. Equal Heating and Cooling Requirements: Both heating and cooling needs are assumed to be equivalent in base value.
c. Pre-installed Building Energy Systems: It is assumed that all buildings already contain internal energy distribution systems.

The simulation employed nPro Tools, focusing on a Fifth Generation District Heating and Cooling (5GDHC) network with a design temperature range of 0–35°C. Notably, the system design omits a dedicated cooling supply, instead relying on heat exchange and storage systems. The building energy consumption parameters used are as follows, and the general number which representing the area of simulation shown in **Table 1**:

Space Heating: 75 kWh/m^2/year
Domestic Hot Water: 2 kWh/m^2/year
Space Cooling: 20 kWh/m^2/year
Process Cooling: 500 kWh/m^2/year

Table 1. Setting parameters for district heating and cooling calculation

Usage type	Number	Floor area (m^2)	Heat (MWh)	Cold (MWh)	Electricity (MWh)	Operating Power (MWh)
Supermarket	40	2000	250	290	114	58
Residental	5000	70	1.5	1000	2.8	200
School	600	1000	21	270	40	54
Hospital	50	3000	250	1250	120	250
Office	50	2000	42	1000	80	200

Based on these inputs, the simulation calculated total annual energy demands for the district as:

Heating: 564 MWh
Cooling: 3,810 MWh
Electricity: 387 MWh

The system is powered by excess heat from a NPPs with a thermal capacity of 40 MWt. To enhance system efficiency, the simulation includes both heat storage and cold storage units, allowing flexibility in thermal energy distribution. The NPP serves as the primary thermal energy source, delivering hot water at approximately 40°C and cold water within a 15–25°C range. The simulation was run with a heat delivery efficiency margin allowing for up to 10% energy loss.

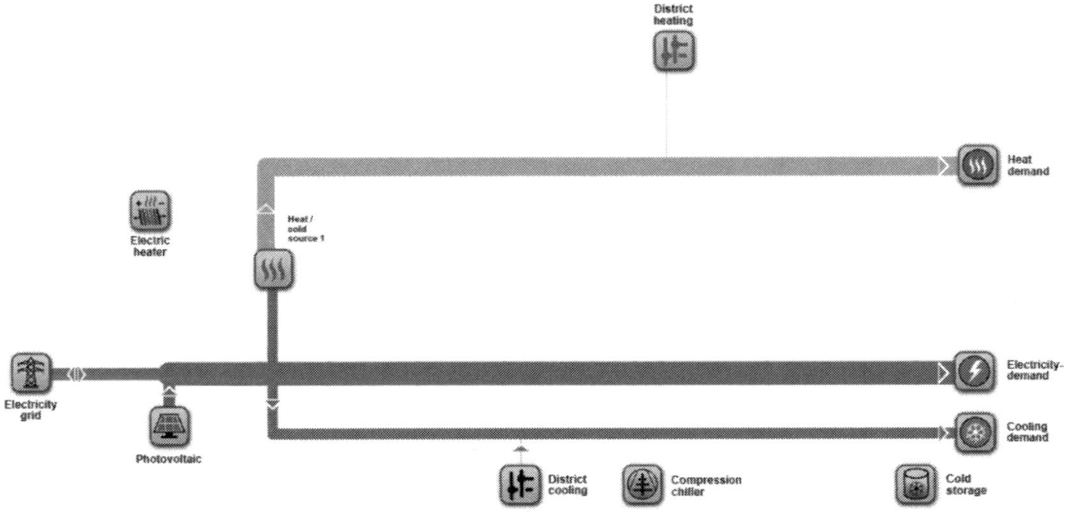

Figure 7.Overview of Annual Operation Simulation

After integrating all parameters including component efficiencies and system losses the simulation identified an optimal scenario for energy performance. **Figure 7** provides a visual representation of the annual energy flow across the system. It was observed that electric heaters, though present, played a minimal role due to the high contribution from thermal storage and the NPP's cogeneration capacity. The simulation confirmed that thermal sources are the dominant contributors to energy provision. Heat demand increases during cooler periods, while cooling loads rise in the hotter months. Integrating an NPP within a DHC network provides a stable, efficient, and low-carbon energy source, especially when paired with renewable energy technologies. In this study, solar photovoltaic (PV) panels were introduced as a complementary source, demonstrating further improvement in energy efficiency and system sustainability.

IOP Conf. Series: Materials Science and Engineering **1330** (2025) 012007 doi:10.1088/1757-899X/1330/1/012007

Table 2. Simulation results for the scenario (1 Euro =Rp 17,400)

Parameter	Value	
Electrical import from the grid	222 MWh	
Renewable electricity generation	284 MWh	
District heating import	0	
Heat from heat source	8760 MWh	
CO_2 emissions	78 t	
CO_2 emissions per heat demand	591 g/kWh	
CO_2 emissions per energy demand	114 g/kWh	
CO_2 emissions per floor area	9.6 kg/m^2	
Primary energy	400 MWh	
Primary energy per heat demand	3 kWh/kWh	
Primary energy per energy demand	0.59 kWh/kWh	
Primary energy per floor area	49.5 kWh/m^2	
System Coefficient of Performance	0.31	
Economic results		
	Scenario 1	Scenario 2
Investment (annuity)	-Rp581,317,226,400/a	-Rp69,038,519,400/a
Energy Cost	-Rp195,763,920,000/a	-Rp772,560,000/a
maintenance cost	-Rp99,925,224,600/a	-Rp10,447,429,800/a
Lump sum cost (annuity)	-Rp69,312,238,800/a	-Rp7,563,414,600/a
Revenues	Rp694,307,154,000/a	Rp17,841,420,600/a
Annual net profits	-Rp252,011,438,400/a	-Rp69,980,503,200/a
Net Present Value	-Rp3,140,619,733,800	-Rp872,111,872,800
Levelized Cost of Heat	Rp466,737.6/kWh	Rp552,832.8/kWh
Levelized Cost of Energy	Rp55,332/kWh	Rp106,940.4/kWh
Energy Cost per Floor Area	Rp32,646,750/m^2	Rp9,006,240/m^2
Monthly Energy Cost per Floor Area	Rp2,720,490/m^2	Rp750,462/ m^2

A comparative economic analysis is presented in **Table 2**, evaluating Scenario 1 and Scenario 2. Interestingly, Scenario 1 yields lower annual energy costs, LCOH, and LCOE, despite its higher total load. This illustrates a core economic principle: larger system loads can justify larger initial investments, improving overall cost-effectiveness over time. The system demonstrates strong potential for low-carbon energy performance, with zero district heating import, substantial renewable electricity generation (284 MWh), and relatively low CO_2 emissions (78 tons). However, a low system COP of 0.31 and high primary energy use per unit of heat (3 kWh/kWh) indicate inefficiencies in energy conversion. Economically, Scenario 1 generates higher revenue but suffers from significantly higher investment and operational costs, leading to a large annual loss (Rp252 billion) and a deeply negative NPV. Scenario 2, while producing less revenue, has much lower costs and a smaller financial deficit (Rp69.9 billion/year), making it more cost-efficient. Despite the challenges, both scenarios highlight the importance of improving system efficiency and optimizing investment strategies to enhance the economic viability of low-carbon DHC systems.

Nonetheless, the implementation of a DHC system represents a substantial initial investment, often leading to negative early cash flows. The integration of NPP cogeneration can mitigate some of these costs by eliminating the need for large-scale standalone heating plants, thus improving economic and environmental performance. However, the simulation does not fully account for real-world carbon emissions at this stage. Successful implementation also depends on community engagement and the development of a supportive policy framework, including promotional tariffs or government subsidies. These initiatives could accelerate adoption and unlock new market opportunities in the regional energy landscape.

4. Conclusion

District Heating and Cooling (DHC) systems offer an efficient, sustainable solution for meeting the thermal energy needs of dense urban areas by centralizing the production and distribution of heating and cooling. As cities strive to meet ambitious climate goals, advanced systems like 5th Generation DHC (5GDHC) are emerging as essential components of low-carbon infrastructure. This study explores the integration of nuclear cogeneration particularly through Small Modular Reactors (SMRs) into DHC networks to enhance reliability, efficiency, and long-term sustainability by providing consistent baseload thermal energy and reducing dependence on fossil fuels. However, successful implementation requires strict safety, security, and regulatory measures, and further real-world validation is needed beyond preliminary simulations. To support this development, the study recommends improving data accuracy, conducting detailed feasibility assessments, establishing strong safety protocols, initiating pilot projects, engaging stakeholders, and fostering ongoing research and innovation. Addressing these priorities will enable the strategic deployment of nuclear-powered DHC systems, contributing to resilient urban energy infrastructures, national energy security, and the global transition to net-zero emissions.

References

[1] Rämä M, Leurent M and Devezeaux de Lavergne J G 2020 Flexible nuclear co-generation plant combined with district heating and a large-scale heat storage *Energy* **193**

[2] Hao Y, Chen P and Li X 2022 Testing the environmental kuznets curve hypothesis: The dynamic impact of nuclear energy on environmental sustainability in the context of economic globalization *Energy Strateg. Rev.* **44**

[3] Saleh Abushamah H A and Skoda R 2022 Nuclear energy for district cooling systems – Novel approach and its eco-environmental assessment method *Energy* **250**

[4] Ali S, Jiang J, Hassan S T and Shah A A 2022 Revolution of nuclear energy efficiency, economic complexity, air transportation and industrial improvement on environmental footprint cost: A novel dynamic simulation approach *Nucl. Eng. Technol.* **54**

[5] Chen X H, Tee K, Elnahass M and Ahmed R 2023 Assessing the environmental impacts of renewable energy sources: A case study on air pollution and carbon emissions in China *J. Environ. Manage.* **345**

[6] Zhang Y, Johansson P and Sasic Kalagasidis A 2022 Assessment of district heating and cooling systems transition with respect to future changes in demand profiles and renewable energy supplies *Energy Convers. Manag.* **268**

[7] Pardo-Bosch F, Blanco A, Mendoza N, Libreros B, Tejedor B and Pujadas P 2023 Sustainable deployment of energy efficient district heating: city business model *Energy Policy* **181**

[8] Mahmood N, Danish, Wang Z and Zhang B 2020 The role of nuclear energy in the correction of environmental pollution: Evidence from Pakistan *Nucl. Eng. Technol.* **52** 1327–33

[9] Beiron J, Normann F and Johnsson F 2023 Carbon capture from combined heat and power plants – Impact on the supply and cost of electricity and district heating in cities *Int. J. Greenh. Gas Control* **129**

[10] Zaw K, Kwik Z Z S, Chang W Q, Islam M R and Poh T K 2023 A techno-commercial decision support framework for optimal district cooling system design in tropical regions *Appl. Therm. Eng.* **220** 1–10
[11] Lipka M and Rajewski A 2020 Regress in nuclear district heating. The need for rethinking cogeneration *Prog. Nucl. Energy* **130**
[12] Galimova T, Ram M and Breyer C 2022 Mitigation of air pollution and corresponding impacts during a global energy transition towards 100% renewable energy system by 2050 *Energy Reports* **8** 14124–43
[13] Wirtz M, Kivilip L, Remmen P and Müller D 2020 5th Generation District Heating : A novel design approach based on mathematical optimization *Appl. Energy* **260** 114158
[14] Mika J, Forgo P, Lakatos L, Olah A B, Rapi S and Utasi Z 2018 Impact of 1.5 K global warming on urban air pollution and heat island with outlook on human health effects *Curr. Opin. Environ. Sustain.* **30** 151–9
[15] Werner S 2017 District heating and cooling in Sweden *Energy* **126** 419–29
[16] Chou M T, Hsu Y P, Hsu C T and Ding J F 2022 Reconstruction mechanism and strategy of global maritime green supply chain against the backdrop of nuclear pollution *Mar. Pollut. Bull.* **185**
[17] Shen J, Cai W, Chen X, Chen X, Zhao Z, Ma Z, Yang F and Zhang S 2022 Synergies of carbon neutrality, air pollution control, and health improvement — a case study of China energy interconnection scenario *Glob. Energy Interconnect.* **5** 531–42
[18] Lyu X, Li K, Guo H, Morawska L, Zhou B, Zeren Y, Jiang F, Chen C, Goldstein A H, Xu X, Wang T, Lu X, Zhu T, Querol X, Chatani S, Latif M T, Schuch D, Sinha V, Kumar P, Mullins B, Seguel R, Shao M, Xue L, Wang N, Chen J, Gao J, Chai F, Simpson I, Sinha B and Blake D R 2023 A synergistic ozone-climate control to address emerging ozone pollution challenges *One Earth* **6** 964–77
[19] Bachtiar V S, Raharjo S, Ruslinda Y, Hayati F and Komala D R 2015 Mapping of ozone gas (O3) concentrations in Padang city *Procedia Eng.* **125** 291–7
[20] Rita E, Chizoo E and Cyril U S 2021 Sustaining COVID-19 pandemic lockdown era air pollution impact through utilization of more renewable energy resources *Heliyon* **7**
[21] Danish, Ud-Din Khan S and Ahmad A 2021 Testing the pollution haven hypothesis on the pathway of sustainable development: Accounting the role of nuclear energy consumption *Nucl. Eng. Technol.* **53** 2746–52
[22] Chen J, Zheng W, Kong Y, Yang X, Liu Z and Xia J 2021 Case study on combined heat and water system for nuclear district heating in Jiaodong Peninsula *Energy* **218**
[23] Maccarini A, Sotnikov A, Sommer T, Wetter M, Sulzer M and Afshari A 2023 Influence of building heat distribution temperatures on the energy performance and sizing of 5th generation district heating and cooling networks *Energy* **275**
[24] Dai W, Xia W, Goh H, Liu Z, Luo C, Wen F and Zhang Z 2023 Reduce operational costs of thermal power units in power systems using the flexibility of district cooling systems *Energy Reports* **9** 1160–8
[25] Xiao X, Cai Q, Chen R, Ding W, Guo K, Zhang K, Tian W, Qiu S and Su G H 2022 An improved MPS-DEM numerical model for fluid–solid coupling problem in nuclear reactor *Nucl. Eng. Des.* **396**
[26] Pursiheimo E, Lindroos T J, Sundell D, Rämä M and Tulkki V 2022 Optimal investment analysis for heat pumps and nuclear heat in decarbonised Helsinki metropolitan district heating system *Energy Storage Sav.* **1** 80–92
[27] Zaw K, Matsuoka K and Poh T K 2023 A real-time operational optimization of commercial district cooling systems: A practical application *Energy Build.* **297**
[28] Ríos-Ocampo J P, Olaya Y, Osorio A, Henao D, Smith R and Arango-Aramburo S 2022 Thermal districts in Colombia: Developing a methodology to estimate the cooling

potential demand *Renew. Sustain. Energy Rev.* **165**
[29] Østergaard P A, Werner S, Dyrelund A, Lund H, Arabkoohsar A, Sorknæs P, Gudmundsson O, Thorsen J E and Mathiesen B V 2022 The four generations of district cooling - A categorization of the development in district cooling from origin to future prospect *Energy* **253**
[30] Neri M, Guelpa E and Verda V 2024 Two-stage stochastic programming for the design optimization of district cooling networks under demand and cost uncertainty *Appl. Therm. Eng.* **236** 121594
[31] Novosel T, Feijoo F, Duić N and Domac J 2022 Impact of district heating and cooling on the potential for the integration of variable renewable energy sources in mild and Mediterranean climates *Energy Convers. Manag.* **272**
[32] Susiati H, Dede M, Widiawaty M A, Ismail A and Udiyani P M 2022 Site suitability-based spatial-weighted multicriteria analysis for nuclear power plants in Indonesia *Heliyon* **8** 1–13
[33] Haq H, Valisuo P, Kumpulainen L and Tuomi V 2020 An economic study of combined heat and power plants in district heat production *Clean. Eng. Technol.* **1** 1–8
[34] Rämä M, Pursiheimo E, Sundell D and Abdurafikov R 2024 Dynamically distributed district heating for an existing system *Renew. Sustain. Energy Rev.* **189**
[35] Abugabbara M, Javed S, Bagge H and Johansson D 2020 Bibliographic analysis of the recent advancements in modeling and co-simulating the fifth-generation district heating and cooling systems *Energy Build.* **224** 1–11
[36] Lyu H, Shen S, Wu Y and Zhou A 2020 Techno-economic analysis of fifth-generation district heating and cooling combined with seasonal borehole thermal energy storage Xiang *Geosci. Front.* 128021
[37] Tai C, Tian G and Lei W 2023 A novel district heating system based on absorption heat exchange and water-heat combined supply *Case Stud. Therm. Eng.* **51**
[38] Hadi Ghazaie S, Sadeghi K, Chebac R, Sokolova E, Fedorovich E, Cammi A, Enrico Ricotti M and Saeed Shirani A 2022 On the use of advanced nuclear cogeneration plant integrated into latent heat storage for district heating *Sustain. Energy Technol. Assessments* **50** 101838
[39] Tai C, Tian G and Lei W 2022 A water-heat combined supply system based on waste heat from a coastal nuclear power plant in northern China *Appl. Therm. Eng.* **200**
[40] Arghand T, Javed S and Dalenbäck J O 2023 Combining direct ground cooling with ground-source heat pumps and district heating: Energy and economic analysis *Energy* **270**
[41] Bilardo M, Sandrone F, Zanzottera G and Fabrizio E 2021 Modelling a fifth-generation bidirectional low temperature district heating and cooling (5GDHC) network for nearly Zero Energy District (nZED) *Energy Reports* **7** 8390–405
[42] Volkova A, Pakere I, Murauskaite L, Huang P, Lepiksaar K and Zhang X 2022 5th generation district heating and cooling (5GDHC) implementation potential in urban areas with existing district heating systems *Energy Reports* **8** 10037–47
[43] Calise F, Cappiello F L, Cimmino L, Dentice d'Accadia M and Vicidomini M 2023 A comparative thermoeconomic analysis of fourth generation and fifth generation district heating and cooling networks *Energy* **284**
[44] Jimenez-Navarro J P, Kavvadias K, Filippidou F, Pavičević M and Quoilin S 2020 Coupling the heating and power sectors: The role of centralised combined heat and power plants and district heat in a European decarbonised power system *Appl. Energy* **270**
[45] Wu Y, An J, Qian M and Yan D 2023 Application-driven level-of-detail modeling framework for occupant air-conditioning behavior in district cooling *J. Build. Eng.* **70**

STUDY OF THE EFFECT OF UTILIZING THORIUM AND PROTACTINIUM-231 IN VODO-VODYANOI ENYERGETICHESKIY REACTOR-1000 FUEL ASSEMBLY USING MONTE CARLO N-PARTICLE (MCNP) SIMULATION

Kevin Wijaya[1] and Dwi Irwanto[1,2]
[1] Nuclear Physics and Biophysics Research Division, Physics Department, Faculty of Mathematics and Natural Science, Bandung Institute of Technology, Jalan Ganesha 10 Bandung 40132, Indonesia.
[2] Nuclear Science and Engineering Department, Faculty of Mathematics and Natural Science, Bandung Institute of Technology, Jalan Ganesha 10 Bandung 40132, Indonesia.

E-mail: kevinwijaya3001@gmail.com

Keywords: Burnable poison, burnup, MCNP, thorium, uranium

Abstract

One type of nuclear reactor is the Vodo-Vodyanoi Enyergeticheskiy Reactor-1000 (VVER-1000). In this work, we will discuss the effect of giving thorium as a fuel and protactinium as a combustible poison in the VVER-1000 fuel assembly using a simulation program based on Monte Carlo N-Particle (MCNP). The purpose of this work was to determine the multiplication factor of the fuel assembly with variations in ThO_2 (without fissile material) and variations in the fractions of gadolinium and protactinium-231, as well as the burnup rate for 5 years. Thorium-232 is predicted to be a substitute for UO_2 fuel in reactors because of its abundant material and can produce uranium-233 which is a better fuel than uranium-235. While protactinium-231 has a function other than absorbing neutrons, generating it can further produce fissile material that aids in chain reactions. Based on the simulation, the multiplication factor with protactinium-231 has a higher value than gadolinium. In addition, the larger the burnable poison fraction, the smaller the multiplication factor value is. In addition, the multiplication factor and burnup rate for 5 years were also obtained. The optimal fuel composition to use is a mixture of ThO_2 of 5% - 15% (from non-burnable poison) and 5% (from burnable poison), as well as the optimal amount of burnable poison used in the case of fuel given 5% ThO_2 is 4% - 6% for Pa-231 and 0.5% - 2.5% for Gd_2O_3.

IOP Conf. Series: Materials Science and Engineering **1330** (2025) 012008 doi:10.1088/1757-899X/1330/1/012008

I. Introduction

Nuclear is part of or related to the atomic nucleus. In nuclear physics, there is a nuclear reaction which is characterized by two nuclei or particles colliding with each other. This reaction is divided into fusion and fission reactions. A fusion reaction is a reaction where two or more atomic nuclei combine and produces energy, while a fission reaction is a reaction where an atom splits into two or more atomic nuclei and releases energy.

At present, many countries utilize fission reactions to produce electrical power by building reactors for Nuclear Power Plants (NPP). Currently, electricity is still one of the energies that humans really need in everyday life. Considering that fossil energy is increasingly depleted, nuclear power plants can be another alternative in terms of supplying electricity needs.

One type of nuclear reactor is the Vodo-Vodyanoi Enyergeticheskiy Reactor-1000 (VVER-1000). For this reason, information is needed in the form of design and analysis of the VVER-1000. To obtain information regarding this matter, in this final research project we will discuss the design and analysis of the VVER-1000. The program used is a Monte Carlo N-Particle (MCNP) based program which is a licensed code released by the Los Alamos National Laboratory (LANL). The basic method of MCNP is Monte Carlo which is statistical in nature.

Several studies have shown the effects of thorium and uranium-233 fuels, including research by Goma et al. in 2021 which discussed the neutronic performance using thorium fuel in the VVER-1000 reactor by varying the mixture of UO_2 and ThO_2 fuels [1], research by Galahom in 2017 which discussed the minimization of fission product waste using thorium fuel by comparing the multiplication factor and burnup rate of UO_2 fuel, ThO_2 with U-235 enrichment, and ThO2 with U-233 enrichment [2], and research by Hassan et al. in 2020 which discussed the comparison of multiplication factors using UO_2 and ThO_2 fuels with uranium-233. From these studies, it was shown that thorium-232 and uranium-233 are the best fertile and fissile fuels [3]. Meanwhile, protactinium-231 is proposed to be used as a burnable poison. In addition to having a cross-section that is not much different from gadolinium-157, research from Kulikov et al. in 2017 discussed the use of protactinium-231 as a burnable poison in UO_2 fuel. From this study, the multiplication factor value of the reactor with protactinium-231 was lower than without burnable poison [4].

The specific objective of this work is to determine how the fuel model of the VVER-1000 works, the relationship between variations in uranium-235 enrichment and variations in the ThO_2 fraction as well as variations in the gadolinium and protactinium-231 fractions with the multiplication factor of the VVER-1000 fuel assembly using Monte Carlo N- Particle (MCNP), as well as the multiplication factor as a function of burnup and burnup rate with variations in thorium and variations in gadolinium and protactinium-231 fractions. The research scope of this work is that the type of reactor analyzed is the Pressurized Water Reactor (PWR) VVER-1000 with variations in uranium enrichment, Gd_2O_3, and Pa-231 fractions of 0.5% to 10% respectively and ThO_2 variations of 0% to 100% without fissile material mixed with UO_2. We don't use fissile material for ThO_2 in this work because we want to know the possibility to use thorium alone without fissile material considering the negative side of U-233 properties and costs. The MCNP program used is MCNPX, the burnup scheme is carried out at intervals of 1 year for 5 years, the number of particles used is 2000. While the limitations and assumptions used in this research are that the thermal-hydraulic analysis aspect is not reviewed in this research, The reactor system reviewed is based on physical conditions only, the simulated system conditions are considered steady state in certain cases, the simulated temperature is adjusted to the operating conditions of the reactor system and only uses material library data in the ENDF nuclear library.

II. Monte Carlo N-Particle (MCNP)

To determine parameters such as multiplication factor and burnup. In this research, Monte Carlo N-Particle (MCNP) simulation was used. Monte Carlo is a statistical numerical method by simulating random numbers to solve problems that are not possible to solve analytically. One computer program based on the Monte Carlo method is Monte Carlo N-Particle (MCNP). MCNP can simulate the travel of neutrons, electrons and photons in 14 three-dimensional materials. This computer program was carried out by the Monte Carlo team at Los Alamos National Laboratory, USA. Important standard features that make MCNP so versatile and easy to use include robust common sources, criticality sources, and surface sources; both geometry and output calculation plots; rich collection of variance reduction techniques; flexible calculation structure; and extensive cross-section data sets.

MCNP contains many flexible calculations: current & surface flux, volume flux (path length), point or ring detectors, particle heating, fission heating, pulse height calculations for energy or charge deposition, mesh calculations, and radiography calculations [5].

MCNP can simulate neutron particles from birth and interact with the material until they reach the dead area. Suppose the neutron particles from the source interact with matter, as shown in Figure 1. The first interaction (indicated by a1) is scattering, then the interaction (indicated by a2) scatters the photons and electrons to the dead zone. Then, the photons formed interact with the material (indicated by b1) to produce scattered neutrons and photons. These photons are absorbed by the material (indicated by b2). The neutrons formed are scattered (indicated by c1) and enter the dead zone [6].

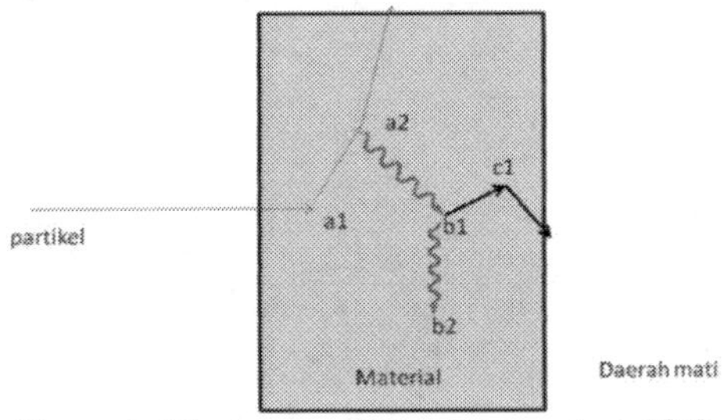

Figure 1. Neutron particles hit a material [6].

The Monte Carlo method uses probability theory to model a system stochastically. There are two probabilities in random sampling, namely the probability of a neutron moving a certain distance without any interaction and the probability of a neutron having the first interaction at a certain distance. Then, the probability density function (PDF) is also used, which is a real-valued function whose integral over any set gives the probability that a random variable has a value in that set. Apart from that, there is also the term cumulative distribution function (CDF) which is the probability that a variable takes a value less than or equal to x' [8]. The MCNP workflow in simulating particle motion and obtaining output can be seen in the flow diagram in Figure 2.

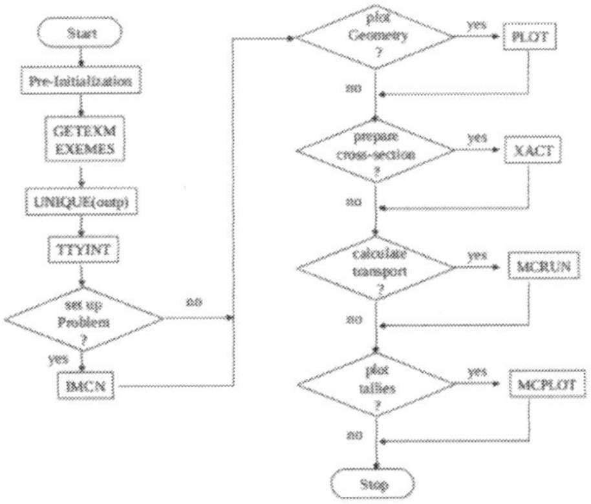

Figure 2. Flowchart in MCNP simulation [7].

To carry out a simulation using MCNP, there are three stages that are followed, namely creating input, running, and interpreting the output. In principle, making MCNP input is simple, namely by filling in what is called a "card". There are three cards in MCNP input, namely cell cards, surface cards, and data cards. The cell card and surface card are input for the geometry of the object to be simulated, while the data card is information about the material of the simulated object, the definition of the particle source, and the tally or physical quantity to be calculated.

In the MCNP input sequence, the first line is filled with "title" in the form of any sentence, it can also be the theme of the simulation to be carried out. For the next row, if it is to be filled with words or sentences, it must be marked "c" in the front column. This is to show that it is just a comment so it will

IOP Conf. Series: Materials Science and Engineering **1330** (2025) 012008 doi:10.1088/1757-899X/1330/1/012008

not be processed. Each card must be separated by a space or blank line. Only 1 empty line, if it is not separated it will be considered one card, while if more than 1 empty line is separated it will not be processed [8]. To visualize the model of the fuel assembly or reactor core, an additional program from MCNP is used, namely MCNPX Visual Editor (Vised).

III. Method

In this section it is broadly divided into 3 main stages, namely literature study, simulation using MCNP, and processing and analysis of the data obtained. The first research carried out was to determine the model of the VVER-1000 fuel assembly and each cell such as the U fuel cell, U-Gd fuel cell, guide tube, and central cell using the MCNPX Visual Editor. The model parameters used in this simulation can be seen in Table 1.

Table 1. VVER-1000 fuel assembly model parameters [9].

No.	Information	Value
1	Number of U fuel cells	300
2	Number of U-Gd fuel cells	12
3	Number of guide tubes	18
4	Number of central tubes	1
5	Inner radius of fuel cell (cm)	0.3860
6	Outer radius of fuel cell (cm)	0.4582
7	Inner radius of central tube cell (cm)	0.4800
8	Outer radius of central tube cell (cm)	0.5626
9	Inner radius of guide tube cell (cm)	0.5450
10	Outer radius of guide tube cell (cm)	0.6323
11	Cell grid length (cm)	1.2750
12	Length of fuel assembly grid (cm)	23.6
13	Fuel temperature (K)	1027
14	Non-fuel temperature (K)	575
15	Gd_2O_3 density (g/cm^3)	7.4
16	Boron fraction (g/kg)	0.6
17	H_3BO_3 density (g/cm^3)	0.7235

`Then, for the U fuel cell, a U-235 enrichment of 3.7% was used and for the U-Gd fuel cell, a U-235 enrichment of 3.5% was used and the Gd_2O_3 fraction was 4%. The first step in the MCNP calculation is to model the U fuel cell in a simple hexagonal lattice (SH). The U fuel cell consists of a 0.3860 cm radius fuel rod surrounded by a 0.0722 cm thick layer of zircalloy-4 without disturbing the air gap between the fuel and the casing. Water occupies the region outside the cladding in the grid. The U-Gd fuel cell model is based on a U fuel cell in an SH lattice with cladding made of zircalloy-4 0.0722 cm thick. The area outside the cladding in the grid is also filled with water. The second step is to model

IOP Conf. Series: Materials Science and Engineering **1330** (2025) 012008 doi:10.1088/1757-899X/1330/1/012008

the guide tube cell on the SH grid using the same technique. The water in the center of the guide tube cell with a radius of 0.5450 cm is surrounded by 0.0873 cm thick cladding and occupies the empty area outside the cladding in the lattice. The model is almost the same as the guide tube, although it has a different function. Small differences exist in the water radius and cladding thickness, namely 0.4800 cm and 0.0826 cm, respectively. The water in both tube cells is used for neutron moderation. All SH grids have the same grid length, namely 1.2750 cm. The final step is to model the fuel assembly of the VVER-1000 by building the fuel cell grid, guide tube and central tube in a hexagonal geometry with a grid side length of 23.6 cm. Then, use the number density equation to determine the isotopic composition of the fuel assembly which can be seen in Tables 2 – 5. The fractions for Gd-152, Gd-154, Gd-155, Gd-156, Gd-157, Gd-158, and Gd-160 are 0.2%; 2.18%; 14.8%; 20.47%; 15.65%; 24.84%; and 21.86% respectively.

Table 2. U isotope composition of fuel cell of VVER-1000 (atom/barn-cm)

Atomic Isotopes	Density
^{235}U	8.6264×10^{-4}
^{238}U	2.2169×10^{-2}
^{16}O	4.6063×10^{-2}

Table 3. U-Gd isotope composition of VVER-1000 fuel cell (atom/barn-cm)

Atomic Isotopes	Density
^{235}U	8.6264×10^{-4}
^{238}U	2.2169×10^{-2}
^{16}O	4.6063×10^{-2}
^{152}Gd	1.0179×10^{-5}
^{154}Gd	1.1095×10^{-5}
^{155}Gd	7.5327×10^{-5}
^{156}Gd	1.0418×10^{-4}
^{157}Gd	7.97×10^{-5}
^{158}Gd	1.26×10^{-4}
^{160}Gd	1.11×10^{-4}

Table 4. Isotope composition of zirconium alloys from VVER-1000 (atoms/barn-cm).

Atomic Isotopes	Density
^{93}Nb	4.2250×10^{-4}
^{90}Zr	2.1913×10^{-2}
^{91}Zr	4.7786×10^{-3}
^{92}Zr	7.3042×10^{-3}
^{94}Zr	7.4021×10^{-3}
^{96}Zr	1.1925×10^{-3}
^{174}Hf	1.0555×10^{-8}
^{176}Hf	3.4700×10^{-7}
^{177}Hf	1.2270×10^{-6}
^{178}Hf	1.7997×10^{-6}
^{179}Hf	8.9851×10^{-7}
^{180}Hf	2.3142×10^{-6}

Table 5. Moderator isotope composition of VVER-1000 (atoms/barn-cm).

Atomic Isotopes	Density

^{1}H	4.843×10^{-2}
^{16}O	2.422×10^{-2}
^{10}B	4.794×10^{-6}
^{11}B	1.942×10^{-5}

After that, a simulation was carried out to determine the multiplication factor value of the fuel assembly using MCNP. The variations used are variations in U-235 enrichment of 0.5%-10% in increments of 0.5% and variations in Gd_2O_3 composition of 0.5% - 10%. After the simulation was carried out, the final k_{eff} data was taken and the multiplication factor values were compared with variations in both U-235 enrichment and Gd_2O_3 fraction variations.

After that, a simulation is carried out to determine burnup. In this method, the "BURN" card is added to the input used. The variations used are variations in U-235 enrichment of 0.5% - 10% in increments of 0.5% and variations in the Gd_2O_3 fraction of 0.5% - 10%. The time used for burnup is 5 years by reviewing conditions every year. Then, the power listed is the thermal power of the fuel assembly which is the result of the VVER-1000 core power divided by the number of fuel assemblies. Then, the volume of each part of the cell (fuel, cladding and moderator) is calculated. The parameters used to calculate power and volume can be seen in Table 6.

After that, a simulation was carried out after the output was produced, data was collected in the form of multiplication factor values each year for 5 years along with standard deviation and 99% confidence intervals, and the burnup rate of each variation for 5 years.

Table 6. Power and volume calculation parameters for burnup [1,9].

No.	Information	Value
1	Thermal power (MW)	3000
2	Number of fuel assemblies	163
3	Tube height (cm)	353
4	Cell grid length (cm)	1.275
5	Length of fuel assembly grid (cm)	23.6
6	Number of U fuel cells	300
7	Number of U-Gd fuel cells	12
8	Number of guide tubes	18
9	Number of central tubes	1
10	Inner radius of fuel cell (cm)	0.3860
11	Outer radius of fuel cell (cm)	0.4582
12	Inner radius of central tube cell (cm)	0.4800
13	Outer radius of central tube cell (cm)	0.5626
14	Inner radius of guide tube cell (cm)	0.5450
15	Outer radius of guide tube cell (cm)	0.6323

Next, a simulation was carried out to determine the multiplication factor value of the fuel assembly by varying the ThO_2 mixture from 0% to 100% in multiples of 5% with 3 different fuel assembly conditions, namely without burnable poison, with 4% Gd_2O_3, and with Pa- 231 as much as 4% too. The U-235 enrichment used in fuel cells with the burnable poison variation is 3.5% while the remaining cells are 3.7%. Information regarding the density used can be seen in the table below.

Table 7. Information on the density of the materials used [1].

No.	Type	Value (gram/cm^3)
1	UO_2	10.53
2	Gd_2O_3	7.4
3	ThO_2	10
4	Pa-231	15.37
5	H_3BO_3	0.7235

Then, a simulation was carried out to determine the multiplication factor value of the fuel assembly with variations in the Gd_2O_3 and Pa-231 fractions of 0.5% - 10% in increments of 0.5%. The fuel sample used was 5% ThO_2 and 95% UO_2. Next, a burnup simulation was carried out to determine the multiplication factor and burnup rate from variations in ThO_2, Gd_2O_3 fraction, and Pa-231 fraction over several years. For the ThO_2 variation used is 0-100% with the burnable poison fraction used being 4% and for the variation of the burnable poison fraction used is 0.5% - 10% with the ThO_2 used being 5%. The time used is 5 years in increments of 1 year.

IV. Results and Discussion

This section shows the results of the VVER-1000 fuel assembly model, multiplication factor and burnup graphs with variations in UO_2 and Gd fraction. First, based on the parameters used (Table 1), a model of the fuel assembly and several cells including U fuel cell, U-Gd fuel cell, guide tube cell, and central tube cell were obtained using the MCNPX Visual Editor.

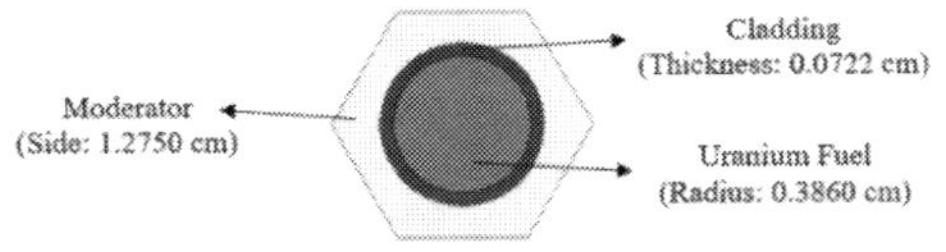

Figure 3. U fuel cell model.

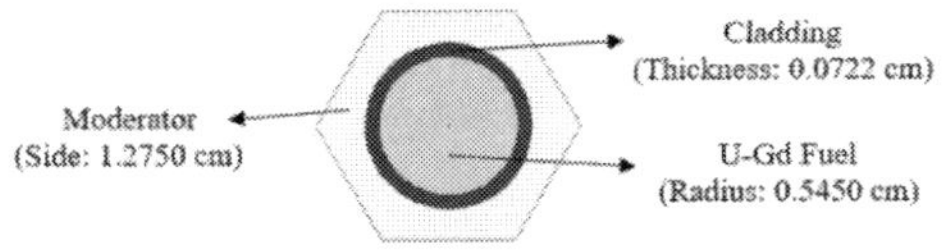

Figure 4. U-Gd fuel cell model.

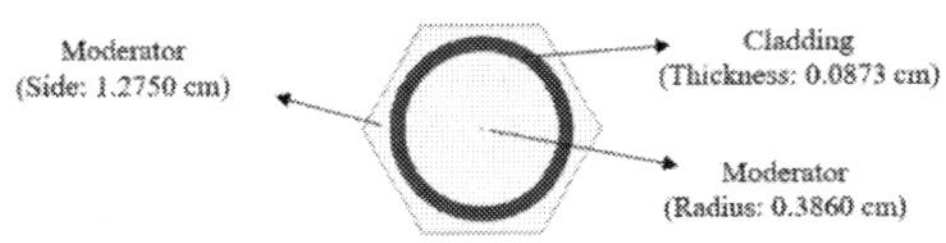

Figure 5. Central tube cell model.

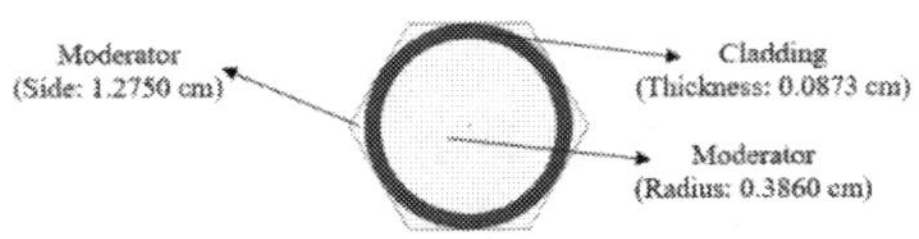

Figure 6. Guide tube cell model.

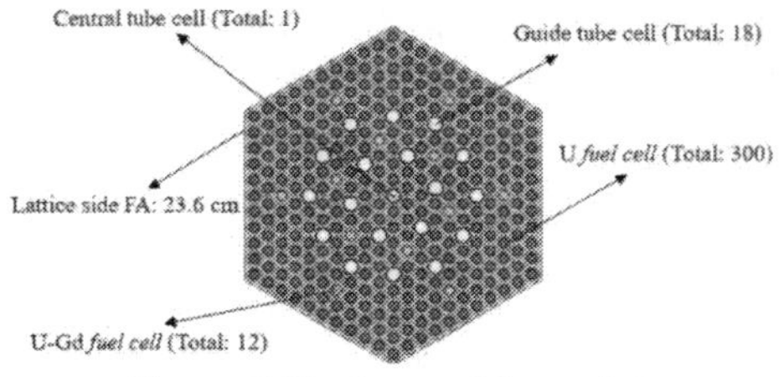

Figure 7. Fuel assembly model.

Then, the multiplication factor (k_{eff}) calculation of the model used was carried out. The variations used are variations in U-235 enrichment and variations in the Gd_2O_3 fraction of 0.5% - 10% each in increments of 0.5%.

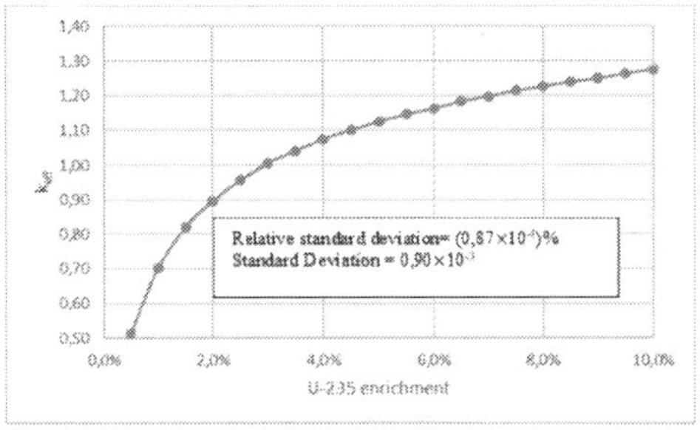

Figure 8. Graph of the multiplication factor of the VVER-1000 fuel assembly with variations in U-235 enrichment.

From this graph, the k_{eff} value increases as the U-235 enrichment value increases. This is caused by the increasing amount of U-235 fissile fuel so that there is more U-235 isotope, and more neutrons are produced. From these values it can be proven that the enrichment of uranium used in commercial reactors is 3% to 5%. Apart from that, relatively small standard deviation and standard deviation values were also obtained. This shows that the distribution of k_{eff} results for each variation obtained tends to have a small difference in value or in other words is close to the k_{eff} obtained. The confidence interval used is 99%.

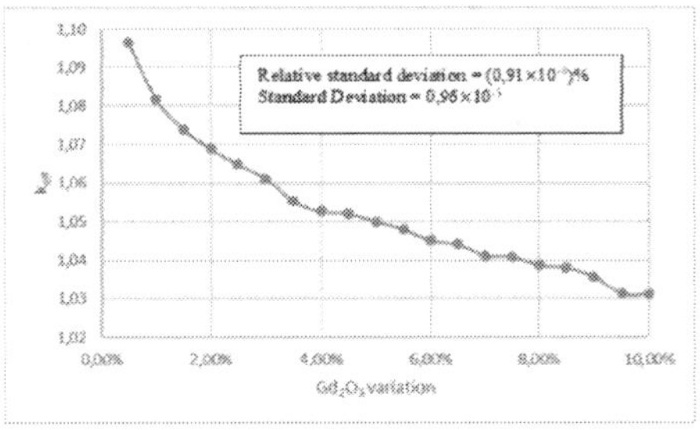

Figure 9. Graph of the multiplication factor of the VVER-1000 fuel assembly with variations in the Gd_2O_3 fraction.

From this graph it can be seen that the keff value decreases as the value of the U-235 fraction increases. This is caused by burnable poison capturing neutrons so that the more burnable poison is used, the more neutrons are captured which will affect the k_{eff} value. Apart from that, relatively small standard deviation and standard deviation values were also obtained. This shows that the distribution of keff results for each variation obtained tends to have a small difference in value or in other words is close to the k_{eff} obtained. The confidence interval used is 99%.

Furthermore, burnup calculations have been carried out with the same variations in U-235 and Gd enrichment, namely 0.5% to 10% respectively with multiples of 0.5%. The results obtained are in the form of multiplication factor values and burnup rate.

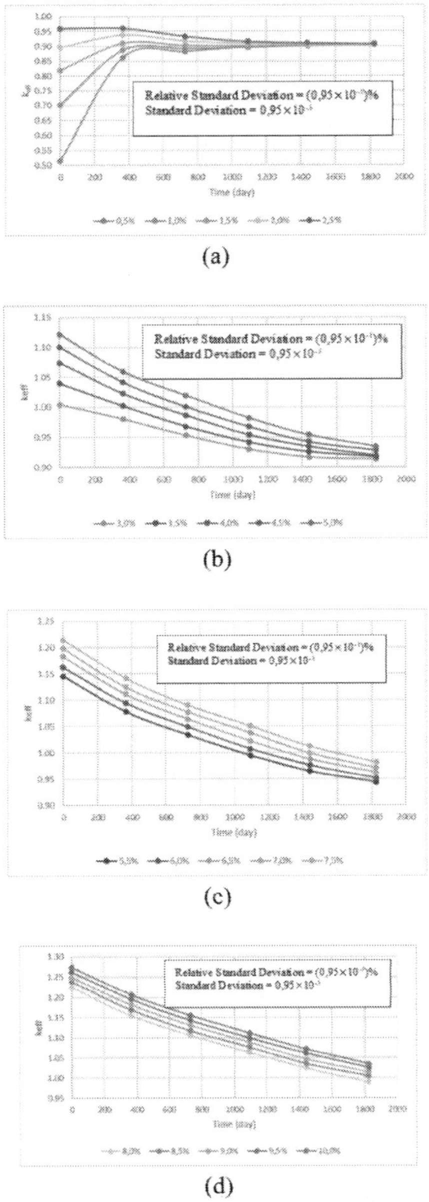

(a) (b) (c) (d)

Figure 10. Graph of multiplication factor against time for VVER-1000 burnup for 5 years with variations in U-235 enrichment: (a) 0,5% - 2,5%; (b) 3% - 5%; (c) 5,5% - 7,5%; and (d) 8% - 10%.

IOP Conf. Series: Materials Science and Engineering **1330** (2025) 012008 doi:10.1088/1757-899X/1330/1/012008

In general, from the graph it can be seen that the value of the multiplication factor will decrease over time. This is because the fuel used will decrease over time so that the number of atomic isotopes, especially U-235, will decrease so that the number of neutrons produced will decrease over time. Apart from that, relatively small standard deviation and standard deviation values were also obtained. This shows that the distribution of k_{eff} results for each variation obtained tends to have a small difference in value or in other words is close to the k_{eff} obtained. The confidence interval used is 99%.

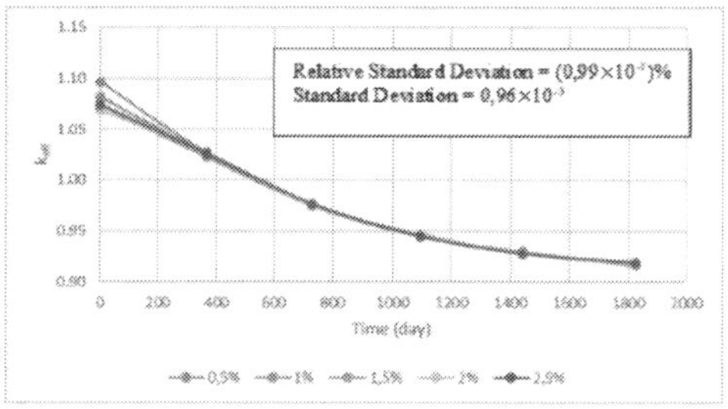

(a)

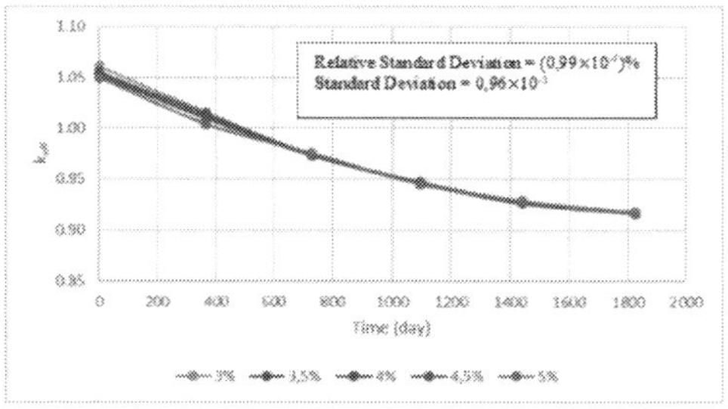

(b)

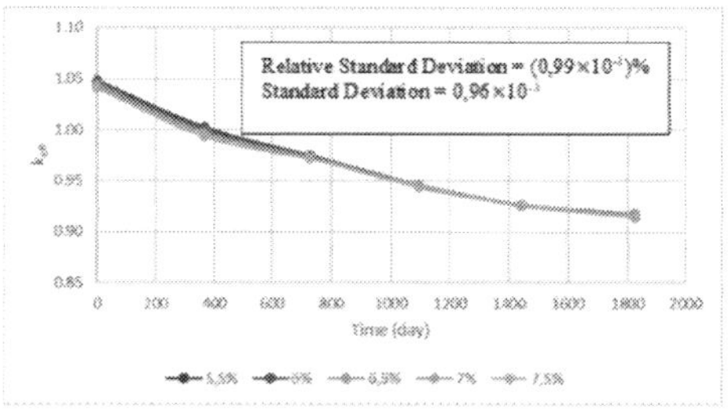

(c)

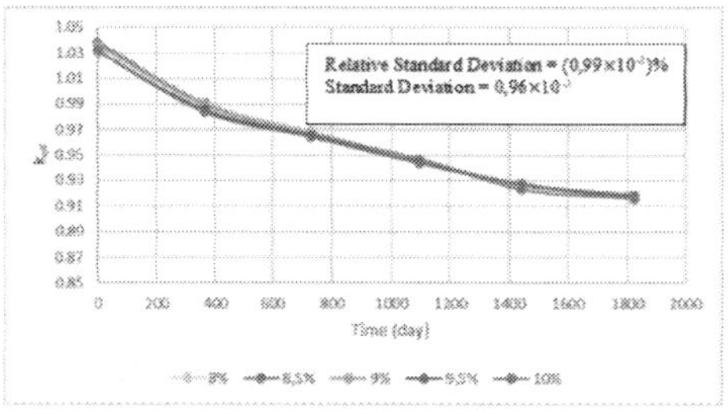

(d)

Figure 11. Graph of multiplication factor against time for VVER-1000 burnup for 5 years with variations in U-235 enrichment: (a) 0.5% - 2.5%; (b) 3% - 5%; (c) 5.5% - 7.5%; and (d) 8% - 10%.

IOP Conf. Series: Materials Science and Engineering **1330** (2025) 012008 doi:10.1088/1757-899X/1330/1/012008

In general, from the graph it can be seen that the value of the multiplication factor will decrease over time. This is because the fuel used will decrease over time so that the number of atomic isotopes, especially U-235, will decrease so that the number of neutrons produced will decrease over time. Then, the low multiplication factor value is caused by gadolinium as a burnable poison. Apart from that, relatively small standard deviation and standard deviation values were also obtained. This shows that the distribution of k_{eff} results for each variation obtained tends to have a small difference in value or in other words is close to the k_{eff} obtained. The confidence interval used is 99%.

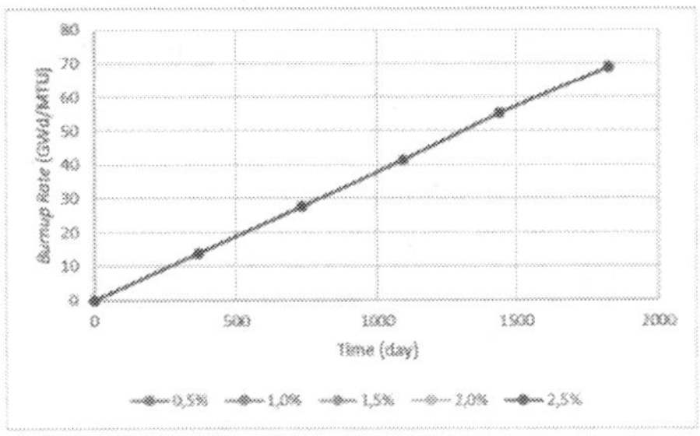

(a)

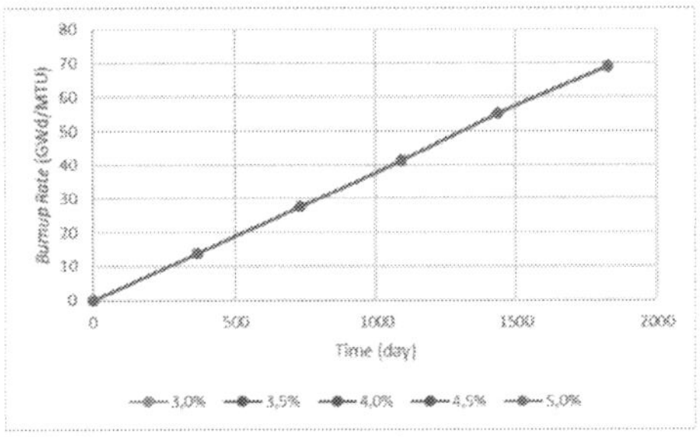

(b)

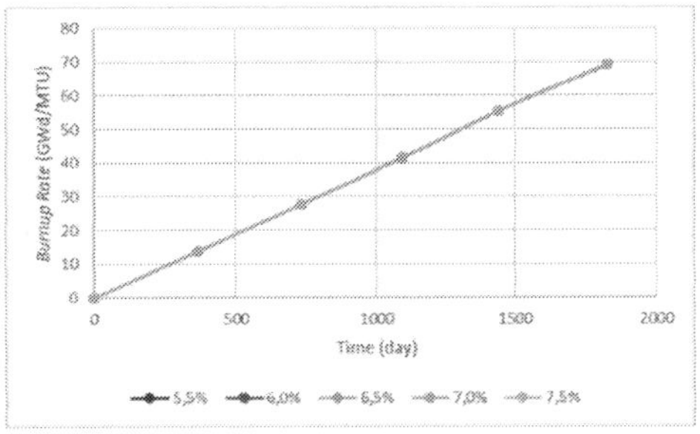

(c)

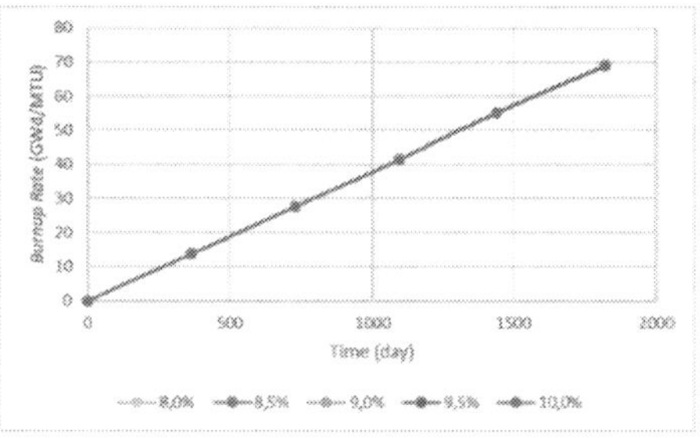

(d)

Figure 12. Graph of the relationship between burnup rate and time with variations in U-235 enrichment: (a) 0.5% - 2.5%; (b) 3% - 5%; (c) 5.5% - 7.5%; and (d) 8% - 10%.

Based on the data obtained, it can be seen in Figure 12 that for each variation of U-235 enrichment, the burnup rate value has a very small difference so it can be said that variations in U-235 enrichment do not affect the burnup rate value. This is because the amount of U-235 fuel used is very small compared to U-238, which means that even though the U-235 enrichment is varied, the difference in U-235 mass is not very significant. Then, it can also be seen that for longer reactor operations, the resulting burnup rate value will also be greater (reaching almost 70 GWd/MTU).

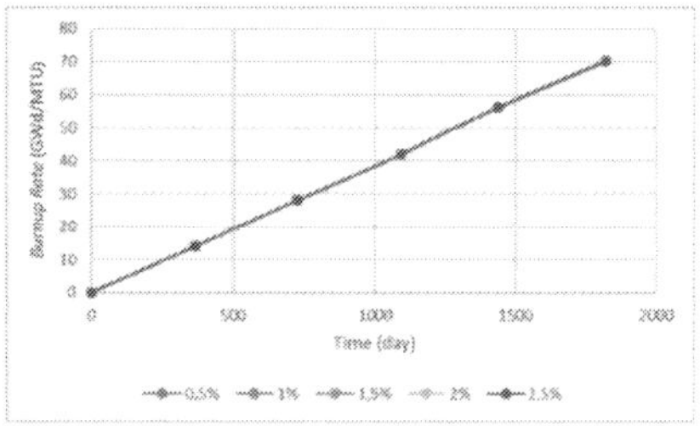

(a)

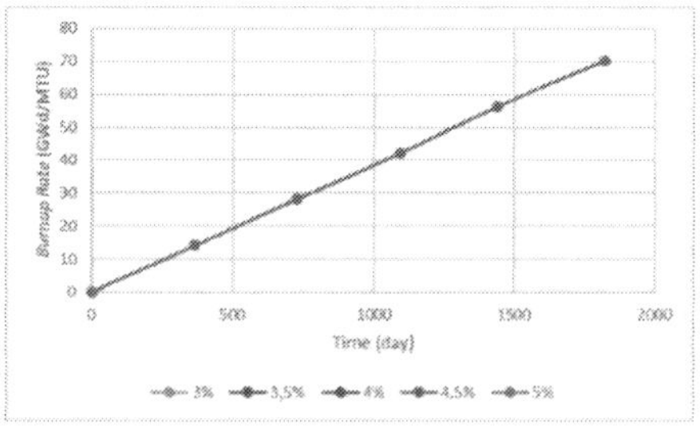

(b)

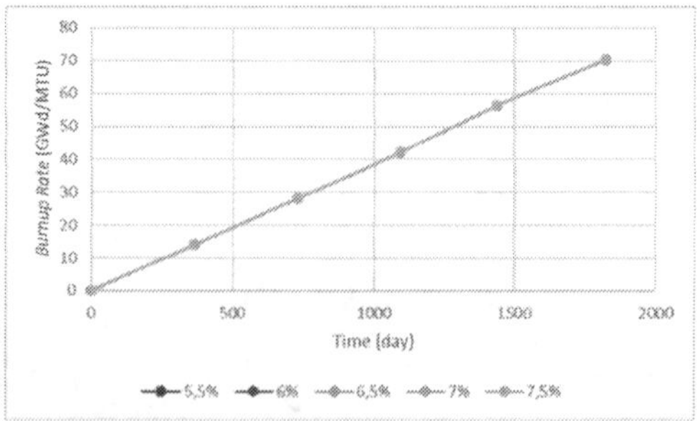

(c)

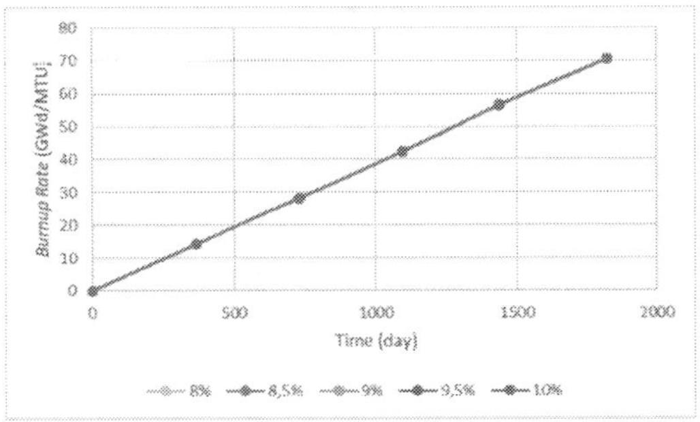

(d)

Figure 13. Graph of the relationship between burnup rate and time with variations in the Gd_2O_3 fraction: (a) 0.5% - 2.5%; (b) 3% - 5%; (c) 5.5% - 7.5%; and (d) 8% - 10%.

Based on the data obtained, it can be seen in Figure 13 that for each variation of U-235 enrichment, the burnup rate value has a very small difference so it can be said that variations in U-235 enrichment do not affect the burnup rate value. This is because the amount of Gd fraction used is compared to the amount of fuel which results in that even though the Gd fraction is varied, the difference in the mass of the Gd fraction is not very significant. Then, it can also be seen that for longer reactor operations, the resulting burnup rate value will also be greater (reaching almost 70 GWd/MTU).

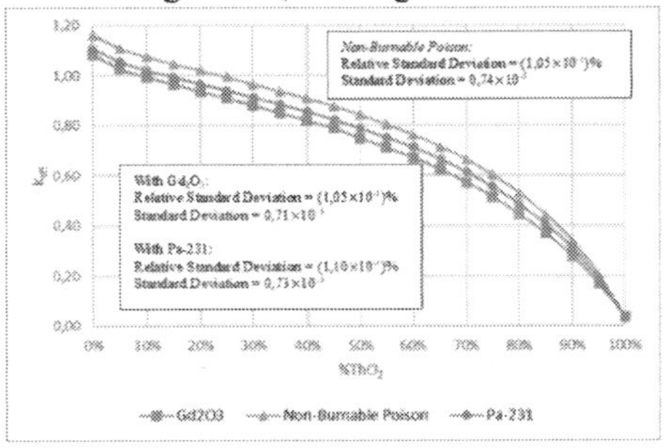

Figure 14. Comparison of multiplication factors with 3 fuel assembly conditions (non-burnable poison, Gd_2O_3, and Pa-231).

The graph in Figure 14 shows the multiplication factor value of the fuel assembly with variations in ThO_2 (without using U-233) and using and without using burnable poisons. The value of the multiplication factor decreases as the ThO_2 fraction increases, this is because thorium is a fertile fuel that cannot trigger a fission reaction, whereas for fissile fuel the presence of ThO_2 causes the quantity of fissile fuel to decrease. Then, it can also be seen that the multiplication factor value of giving Pa-231 is greater than Gd. This is because apart from the neutron absorption cross section, the next generation of Pa-231 produces U-232 and U-233 which are fissile fuels. This is proven in the graph below:

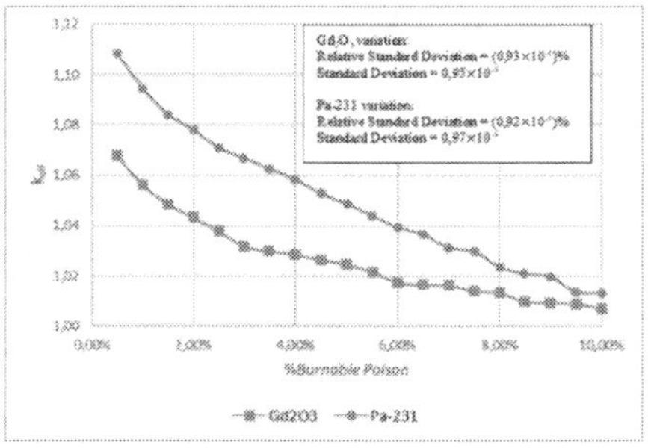

Figure 15. Comparison of multiplication factors with variations in the burnable fraction of poison gadolinium with protactinium-231.

Then, a simulation was carried out to determine the multiplication factor as a burnup function. For variations without burnable poison, Gd_2O_3, and Pa-231, the results can be seen in the graph below:

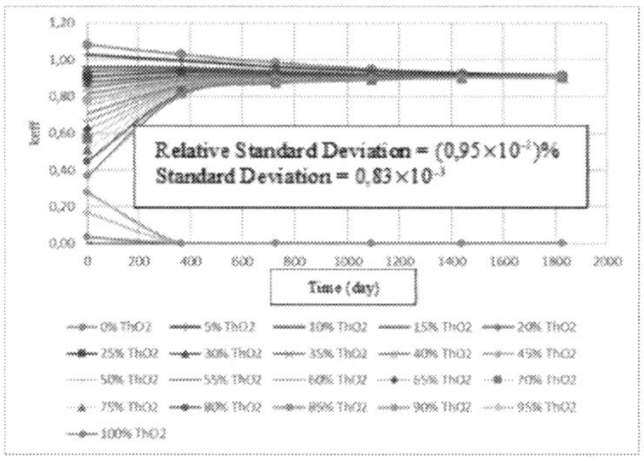

(a)

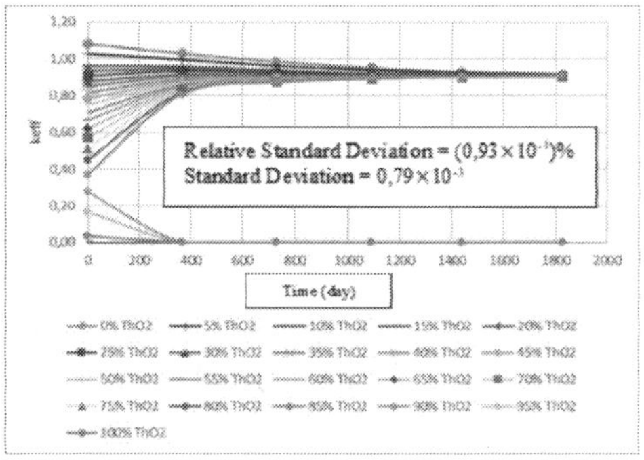

(b)

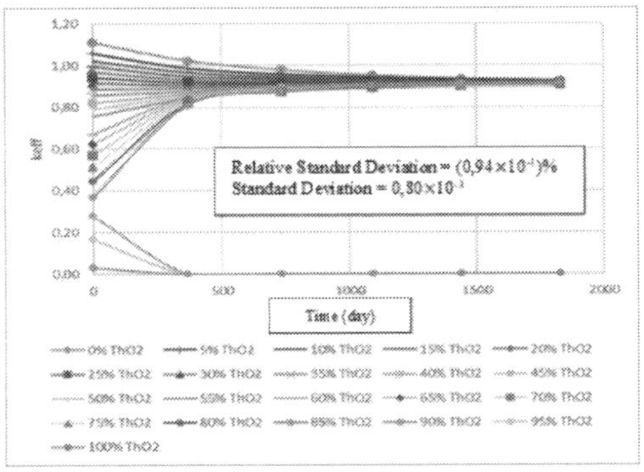

(c)

Figure 16. Graph of the multiplication factor against time for VVER-1000 burnup for 5 years with variations in ThO_2 (a) without burnable poison, (b) with Gd_2O_3, and (c) with Pa-231.

Like the previous explanation about burnup, the longer the reactor is in operation, the less fuel will be used, which will affect the value of the multiplication factor. From the data obtained, looking at the resulting effective multiplication factor value, the optimal amount of ThO_2 used as a UO_2 mixture is 5% - 15% in the case of not being given burnable poison and 5% ThO_2 when given 4% burnable poison. Likewise, if burnable poison varies by using ThO_2 of 5%, both Gd and Pa-231, the results can be seen in the following graph:

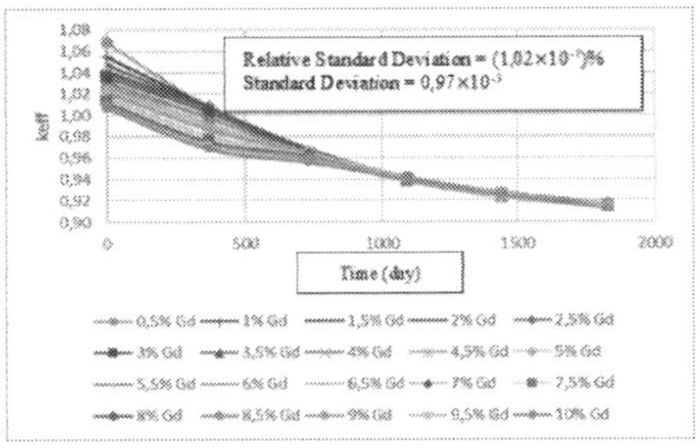

(a)

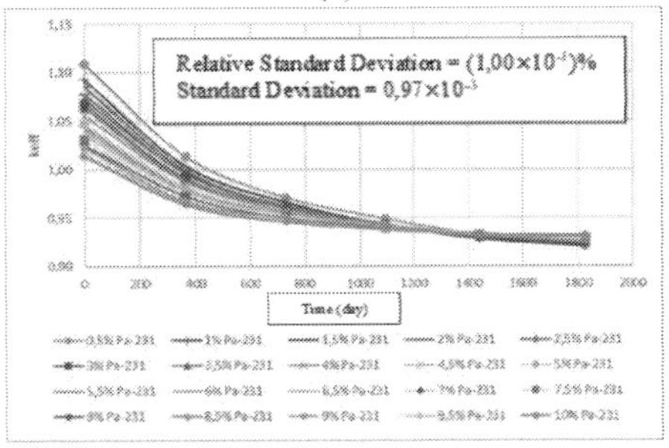

(b)

Figure 17. Graph of multiplication factor against time for VVER-1000 burnup for 5 years with ThO_2 and (a) Gd_2O_3 variations, (b) Pa-231 variations.

From this graph, it is found that the optimal amount of burnable poison used is 4% - 6% for Pa-231 and 2% - 4% for Gd_2O_3. Then, the burnup rate value for each variation was also obtained with the results which can be seen in the following graph:

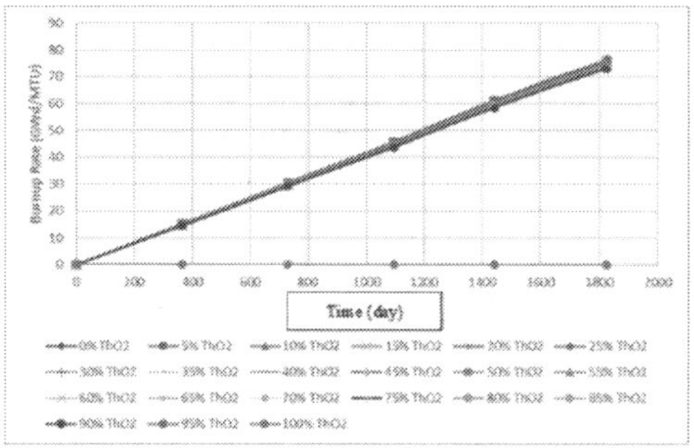

(a)

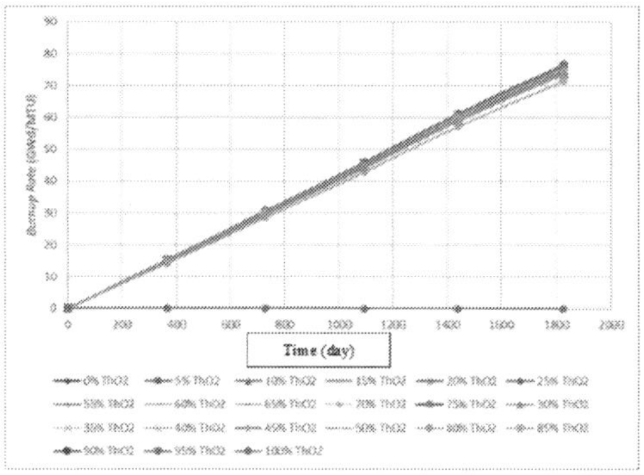

(b)

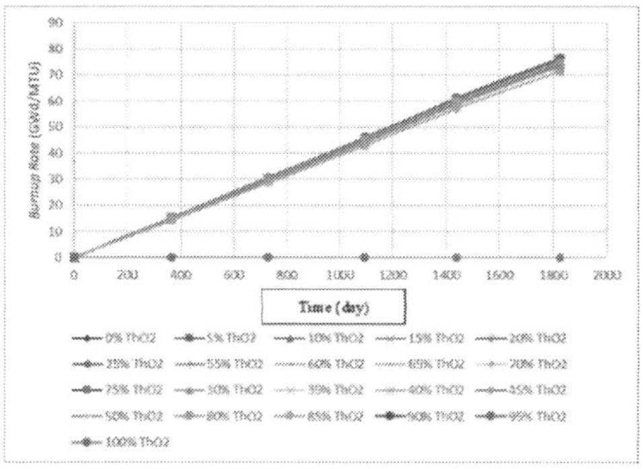

(c)

Figure 18. Graph of burnup rate versus time for VVER-1000 burnup for 5 years with (a) without burnable poison, (b) with Gd_2O_3, and (c) with Pa-231.

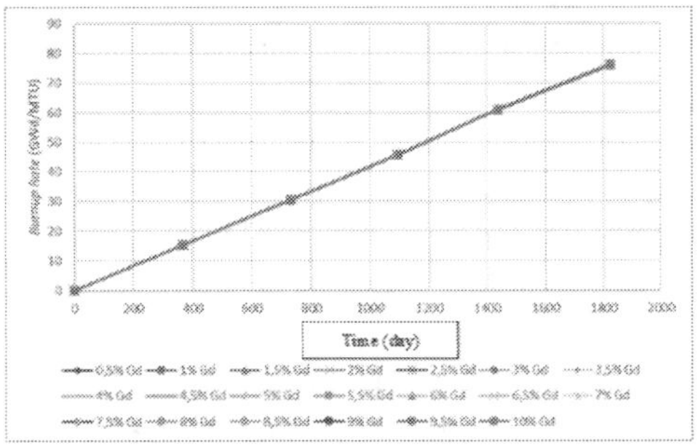

(a)

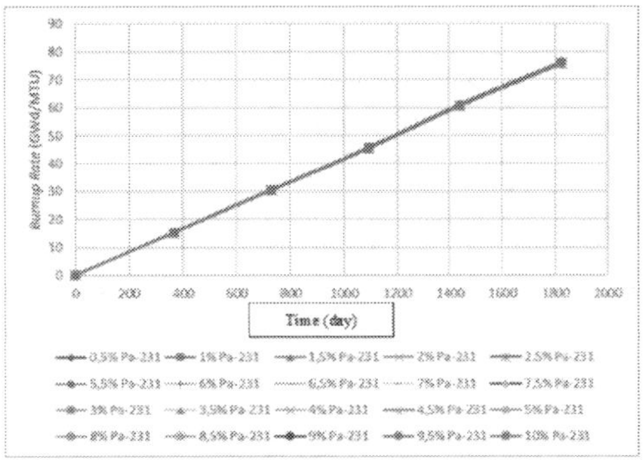

(b)

Figure 19. Graph of burnup rate against time for VVER-1000 burnup for 5 years with ThO_2 and (a) Gd_2O_3 variations, (b) Pa-231 variations.

While Pa-231 exhibits promising neutronic characteristics in terms of reactivity control and delayed neutron absorption, its practical implementation within a thorium-based fuel cycle remains subject to several critical challenges. Unlike Gd_2O_3, which is widely used as a burnable poison due to its well-characterized behavior and availability, Pa-231 is a rare and radioactive isotope, primarily derived from the decay of U-235 or through neutron capture in Th-232, making its procurement and handling more complex. Additionally, its radiotoxicity and long half-life pose storage and waste management issues that are not as prominent in Gd-based systems. From a reactor design standpoint, the inclusion of Pa-231 could offer more gradual reactivity suppression compared to Gd_2O_3; however, its integration would necessitate substantial modifications to fuel fabrication and regulation protocols. Therefore, while the neutronic performance of Pa-231 may seem favorable in isolation, demands a careful balance between performance benefits and operational feasibility.

Note that MCNP has several weaknesses. First is approximation in cross-section libraries. Accuracy of simulations is limited by the quality of cross-section data. For some isotopes, particularly minor actinides, or at high temperatures, available data may be sparse or approximate. Second is because the basis of the Monte Carlo method is statistics, the value of parameters obtained aren't 100% accurate or in other words the value is in a certain interval. For example, the value of parameters depends on the number of particles used and confidence interval chosen. Also, the number of particles affects the standard deviation value and computation time.

V. Conclusion

Based on the design and analysis of the VVER-1000 reactor using an MCNP-based simulation, several important conclusions can be drawn. First, the reactor's fuel assembly was successfully modeled using key cell parameters, including uranium fuel cells, U-Gd fuel cells, central tube cells, and guide tube cells. These models served as the basis for evaluating the neutronic performance of the system. Second, the simulation showed that variations in U-235 enrichment and Gd_2O_3 fraction significantly influence the multiplication factor. A similar trend was observed when ThO_2 was used as part of the fuel, with or without a burnable poison. Third, it was found that the addition of Th-232 reduced the multiplication factor due to its nature as a fertile fuel, especially when U-233 was not present as a fissile component. However, Pa-231 provided higher reactivity than Gd_2O_3 under similar conditions. This is attributed to Pa-231's ability to breed fissile isotopes U-232 and U-233 during reactor operation. Fourth, the behavior of the multiplication factor over 5 years of burnup indicated that reactivity trends depend on both fissile enrichment and burnable poison content. Gadolinium's effectiveness as a burnable poison was especially evident in higher enrichment levels, while lower enrichments showed less reactivity, keeping the reactor subcritical. Fifth, burnup rate analysis for different enrichment and poison fractions showed relatively minor differences across cases, suggesting that fuel consumption behavior is influenced more by the presence or absence of fissile isotopes than by enrichment alone. Last, the analysis indicates that the optimal fuel composition for the VVER-1000 system includes 5%–15% ThO_2 without burnable poison, or 5% ThO_2 with an optimal burnable poison fraction of 4%–6% for Pa-231 and 0.5%–2.5% for Gd_2O_3. The presence and amount of fissile fuel directly affect the reactivity and burnup characteristics of the system. Although Pa-231 showed favorable neutronic characteristics compared to Gd_2O_3 in thorium fuel, its practical application would require further study due to availability and radiological concerns.

The author's recommendation for this work is to use fissile materials U-235 and U-233 for ThO_2 in simulation for VVER-1000 although U-233 is expensive and dangerous. From that, the multiplication factor and burnup rate can be compared between those fissile materials and from that can be obtained the optimal amount of mixed thorium used for VVER-1000. Furthermore, this work can be continued by reviewing the reactor core with various amounts of fuel and burnable poison.

VI. References

[1] Goma I M 2021 Neutronic Performance of the VVER-1000 Reactor Using Thorium Fuel with ENDF Library *Adv. High Energy Phys.* https://doi.org/10.1155/2021/8838097

[2] Galahom, A. Abdelghafar. (2017). Minimization of the fission product waste by using thorium-based fuel instead of uranium dioxide. https://doi.org/10.1016/j.nucengdes.2017.01.024.

[3] Hassan A A *et al* 2020 Investigation of using U-233 in thorium base instead of conventional fuel in Russian PWR by SERPENT Code *J. Phys.: Conf. Ser.* **1689** 012031

[4] Kulikov G G 2017 Protactinium-231 – New burnable neutron absorber *Nucl. Eng. Des.* **256–257** 256–60

[5] Los Alamos National Laboratory 2018 *A General Monte Carlo N-Particle (MCNP) Transport Code* https://mcnp.lanl.gov

[6] Tursinah R 2019 *Simulasi Monte Carlo untuk Dosimetri Neutron*. Pelatihan Simulasi MCNP. PTKMR-BATAN, 19–20 Desember 2019

[7] Goorley J T 2013 *MCNP6 User's Manual Version 1.0 (Manual Rev)* Los Alamos National Laboratory, USA

[8] Brewer R 2009 *Criticality Calculations with MCNP5: A Primer* http://www.mcnp.ir/imgs/1354172977.Criticality Calculations with MCNP5 A Primer.pdf

[9] Tran H-N, Hoang V-K, Liem P H, Hoang H T P 2019 Neutronics design of VVER-1000 fuel assembly with burnable poison particles *Nucl. Eng. Technol.* **51** 1730–31 https://doi.org/10.1016/j.net.2019.05.026

Neutronic and Thermal Hydraulic Analysis of Gas Cooled Fast Reactor in Indonesia – An Overview

Kevin Wijaya[1] and Dwi Irwanto[1,2]

[1] Nuclear Physics and Biophysics Research Division, Physics Department, Faculty of Mathematics and Natural Science, Bandung Institute of Technology, Jalan Ganesha 10 Bandung 40132, Indonesia

[2] Nuclear Science and Engineering Department, Faculty of Mathematics and Natural Science, Bandung Institute of Technology, Jalan Ganesha 10 Bandung 40132, Indonesia

E-mail: kevinwijaya3001@gmail.com

Keywords: Burn up, fast reactor, GFR, hydraulic thermal, multiplication factor

Abstract

Gas-cooled Fast Reactor (GFR) is one of the fourth-generation reactors that is still being developed. The GFR system combines the advantages of a fast spectrum system for long-term sustainability of uranium resources and waste reduction (through fuel reprocessing and long-lived actinide fission) with a high-temperature system (industrial use and high thermal cycle efficiency). Several types of reactors, research methods have been carried out such as the type of output power produced, fuel variations, fuel composition variations, ring geometry variations, and core configuration variations have been carried out. Research has been carried out related to the thermal hydraulic analysis of GFR using various numerical methods such as Computation Fluid Dynamics, Runge Kutta method up to Order 5, a genetic algorithm, and MATLAB code to find some parameters i.e. d. This article discusses the results of several neutronic and thermal hydraulic analyzes from research that has been carried out and discusses things that can be developed from previous GFR research.

1. Introduction

High-temperature, helium-cooled fast-spectrum reactors with a closed fuel cycle, GFR systems combine the advantages of fast-spectrum systems for long-term uranium resource sustainability and waste reduction (through fuel reprocessing and long-lived actinide fission) with high-temperature systems (industrial use and high thermal cycle efficiency) [1]. GFRs do not have significant technical maturity among the advanced reactors being considered by vendors and governments, and none are operating at a rigorous experimental level. However, laboratory-scale design studies and testing of materials and fuels continue to be conducted in Europe and the United States at modest levels. GFRs promise high fuel utilization from fast-spectrum reactors with high outlet temperatures. This supports high-efficiency generation of electricity and process heat for industrial applications [2].

Gougar et al (2015) published a technology maturity assessment report that identified critical subsystems within a reactor system to achieve performance and safety objectives. Some advanced reactor concepts may have common systems and subsystems. Common systems or subsystems in multiple reactor concepts (particularly GFRs) include:

- Core instrumentation
- High temperature structural alloys and joining techniques

- Brayton Gas Cycle
- Multi-module control

This article contains a brief discussion of the GFR, the development of study results on neutronic and thermal hydraulic analysis working in the GFR [3].

2. GFR explanation

Then, based on the power generated, GFR consists of GFR600 and 2400 which each generate power of 600 MWth and 2400 MWth. GFR600 was developed by Europe and Japan, but because it did not meet the utilization target, GFR2400 was developed.

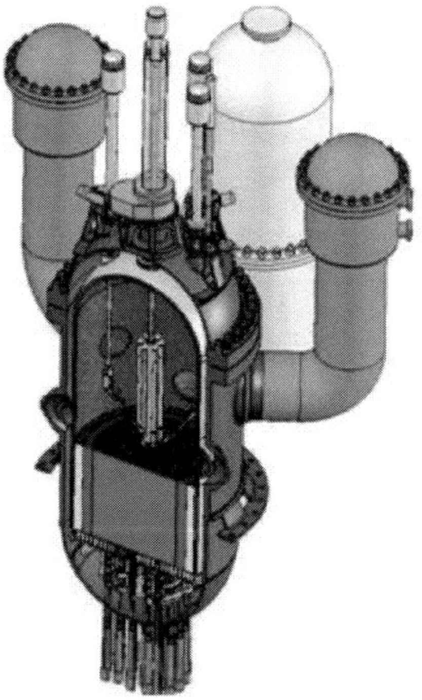

Figure 1. GFR2400 schematic [1]

3. GFR neutronic analysis

Existing nuclear calculation tools and data libraries must be validated for gas-cooled fast reactor designs to meet neutronic and zero-power requirements. Validation studies on sodium-cooled fast reactors must be complemented by special experiments that incorporate special elements of gas-cooled designs, such as slightly different spectral conditions, innovative materials, various ceramic materials (UC, PuC, SiC, ZrC, and Zr_3Si_2), and unique abnormal conditions (depressurization, steam ingress) [1]. Several studies have been conducted to perform neutronic analysis of various variations of GFR output power, fuel, and fuel volume fraction using various codes such as Monte Carlo N Particle (MCNP), openMC, and SRAC.

Raflis et al (2020) varied the Plutonium fuel by 5-25% and used the MCNP6 code. The fuel cell, fuel assembly, and GFR core schematics can be seen in Figures 1, 2, and 3 [4].

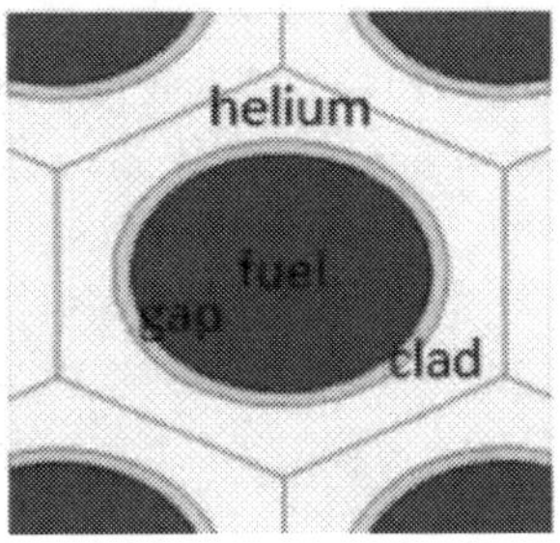

Figure 2. Fuel cell [4].

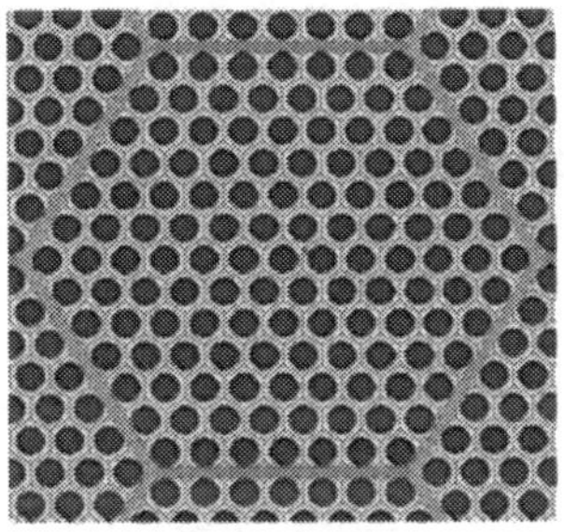

Figure 3. Fuel assembly with 127 cells [4].

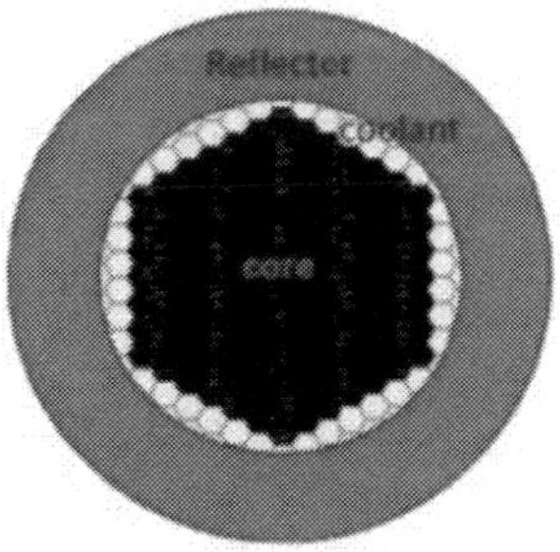

Figure 4. GFR core with 127 fuel assemblies [4].

Variations in the type of fuel used adalah Carbide (U-Pu)C, Nitride (U-Pu)N, Oxide (U-Pu)O_2 and Metallic (U-Pu)Zr. The parameters of each type of fuel can be seen in The GEN IV GCFR: Status of Studies, Pioneering Science and Technology (2005). In general, the effective multiplication factor value (k_{eff}) is more than 1 when the amount of plutonium reaches 14%

and for the variation of fuel types, (U-Pu)C has the highest k_{eff} compared to other types of fuel. Then, the burn up calculation for 8000 days was also obtained.

Furthermore, Ilham et al (2020) conducted a study similar to Raflis et al (2020) except that the plutonium variation used was 10-20% and the fuel fraction used was 40%, 50%, and 60%. Details regarding the parameters used can be seen in Ilham et al (2020). From the results obtained, there are differences in k_{eff} from fuel variations, both in terms of fractions and types used. In addition, flux distributions from various models with variations in %Pu were also obtained [5]. Furthermore, Angelina et al (2024) conducted a study focusing on heterogeneous radial directions with thorium fuel and using a comparison of SRAC and openMC codes. The ring variations used can be seen in Table 1 and Figure 4 [6].

Tabel 1. Geometric ring variations [6].

Variation	Fuel 1 (F1)	Fuel 2 (F2)	Fuel 3 (F3)
1	1 ring	4 rings	1 ring
2	2 rings	2 rings	2 rings
3	1 ring	2 rings	3 rings
4	1 ring	3 rings	2 rings
5	1 ring	1 ring	4 rings

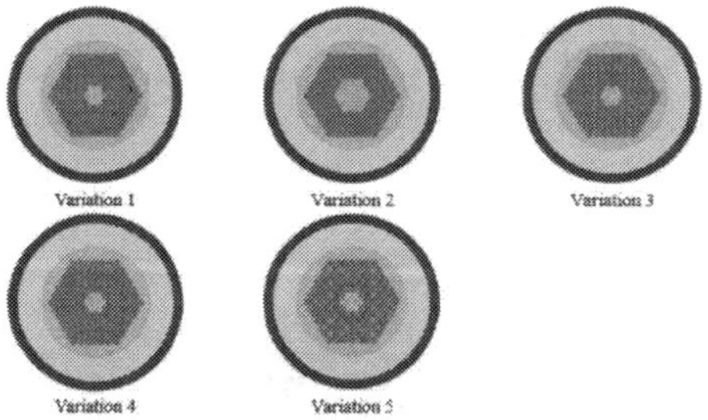

Figure 5. Design variations of ring geometry [6].

The results of the study were obtained in the form of a comparison of k_{eff} openMC with SRAC, as well as a comparison of k_{eff} to variations in ring geometry. There is a difference in k_{eff} values between openMC and SRAC and the most optimal design is the second variation. Syarifah et al (2023) conducted a study using UN-PuN fuel and varying the fuel fraction, plutonium percentage, and coolant volume. Several results were obtained in the form of k_{eff} values for 10 years from each variation and a power density distribution graph from the X, Y, and Z directions. A power density of zero indicates that the area is a reflector that causes no chain fission reaction [7].

Syarifah et al (2017) conducted a study by comparing homogeneous and heterogeneous core configurations using thorium and U-233 fuels. By varying the enrichment of U-233, the maximum value of k_{eff} and power density of the heterogeneous core configuration were greater than those of the homogeneous core configuration [8]. Irka et al (2019) investigated the GFR using the Modified CANDLE combustion scheme in the radial direction with reactor output power varying from 300 to 600 MWth. The area distribution scheme in the radial direction and randomization can be seen in Figures 5 and 6 [9].

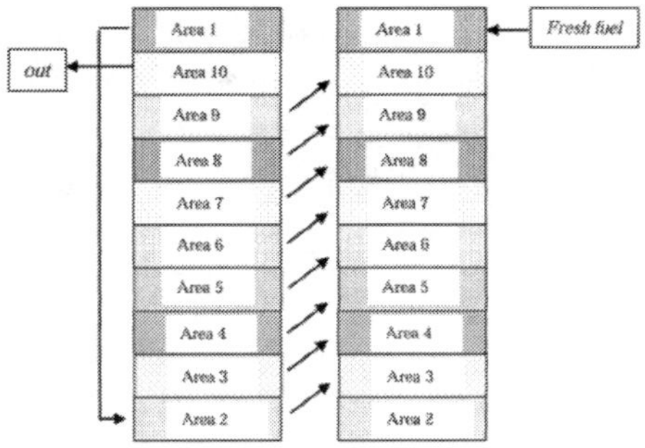

Figure 6. Schematic of area distribution in radial and random directions [9].

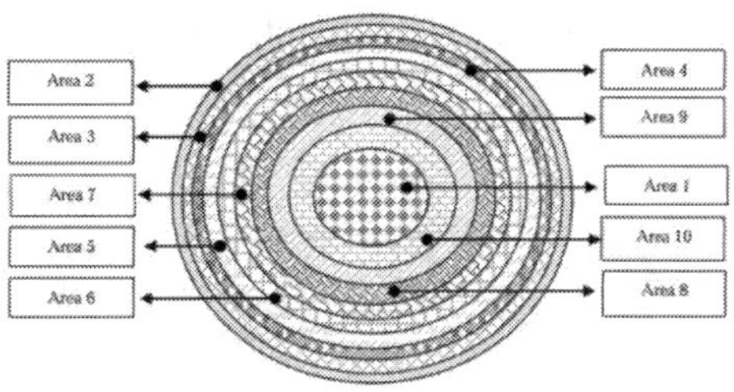

Figure 7. Schematic of area distribution in radial and random directions [9].

Variations in the height power and core diameter were carried out. Data were obtained in the form of k_{eff} and k_{inf} curves against time and conversion ratio values. The k_{eff} value increased over time while the conversion ratio value decreased over time. The density of U-238 atoms was also obtained which decreased over time and the density of Pu-239 atoms which increased and eventually began to decrease slightly over time. Several other studies related to GFR have been conducted.

4. GFR thermal hydraulic analysis

Air and helium tests on sub-assembly mock-ups are required for core thermal hydraulics to evaluate the heat transfer and pressure drop uncertainties of a particular GFR technology. The ALLEGRO licensing process requires a large-scale demonstration of passive DHR functionality to assess the transient behavior of the system. This will be done through air and then helium tests [1].

From research conducted by Han et al (2005), a thermal hydraulic analysis code for GFR was created using a block element heat transfer model. The helium energy equation was solved implicitly. In normal operation, annular fuel showed a peak temperature 80°C lower than solid fuel for the same power in a Japanese high-temperature engineering test reactor (HTTR), although the pressure drop was higher in annular fuel [10].

Hutli and Kridan (2023) performed numerical calculations using the MIGHT code that can perform sub-channel thermal-hydraulic analysis of (GCFBR) cooled by helium (He). Calculations and comparisons of temperature and heat flux distribution profiles were performed for various types of power distribution. For the cosine power profile, the coolant mass flow rate and power level affect the thermal-hydraulic performance of the tested GCFBR. The results show that, at nominal operating power (2530 MWt), the lowest flow rate for the tested reactor is 80% of the nominal flow rate (10×10^6 kg/hour). The cladding temperature is more sensitive to power level changes than mass flow rate. The results show that the maximum cladding temperature remains below the recommended GCFR design limit (700–750°C) while the power is increased by 10% and 15%. The GCFBR reactor data is commonly used for input data and code validation [11].

For GFR2400, Osusky et al. (2018) conducted a study on the development of a thermal-hydraulic approach that can be used for the combined calculation of GFR2400. The CFD analysis performed the supporting analysis; the axial distribution of fission reaction rates used in the CFD analysis was calculated with the Monte Carlo module KENO VI in the SCALE code. The FLUENT code in the ANSYS software package was used to investigate the effective heat transfer coefficient. The estimated effective heat transfer coefficient was taken from data published in the GoFastR project and fed into the NESTLE thermal-hydraulic subchannel code. The FLUENT calculation and the previous thermal-hydraulic calculation were performed with the FlowVision CFD code. One of the important innovations in the current design is the application of a W14Re layer as a refractory metal coating. This layer improves the retention of fission products on the cladding, which results in a significant neutronic penalty during normal operation. In addition, the W14Re layer prevents cladding carbonization caused by the interaction of the cladding with the carbide fuel. Consisting of reinforced silicon carbide (SiCf/SiC), its slack. The neutron spectrum of GFR is one of the loudest among fast reactors because there is practically no moderation. As a result, recycling all actinides, including minor actinides (MA), is an ideal choice. The results of the study showed that there were differences in the results between types of codes, including differences in the effective heat transfer coefficient values as a function of temperature, coolant temperature and fuel temperature as a function of axial position [12].

Indonesia has conducted various studies focusing on the thermal-hydraulic analysis of GFR. Anggraeni (2018) conducted a study using the Runge Kutta method up to Order 5 and obtained results in the form of maximum temperature values at the fuel centerline and cladding surface temperature. In addition, other hydraulic thermal parameter values were also obtained such as the Reynolds number, Prandtl number, Nusselt number, coolant diffusion coefficient, friction pressure drop, pressure drop due to shape, pressure drop due to gravity, total pressure drop, and axial coolant temperature distribution to the length of the channel in the form of inlet and outlet temperatures [13]. Then, Drajat et al (2012) conducted a multi-channel GFR study using a genetic algorithm and obtained several results such as minimum pressure drop and maximum outlet temperature with variations in mass flow rate [14]. Saro (2019) conducted a study on power distribution and maximum temperature using MATLAB code and in a steady state. The results showed that the power distribution and maximum temperature values were still within safe limits.

5. Discussion

This article has reviewed various studies on the neutronic and thermal-hydraulic analysis of Gas-Cooled Fast Reactors (GFR). In general, GFRs can utilize different types and compositions of nuclear fuel, with heterogeneous radial and core configurations proving superior to homogeneous arrangements. Most of the studies employ the Monte Carlo method for simulations, which is widely used in numerical analysis involving random sampling from a large sample space to predict system behavior. In nuclear applications, due to the stochastic nature of neutron behavior and the large number of particles involved, the Monte Carlo method is ideal for estimating parameters such as the neutron multiplication factor (k).

Several studies also incorporate the CANDLE burnup method, particularly modifications developed by Prof. Zaki Su'ud. The CANDLE method offers distinct advantages in terms of fuel sustainability and long-life reactor operation, as detailed in the work of Sekimoto and Tomita (2004) [16]. Given these benefits, future research could explore comparative studies of burnup performance using the CANDLE method versus traditional methods, as well as comparisons with other approaches, including small modular reactor (SMR) configurations. Additionally, current studies focus primarily on reactor powers in the 300–600 MWth range; limited work exists on the 2400 MWth scale, presenting opportunities for further exploration.

In terms of coolants, supercritical CO_2 (s-CO_2) is a promising alternative to helium and warrants further investigation due to its thermophysical properties. For thermal-hydraulic analysis, studies have employed various numerical methods, with the Runge-Kutta method favored for its improved accuracy compared to simpler approaches like Euler's method. Future development in this field could include studies under unsteady-state conditions, application of alternative numerical methods, implementation of multi-channel models, and determination of critical operating parameters when operating near material limits. Comparative studies across different power levels and coolant types also represent valuable research directions.

6. Conclusion

The neutronic and thermal-hydraulic behavior of GFRs has been widely studied using numerical methods, with the Monte Carlo and Runge-Kutta methods being particularly prominent. GFRs demonstrate flexibility in fuel usage and core configuration, with heterogeneous designs offering performance advantages. The CANDLE burnup strategy, especially its modified forms, offers a promising avenue for sustainable reactor operation and warrants further comparative evaluation.

Research opportunities remain, particularly in exploring high-power GFR configurations (e.g., 2400 MWth), evaluating alternative coolants such as s-CO_2, and expanding numerical modeling approaches to include transient conditions and multi-channel analyses. These directions can significantly contribute to advancing the design and performance of future GFR systems.

This article has discussed several studies on neutronic and thermal hydraulic analysis of GFR. In general, GFR can use various types of different fuels and with different fuel compositions and obtain heterogeneous radial directions and heterogeneous core configurations that are better than homogeneous ones. In general, this study uses the Monte Carlo method in conducting simulations or numerical analysis, the Monte Carlo method is generally used in numerical simulations where repeated sampling of a large random sample space is used to obtain results. In the nuclear case, the neutron particles produced from the fission reaction are very large in number and the movement of the neutron particles cannot be determined with certainty so that a random sampling technique is carried out in order to provide an overview of the movement of neutron particles and the number of neutrons produced from the fission reaction (in this case the value of the multiplication factor (k). In addition, there are several studies that use the CANDLE burn up method, especially the CANDLE modification initiated by Prof. Zaki Su'ud. For an explanation of the advantages of the CANDLE method, see the article by Sekimoto and Tomita (2004) [16]. Considering the advantages offered by the CANDLE method, this topic presents potential for further research, such as a comparative study of burnup performance with and without the CANDLE approach, as well as with other methods or the small modular reactor (SMR) design, where applicable. Then, the type of reactor power studied is worth 300-600 MWth while for the 2400 MWth version there is still little research found so this can open up new opportunities to make the GFR 2400 MWth and the CANDLE method a research topic. For coolants such as s-CO_2 can be considered as an alternative to Helium to be studied.

For the case of hydraulic thermal analysis, analysis has been carried out using various numerical methods. Runge-Kutta is used because it has higher accuracy compared to other methods such as Euler for example. Then, from the research that has been done, research can be developed by considering unsteady conditions, using other numerical methods, using multi-channel assumptions, and determining maximum parameters if the desired maximum temperature is close to the melting point of the material. In addition, research can be carried out with the aim of comparing different power and coolants.

References

[1] Generation IV International Forum (GIF) 2014 *Technology Roadmap Update for Generation IV Nuclear Energy Systems* (France: OECD).

[2] Idaho National Laboratory 2017 *Gas-Cooled Fast Reactor Research and Development Roadmap*

[3] Gougar H, Bari R, Kim T K, Sowinski T and Worrall A 2015 *Assessment of the Technical Maturity of Generation IV Concepts for Test or Demonstration Reactor Applications* INL/EXT-15-36427 Rev. 1 (Idaho Falls, ID: Idaho National Laboratory)

[4] Helen Raflis et al 2020 *J. Phys.: Conf. Ser.* **1493** 012008.

[5] Muhammad Ilham *et al* 2020 *J. Phys.: Conf. Ser.* **1493** 012007.

[6] Tsania Eksa Angelina *et al* 2024 *J. Phys.: Conf. Ser.* **2734** 012066.

[7] Syarifah R D, Hidayatul A M, Arkundato A and Nugroho A T 2023 Design study of 300 MWth GFR with UN–PuN fuel using SRAC-COREBN code *ARPN J. Eng. Appl. Sci.* **18** 269 ISSN 1819-6608.

[8] Syarifah R D, Yulianto Y, Su'ud Z and Irwanto D 2017 Neutronic analysis of thorium nitride (Th, U-233)N fuel for 500 MWth gas cooled fast reactor (GFR) long life without refueling Key Eng. Mater. 733 47–52 https://doi.org/10.4028/www.scientific.net/KEM.733.47.

[9] F H Irka *et al* 2021 *J. Phys.: Conf. Ser.* **2072** 012013

[10] Han K H, Seo K W, Hwang D H and Chang S H 2005 Development of a thermal hydraulic analysis code for gas-cooled reactors with annular fuels *Nucl. Eng. Des.* **235** 1947–1957 https://doi.org/10.1016/j.nucengdes.2005.07.006.

[11] Hutli E and Ramadan K 2023 Thermal-hydraulic analysis of a high-temperature helium-gas-cooled reactor under various steady-state operating conditions *Heat Transf. Res.* **54** (DOI: 10.1615/HeatTransRes.2023044244).

[12] Osusky F, Vrban B, Luley J, Cerba S and Hascik J 2018 Definition of the thermal-hydraulic model of the gas-cooled fast reactor *J. Nucl. Res. Dev.* **15** https://inis.iaea.org/search/search.aspx?orig_q=RN:51083153

[13] Anggreani A A, Yulianti Y and Maurung P 2018 *Analisis termal-hidrolik reaktor cepat berpendingin gas (Gas Cooled Fast Reactor) menggunakan metode Runge Kutta J. Teor. Apl. Fis.* **6** 141–6 https://doi.org/10.23960/jtaf.v6i2.1913.

IOP Conf. Series: Materials Science and Engineering **1330** (2025) 012009 doi:10.1088/1757-899X/1330/1/012009

[14] Drajat R Z, Su'ud Z, Soewono E and Gunawan A Y 2012 *Multi-channel thermal hydraulic analysis of gas cooled fast reactor using genetic algorithm AIP Conf. Proc.* **1450** 246–53 https://doi.org/10.1063/1.4724148.

[15] Saro B L L 2019 *Studi awal distribusi temperatur elemen bahan bakar reaktor cepat berpendingin gas* (Undergraduate thesis, Universitas Sriwijaya, Indonesia).

[16] Sekimoto H and Tomita S 2004 *Application of CANDLE burn up on small fast reactor Proc. 5th Int. Conf. on Nuclear Option in Countries with Small and Medium Electricity Grids* (Croatia).

IOP Conf. Series: Materials Science and Engineering **1330** (2025) 012010 doi:10.1088/1757-899X/1330/1/012010

Effect of the Number of Freeze Valve Channels on Core Discharge Rate and Pressure in Core and Drain Tank

Ahmad Muzaki Mabruri[1*], Amna Yasya Mubarok[1], Bagja M B Kertasafari[1], Asril Pramutadi Andi Mustari[1,2]

[1] Department of Nuclear Science and Engineering. Bandung Institute of Technology, Bandung, Indonesia
[2] Department of Physics, Bandung Institute of Technology, Bandung, Indonesia

*E-mail: amuzaki073@gmail.com

Abstract. Abstract. Analysis of the Influence of the Number of Freeze Valve Channels on Core Discharge Rate and Pressure in Core and Drain Tank has been conducted. This research employed a 2D freeze valve design representation using the Moving Particle Semi-implicit Method. This method utilizes the behavior and movement of particles within a predefined geometry, influenced by various physical parameters within the system. The aim of this study is to observe velocity profiles and pressure profiles during the core discharge process with variations in freeze valve geometries V1, V2, and V3. The results indicate that the core discharge times for V1, V2, and V3 are 4.75 s, 6.40 s, and 6.15 s respectively. V3 exhibits a lower average system pressure compared to V1 and V3, suggesting that V3 is more effective in managing pressure during the core discharge process.

1. Introduction

A molten-salt reactor (MSR) is a type of nuclear fission reactor where the primary nuclear reactor coolant and/or the fuel is a mixture of molten salt with a fissionable material [1]. This design offers several advantages over traditional water-cooled reactors, including the elimination of the nuclear meltdown scenario and the potential for refueling while operating [2][3]. The MSR technology also offers improved safety features, such as the ability to drain the fuel mixture in an emergency, where it solidifies, quenching the reaction, and the elimination of hydrogen evolution, which reduces the risk of hydrogen explosions [4]. Furthermore, the operating temperatures of MSR reactors are significantly higher than traditional light-water reactors, typically around 700°C (1,292°F), which increases electricity-generation efficiency and process-heat opportunities. This higher temperature also presents design challenges related to the corrosivity of hot salts and the changing chemical composition of the salt as it is transmuted by the neutron flux. Additionally, MSR reactors have the potential to operate on a variety of fuels, including uranium, thorium, and even waste materials, making them a promising option for nuclear waste management and disposal [5].

The freeze valve is a crucial passive safety system in Molten Salt Reactors (MSRs), designed to ensure the safe operation and shutdown of the reactor in the event of an emergency [6]. This innovative system is particularly significant in MSRs, where the nuclear fuel is a molten salt base. The freeze valve functions by freezing the salt in the pipe connecting the reactor core and

IOP Conf. Series: Materials Science and Engineering **1330** (2025) 012010 doi:10.1088/1757-899X/1330/1/012010

the drain tanks using external cooling. In the case of a station blackout, the cooling system stops, and the frozen salt melts automatically, opening the freeze valve and allowing the liquid fuel to drain out of the reactor core to the drain tanks, thereby stopping the nuclear reaction passively [7]. The valve is made of heat-resistant materials, designed to be highly reliable and maintain structural integrity under extreme conditions, ensuring efficient and safe operation [8]. Its adaptability to various operating conditions and ability to quickly and reliably open make it an essential safety feature in MSR reactors, providing an added layer of protection against potential accidents and ensuring the safe operation of these reactors [9].

The development of freeze valves in Molten Salt Reactors (MSRs) has been a crucial aspect of enhancing the safety systems of these reactors. The concept of freeze valves in MSRs dates back to the Molten Salt Reactor Experiment (MSRE), which was the first MSR to implement the use of freeze valves. Since then, significant research has been conducted to improve the design and performance of freeze valves, focusing on enhancing their reliability and efficiency in ensuring the safe operation of MSRs [10]. The freeze valve system in MSRs is particularly significant due to the unique characteristics of molten salt fuel. In the event of a station blackout, the cooling system fails, and the reactor overheats. The freeze valve is designed to automatically open, allowing the molten salt to drain out of the reactor core and preventing a meltdown. This passive safety feature ensures that the reactor is safely shut down, even in the absence of active safety control systems. Recent studies have focused on optimizing the design of freeze valves to improve their performance, including investigating the effects of fundamental design parameters, such as thickness and orientation, on the melting behavior of the freeze valve. Additionally, researchers have explored the use of advanced materials and simulation techniques to enhance the reliability and efficiency of freeze valves. The development of freeze valves in MSRs has significant implications for the safety and efficiency of these reactors, ensuring the safe operation of MSRs and minimizing the risk of nuclear waste disposal [11].

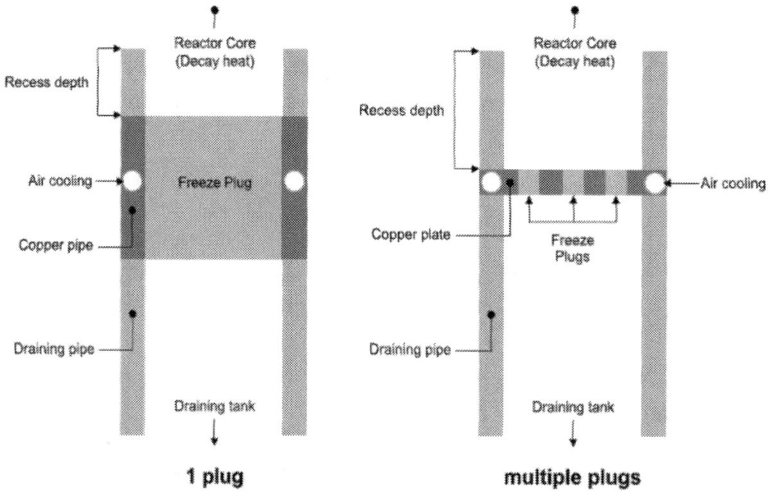

Figure 1. Single-channel and multiple-channel freeze plug design [10].

Previous study by Ilham [12] conducted systematic numerical simulations to investigate the impact of wall thickness and inner diameter of freeze plugs used in molten salt reactors (MSRs). These simulations aimed to understand how these factors influence the opening time of the freeze plug, which is crucial for ensuring the passive safety of MSRs. The research involved two main simulations: the solidification process of the frozen salt forming in the freeze plug tube, followed by the melting process. The study indicates that understanding the processes of solidification, melting, and drainage is essential to effectively use freeze plugs as a passive safety system in MSRs [12]. Another study by Tiberga [13] analyzes an unconventional freeze valve design featuring multiple salt channels, as shown in **Figure 1**. The purpose of this design is to achieve more uniform and rapid heating, thereby facilitating faster melting of the freeze valve [13].

The analysis of Freeze Plug shape by [13], focuses on the melting behavior of the freeze-plug in MSFR [13]. The main objective of this research is to investigate the influence of various design configurations and operational parameters on the freeze-plug melting process. The research findings indicate that the single-plug design is generally more effective in the melting process compared to the multi-plug design. Melting in the single-plug design occurs faster and more uniformly. In contrast, the multi-plug design is found to form a frozen layer above the metal grid that holds the plug, significantly impeding the melting process. This frozen layer forms due to uneven heat flow, causing some parts of the plug to remain frozen for a longer time [12]. The design conducted by [12], is limited to the state before the freeze valve melts. Based on this limitation, this paper attempts to analyze the fluid flow conditions after the freeze valve melts and the fluid begins to flow out from the core to the drain tank according to the freeze valve design provided by [12], to determine its impact on the core emptying time. Additionally, the flow velocity profile and the pressure profile received by the drain tank are observed in more detail.

2. Method

2.1 Moving Particle Semi-Implicit

The MPS method is used for modeling the fragmentation of incompressible fluids. MPS stands out as one of the numerical methods due to its notable numerical stability and significantly increased computational speed. The equations commonly used to describe incompressible flow are the Navier-Stokes equations, which are mathematically represented as follows:

$$\frac{D\rho}{Dt} = 0 \quad (1)$$

$$\frac{Du}{Dt} = -\frac{1}{\rho}\nabla \mathrm{P} + v\Delta^2 u + g \quad (2)$$

$$\frac{Dh}{Dt} = k\nabla^2 T + Q \quad (3)$$

With ρ, u, P, v, g, h, j, and Q representing density, velocity, pressure, kinematic viscosity, gravity, enthalpy, thermal conductivity, and heat source respectively. Eq. (1) represents the

IOP Conf. Series: Materials Science and Engineering **1330** (2025) 012010 doi:10.1088/1757-899X/1330/1/012010

conservation of mass, while Eq. (2) is the Navier-Stokes equation accounting for particle movement. Eq. (3) describes the conservation of energy. In the MPS method, interactions between particles within a certain radius are considered, which are mathematically represented in Eq. (4) [14]:

$$w(r)=\begin{cases}1-\dfrac{r}{r_e} & 0\le r\le r_e\\ 0 & r_e\le r\end{cases} \quad (4)$$

$$n_i=\sum_{j\ne i} w(|\vec{r_j}-\vec{r_i}|) \quad (5)$$

With $w(r)$ representing the weighting function that describes the interaction between particles, Eq. (5) expresses the particle density as the total sum of the weighting functions of the particles. Here, $\vec{r_i}$ and $\vec{r_j}$ are the position vectors of particles i and j respectively. The particle interactions are described by the gradient, divergence, and Laplacian operators, as shown in the following equations:

$$\langle\nabla\phi\rangle_i=\frac{d}{n^0}\sum_{j\ne i}\left[\frac{\phi_j-\hat{\phi}_i}{|\vec{r_j}-\vec{r_i}|^2}(\vec{r_j}-\vec{r_i})\,\omega(|\vec{r_j}-\vec{r_i}|)\right] \quad (6)$$

$$\langle\nabla .u\rangle=\frac{d}{n^0}\sum_{j\ne i}\left[\frac{(\vec{u_j}-\vec{u_i}).(\vec{r_j}-\vec{r_i})}{|\vec{r_j}-\vec{r_i}|^2}\,\omega(|\vec{r_j}-\vec{r_i}|)\right] \quad (7)$$

$$\langle\nabla^2\phi\rangle_i=\frac{d2}{\lambda n^0}\sum_{j\ne i}\left[(\phi_j-\phi_i)\,\omega(|\vec{r_j}-\vec{r_i}|)\right] \quad (8)$$

Where $d, n^0, \phi_j, \hat{\phi}_i$ represent the spatial dimensionality, the initial number density of particles, the scalar value of particle j at position $\vec{r_j}$, and the minimum scalar value at the target particle i respectively. The value of λ is indicated in Eq. (9).

$$\lambda=\frac{\sum_{j\ne i} w|\vec{r_j}-\vec{r_i}|(|\vec{r_j}-\vec{r_i}|)^2}{\sum_{j\ne i} w|\vec{r_j}-\vec{r_i}|}\cong\frac{\int w(r)r^2dV}{\int w(r)dV} \quad (9)$$

The MPS method employs a semi-implicit algorithm where pressure gradient is implicitly computed, while viscosity, gravity, and temperature are explicitly calculated. Momentum equation is explicitly evaluated at each time step to derive particle velocity and temporal position, followed by implicit pressure computation using Eq. (10).

$$\nabla^2 P^{k+1} = -\frac{\rho^0}{\Delta t^2}\frac{n^* - n^0}{n^0}$$

$$\boldsymbol{u}^{k+1} = \boldsymbol{u}^* - \frac{\Delta t}{\rho^0}\nabla P$$

$$\boldsymbol{r}^{k+1} = \boldsymbol{r}^* - \frac{\Delta t^2}{\rho^0}\nabla P$$

Eq. (11) and Eq. (12) indicate the position and velocity of particles, taking into account pressure gradients [15]

2.2 Specifications and Design

The effect of the number of channels in the freeze valve was evaluated with variations in channel geometry as shown in **Figure 2**. The geometries from left to right are referred to as V1, V2, and V3. Geometry V1 is a freeze valve with a single fuel discharge channel. Geometry V2 has four fuel discharge channels, which are reduced to two channels in a 2D simulation. Geometry V3 is a 2D cross-section of a freeze valve with seven channels: one central channel and six surrounding channels arranged in a hexagonal pattern, resulting in three channels in the 2D representation. The inner diameter of the fuel discharge pipes is kept constant for each variation, allowing for an analysis of the effect of the number of channels on the flow rate and pressure in the drain tank.

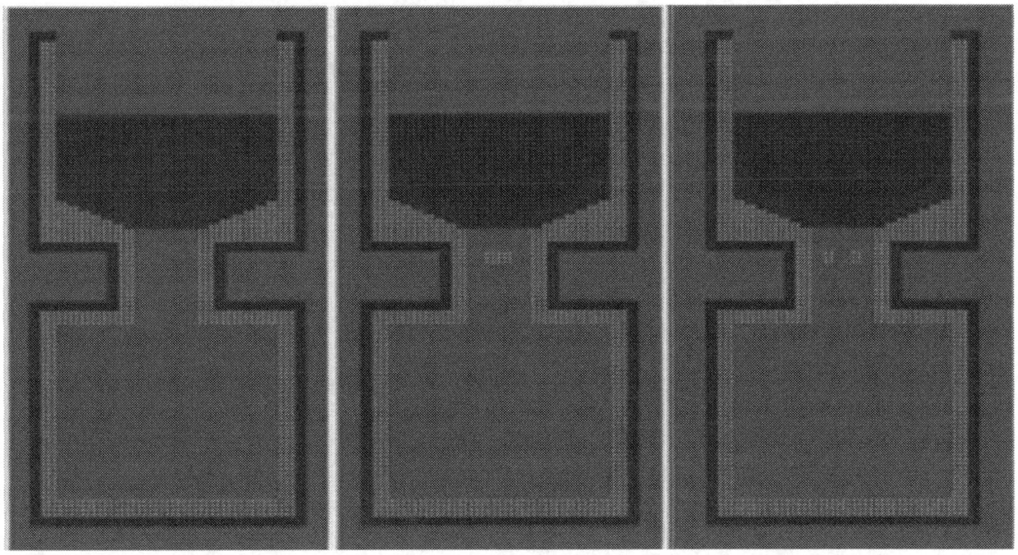

Figure 2. Geometry of MSR Terrace with 3 variations of freeze valve channels.

The blue-colored particles in **Figure 2** represent the type of fluid particles simulated using MPS. These particles represent a mixture of LiF-ThF4, which is the liquid fuel of the MSR reactor.

IOP Conf. Series: Materials Science and Engineering **1330** (2025) 012010 doi:10.1088/1757-899X/1330/1/012010

To ensure that the LiF-ThF4 mixture accurately represents the state of freeze valve melting, several parameters need to be adjusted, primarily the temperature, indicating that the reactor must be shut down immediately due to an uncontrollable temperature surge. The parameters listed in Table 1 represent the condition where the freeze valve has already melted.

Table 1. Fluid Material Specifications [16]

Parameter	Specification
Fluid	Lif-UF_4
Temperature	900 °C
Density	3948 kg/m^3
Kinematic Viscosity	1.29 x 10^{-6} m^2.s

At a temperature of 900°C, the fluid represents an overheated condition in MSR reactors, which typically operate at temperatures of 700-750°C. This state is then represented in the MPS method by describing the fluid characteristics under such conditions. The values entered include Density and Kinematic Viscosity as shown in Table 1.

3. Results and discussion

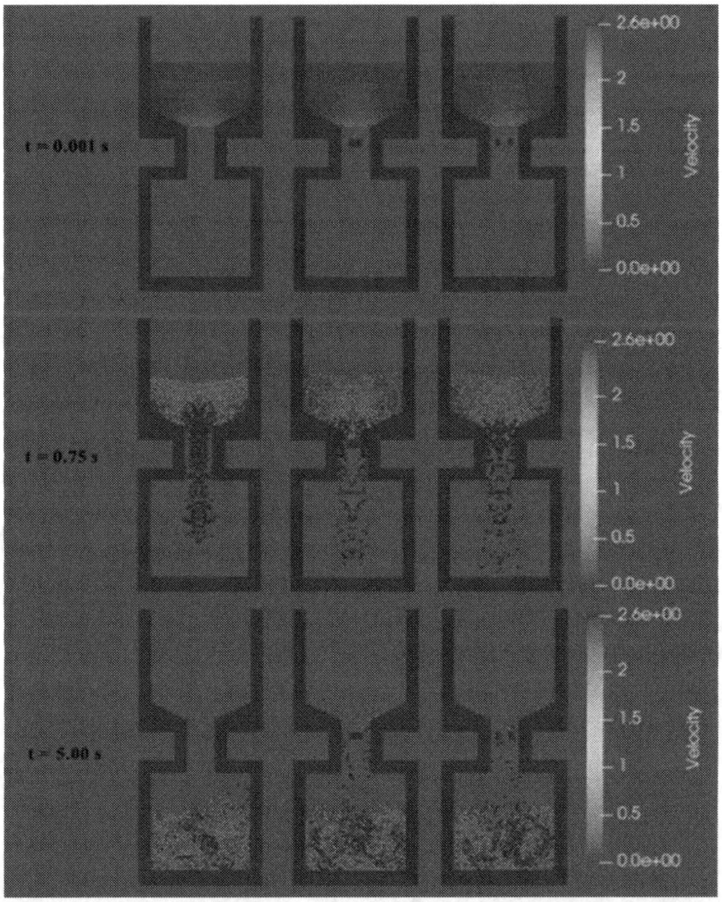

Figure 3. Particle flow velocity profile during fluid discharge process.

The effect of varying the number of channels in the freeze valve on the fuel discharge rate from the core and the pressure in the drain tank has been investigated. The impact on the discharge rate is shown in **Figure 3**, based on the time required for fluid particles to travel from the core to the drain tank. **Figure 3** illustrates the particle velocity profiles from left to right for V1, V2, and V3.

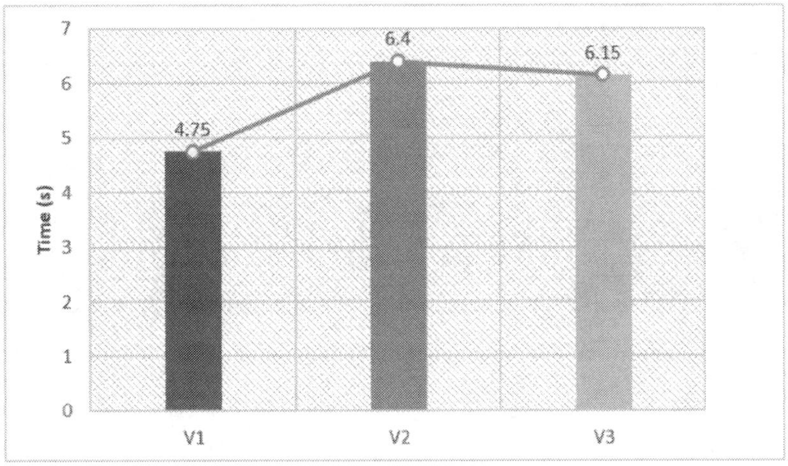

Figure 4. Core discharge time

Based on **Figure 3**, at the first-time step (t=0.001 s), particles in each variation start moving towards the drain tank with the same velocity profile. The velocity profile at the subsequent time step (t=0.75 s) indicates that V3 reaches the bottom wall of the drain tank first, followed by V2 and then V1. However, at this time step (t=0.75 s), it is evident that the particle discharge rate in V1 is significantly higher compared to V2 and V3. At the next time step (t=5.0 s), V1 is observed to be empty first, compared to V2 and V3. This is due to the absence of obstacles that would impede the particles from falling or flowing downward, resulting in a higher particle discharge rate for the same time interval. In contrast, V2 and V3 have obstacles that reduce the diameter of the channels through which the particles flow. It is observed that the particle discharge rate in V3 is higher than in V2, which is also influenced by the total channel diameter, significantly affecting the core emptying time, as shown in **Figure 4**.

Figure 4 provides a clearer explanation of the influence of the fluid particle flow rate shown in **Figure 3**. It is observed that the draining time for V1 is 4.75 seconds, corresponding to the higher particle flow rate in V1 compared to V2 and V3. The characteristics for V2 and V3 also align with the particle flow rate analysis in Figure B, resulting in core draining times for V2 and V3 of 6.4 seconds and 6.15 seconds, respectively. This difference is not significant due to the relatively small variation in total diameter between the V2 and V3 configurations, with a diameter ratio of 6:7 in this simulation.

The pressure profiles when fluid particles from V1, V2, and V3 make contact with the walls of the drain tank occur at t = 0.85 s, as shown in **Figure 5**. The pressure patterns observed in V1, V2, and V3 have been normalized within the pressure range of $0 - 10^5$ atm to illustrate the comparative pressure distribution within the same magnitude range. It is evident that there are variations in pressure profiles both within the core and the drain tank for all three configurations. In the core region, fluid exerts pressure on the lower walls of the core (cone section) as it flows towards the drain tank. In V1, the pressure in this region is not evenly distributed due to the large number of particle discharges, resulting in reduced pressure on the lower core walls. Conversely, in V2 and V3, the particle discharge is lower compared to V1. In V3, particles exert pressure on the cone tip, whereas in V2, the pressure is almost uniform across the entire cone section. This observation suggests that the amount of fluid retained in the core during the fluid transfer process towards the drain tank imparts significant pressure on the core.

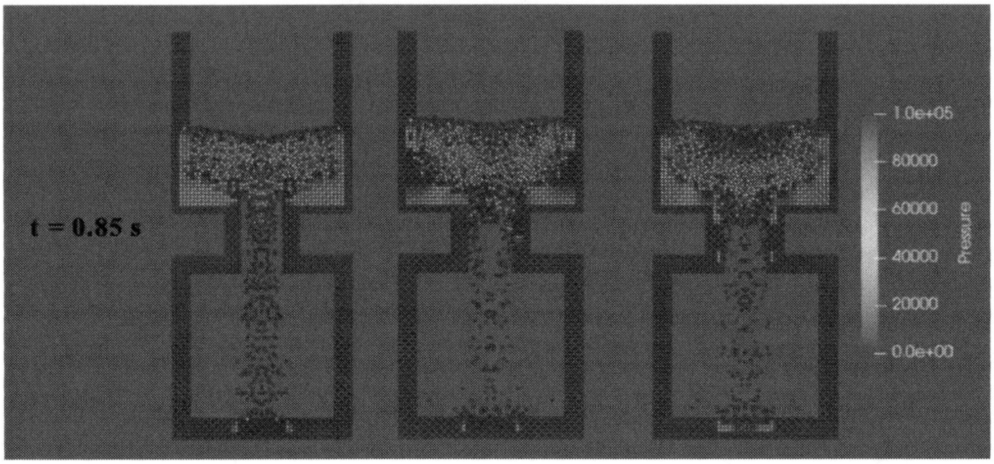

Figure 5. Pressure profile at t=0.85 s

Figure 5 also illustrates the pressure profile when the fluid falls and contacts the bottom wall of the drain tank. Based on **Figure 5**, the fluid pressure patterns for V1 and V2 exhibit similarity, where the fluid exerts maximum pressure on the drain tank over a wide area. In contrast, for V3, the maximum fluid pressure is dispersed over a smaller area. The clearer distribution pattern is depicted in **Figure 6**, representing the magnitude of pressure experienced by particles based on their particle IDs. It is evident that for V1 and V2, the pressure peaks are centralized with one highest peak, while V3 exhibits multiple peaks with a more uniform distribution. It is also observed that a decrease in fluid flow rate or particle flow rate, due to obstruction by the freeze valve channels, results in a significant reduction in pressure on the drain tank. V1 with 1 channel yields the highest pressure, V2 with 2 channels provides lower and more uniform pressure compared to V1. In V3, the pressure reduction is notably clear, with maximum values much lower than those of V1 and V2. This is because V3 is capable of dispersing fluid flow, causing particles to scatter and not exert direct pressure on the same area of the drain tank.

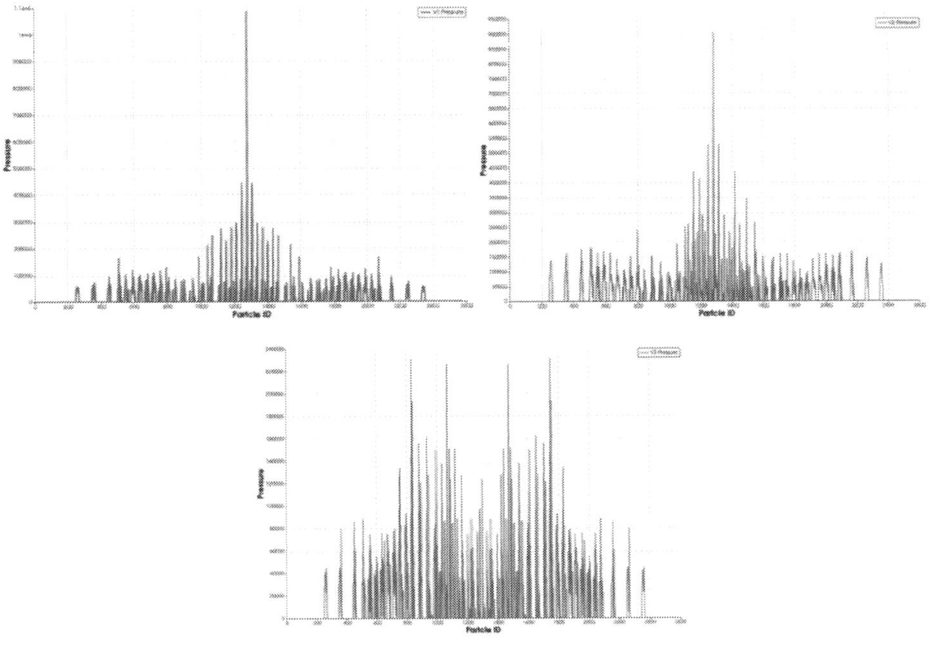

Figure 6. Pressure distribution at time t = 0.85 s

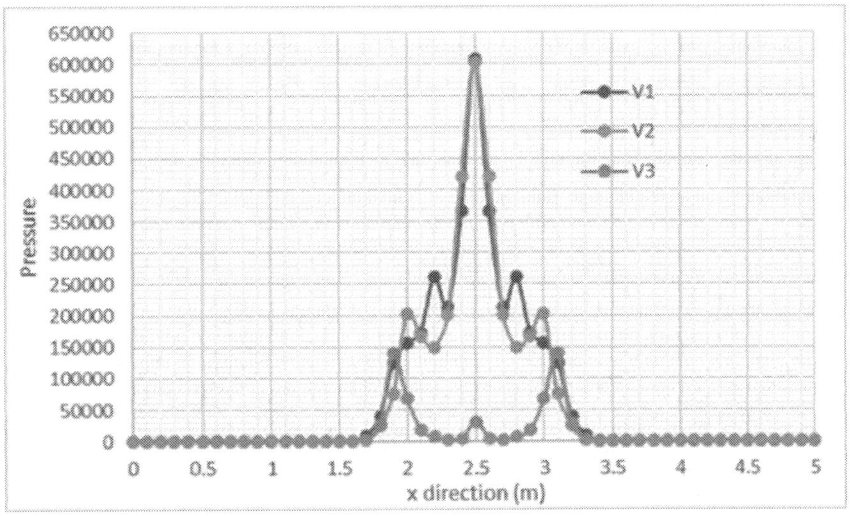

Figure 7. Average pressure at the bottom of the drain tank.

Figure 7 further represents **Figure 6** by focusing on the pressure at the bottom part of the drain tank, considering 3 layers of wall particles in that region. It is observed that the pressure results from V1 and V2 exhibit a similar profile with one peak at the center and two smaller peaks on either side. The peak pressure values are nearly identical, 607053.75 Pa/s for V1 and 603115.75 Pa/s for V2. In contrast, V3 reduces the peak pressure at the center of the drain tank and produces two smaller peaks with a pressure value of 140495.2 Pa/s.

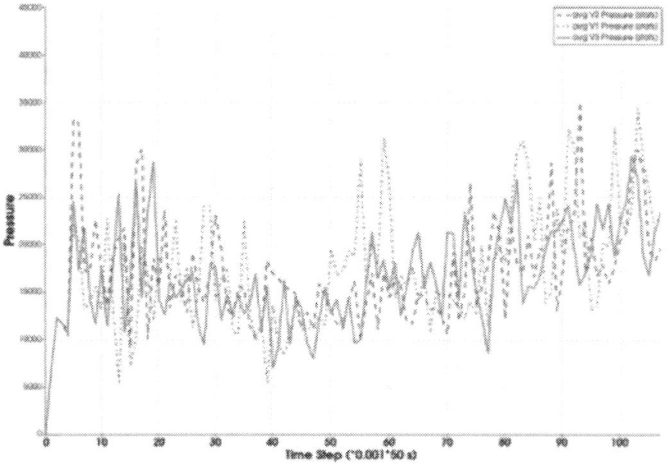

Figure 8. Average pressure in the system

Figure 8 shows a comparison of the average pressure magnitude experienced by the system in V1, V2, and V3 configurations. At time step 17 (t = 0.85 s), it is observed that the average pressure in the V2 system is greater, consistent with **Figure 5** and **Figure 6** , which show a uniformly higher maximum pressure in the V2 system. Across all time steps shown in **Figure 8**, V3 exhibits a lower average pressure magnitude compared to V1 and V2. This finding suggests that V3 is more effective in mitigating pressure during fluid discharge in this study.

4. Conclusion

The research results described previously show that variations in the number of channels in the freeze valve channel greatly influence the fluid rate profile in the core emptying process and the pressure profile exerted by the fluid on the core and drain tank. V2 and V3 respectively can provide an increase in fluid discharge time of 1.65 s and 1.40 s compared to V1. Based on the presented pressure profiles, V3 is more effective than V2 in mitigating the pressure impact exerted by the fluid on both the core and drain tank.

References

[1] R. D. Syarifah, B. A. Putri, I. K. Aji, and A. M. Mabruri, "Neutronic Analysis on Molten Salt Reactor (MSR) Using OpenMC Code With Variations of Geometry Core Fueled By LiF-BeF2-UF4," El-Cezeri Journal of Science and Engineering, vol. 11, no. 2, pp. 152–159, Jul. 2024, doi: 10.31202/ecjse.1364028.

[2] A. M. Mabruri, R. D. Syarifah, I. K. Aji, A. Arkundato, and N. Trianti, "Validation of OpenMC Code for Low-cycle and Low-particle Simulations in the Neutronic Calculation Criticality k-eff Low-cycle MSR FUJI-12 OpenMC

IOP Conf. Series: Materials Science and Engineering **1330** (2025) 012010 doi:10.1088/1757-899X/1330/1/012010

Corresponding Author," Jurnal Ilmu Fisika, vol. 16, no. 2, pp. 107–117, 2024, doi: 10.25077/10.25077/jif.16.2.107-117.2024.

[3] A. M. Mabruri, R. D. Syarifah, I. K. Aji, Z. Hanifah, A. Arkundato, and G. Jatisukamto, "Neutronic analysis on molten salt reactor FUJI-12 using 235U as fissile material in LiF-BeF2-UF4 fuel," Eastern-European Journal of Enterprise Technologies, vol. 5, no. 8(119), pp. 6–12, 2022, doi: 10.15587/1729-4061.2022.265798.

[4] Krepel, J., Hombourger, B., Fiorina, C., Nichenko, S., Kalilainen, J., Burgherr, P., Prasser, H.M. and Pautz, A., 2019. Molten Salt Reactor Research in Switzerland. In Thorium—Energy for the Future: Select Papers from ThEC15 (pp. 455-468). Springer Singapore.

[5] Merle-Lucotte, E., Heuer, D., Allibert, M., Ghetta, V. and Le Brun, C., 2008. Introduction to the physics of molten salt reactors. In Materials Issues for Generation IV Systems: Status, Open Questions and Challenges (pp. 501-521). Springer Netherlands.

[6] Riupassa, R.D., Basar, K. and Waris, A., 2022, June. Numerical Investigation of Freeze Valve Melting Behavior in Molten Salt Reactor using OpenFOAM. In Journal of Physics: Conference Series (Vol. 2243, No. 1, p. 012119). IOP Publishing.

[7] Aji, I.K., Tatsuya, T., Kinoshita, M. and Okawa, T., 2018, July. An experimental study on freeze valve performance in a molten salt reactor. In International Conference on Nuclear Engineering (Vol. 51531, p. V009T16A044). American Society of Mechanical Engineers.

[8] Chisholm, B.M., Krahn, S.L. and Sowder, A.G., 2020. A unique molten salt reactor feature–The freeze valve system: Design, operating experience, and reliability. Nuclear Engineering and Design, 368, p.110803.

[9] Mustari, A.P.A., Irwanto, D., Permana, S. and Pramuditya, S., 2021. Studi Pengaruh Perbedaan Ketebalan Freeze-valve di Msr (Molten Salt Reactor) dalam Pengendalian Kecelakaan. GANENDRA Majalah IPTEK Nuklir, 24(1), pp.29-40.

[10] Aji, I.K., 2020. Investigation of Basic Parameters in Developing High Performance Freeze Valve for Molten Salt Reactor. University of Electro-Communications.

[11] Mustari, A.P.A. and Eben, V., 2023, November. Study of Thickness Effect on Freeze-Valve Scheme for Molten Salt Reactor Safety System. In 4th Borobudur International Symposium on Science and Technology 2022 (BIS-STE 2022) (pp. 661-671). Atlantis Press.

[12] Ilham, M., Aji, I.K. and Okawa, T., 2023. Numerical investigation on the effects of fundamental design parameters on freeze plug performance in molten salt reactors. Nuclear Engineering and Design, 403, p.112144.

[13] Tiberga, M., Shafer, D., Lathouwers, D., Rohde, M. and Kloosterman, J.L., 2019. Preliminary investigation on the melting behavior of a freeze-valve for the Molten Salt Fast Reactor. Annals of Nuclear Energy, 132, pp.544-554

[14] Hidayati A. N., A. Waris, A. P. Andi Mustari, N. A. Apriani, M. Ifthacharo, and R. Wulandari. 2020. The Effect of Temeperature Variations on Wood's Metal Plate Melting Simulation by Using MPS. Journal of Physics: Conference Series, 1493, no. 1.

[15] Yamaji A. and X. Li. 2016. Development of MPS Method for Analyzing Melt Spreading Behavior and MCCI in Severe Accidents. Journal of Physics: Conference Series, 739, no. 1.

[16] Rubiolo, P. R., Retamales, M. T., Ghetta, V., & Giraud, J. 2017. High temperature thermal hydraulics modeling of a molten salt: application to a molten salt fast reactor (MSFR). ESAIM: Proceedings and Surveys, 58, 98–117. https://doi.org/10.1051/proc/20175809

IOP Conf. Series: Materials Science and Engineering **1330** (2025) 012011 doi:10.1088/1757-899X/1330/1/012011

Computational Fluid Dynamics for Irregular Pentagon Natural Circulation Loop

Rahmad Dahana Saputra[1], **Duwi Hariyanto**[3], **Syeilendra Pramuditya**[1,2], **Syaiful Bakhri**[4]

[1]Nuclear Physics and Biophysics Research Division, Physics Department, Faculty of Mathematics and Natural Science, Bandung Institute of Technology, Jalan Ganesha 10 Bandung 40132, Indonesia
[2]Nuclear Science and Engineering Department, Faculty of Mathematics and Natural Science, Bandung Institute of Technology, Jalan Ganesha 10 Bandung 40132, Indonesia
[3]Industrial Technology Faculty, Institut Teknologi Sumatera, Jl. Terusan Ryacudu, Lampung, Indonesia
[4]Research Center for Nuclear Material and Radioactive Waste Technology (PRTBNLR), Research Organization for Nuclear Energy, National Research and Innovation Agency (BRIN), South Tangerang 15314, Indonesia

E-mail: rahmad.dahana@gmail.com

Abstract. Electricity is an important needs for everyday life. Nuclear Reactor is a power plant that have low CO_2 emission and the energy is available at any time. Nuclear reactor needs a passive safety system that allow it to run without external power source or any human intervention. One of the passive safety system in nuclear reactor is natural circulation. A study of irregular pentagon natural circulation loop using water as its coolant has been done previously. In this study the natural circulation loop is simulated using ANSYS Fluent. A mesh independence study is done for obtaining mesh with accurate result but at effective computational cost. After that some flow parameters is obtained, namely cold leg temperature, hot leg temperature, and the flow Reynolds number with a value of 34.29 C, 39.93 C, and 86.8-132.6 respectively . Those parameters is compared to a reference study and an error less than 3.5% are obtained. This result means that this simulation model can be used for further study.

1 Introduction

Burning coal and gas contribute to the 62,5% electricity in the world [1]. Unfortunately, this method produce CO_2 emission that in the long term will cause climate issue. On the other hand renewable source like solar, water, and wind power is an intermittent source of energy. We cannot depend only on renewables because it is not available at any time we need it. The solution for reliable source of electricity with low CO_2 emission is nuclear power. Nuclear reactor needs a safety mechanism that does not depends on external power or human intervention called passive safety system. This passive safety can prevent another accident like fukushima nuclear disaster that happen because the lost of power for the cooling pump [2].

Natural circulation is an example of passive safety system. It uses the temperature gradient available in the system to circulate the cooling fluid. Research in natural circulation phenomenon usually done in a pipe loop called Natural Circulation Loop (NCL). One of the research in NCL was done by [3] using pentagon shaped NCL. In that study water is used as the fluid and circulated around the loop using

IOP Conf. Series: Materials Science and Engineering **1330** (2025) 012011 doi:10.1088/1757-899X/1330/1/012011

enamel wire as heater and wind blower as cooler. A temperature sensor is placed around the loop to measure temperature in the hot leg and cold leg part of the loop. The loop is also simulated to obtain the flow Reynolds number which is hard to measure in the experiment.

The NCL from [3] can be further developed to use molten salt as its coolant that represent a Molten Salt Reactor (MSR). Due to the difficulty in obtaining molten salt, a simulation study is chosen. But before simulationg the loop using molten salt, a validation of the simulation model needs to be done by comparing it to the previous study. In this study, the validation will be done by comparing the flow parameters obtained from simulation with the reference. Additionally, an optimum mesh size, which is not yet discussed in [3], will also be determined.

2 Methodology

2.1 Geometry and Mesh Creation

A 2D NCL geometry from [3] was created using ANSYS Discovery which is a Computer Aided Design (CAD) software. The loop has 7.4 mm inner pipe diameter and 9.8 mm outer pipe diameter. Other detailed dimension of the loop is showed in Figure 1. The geometry that was created then divided into a small elements called mesh. In this study, four mesh with different elements size was tested to obtain the most efficient one. Smaller mesh will results in more accurate calculation, but the computational cost will be higher. The mesh size used in this study is 2, 1, 0.7, and 0.5 mm which is displayed in figure 2. Detailed properties of the mesh is tabulated in table 1.

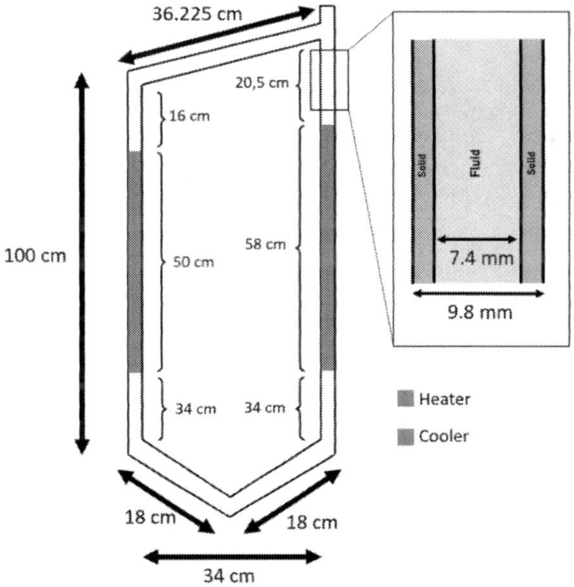

Figure 1: Natural circulation loop dimensions

Table 1: Mesh configuration for mesh independence test

Number	Solid Size	Fluid Size	Inflation layer	Number of elements
1	2 mm	2 mm	5	20198
2	1 mm	1 mm	7	61306
3	0.7 mm	1 mm	10	109404
4	0.5 mm	1 mm	10	183504

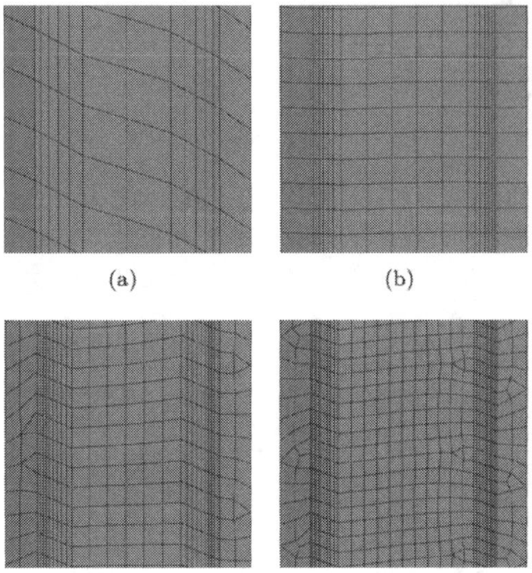

(a) (b) (c) (d)

Figure 2: Mesh with various size (a) 2 mm, (b) 1 mm, (c) 0.7 mm, dan (d) 0.5 mm

2.2 CFD Calculation

The CFD calculation was done using ANSYS Fluent which is a computational fluid dynamics (CFD) software based on finite volume method [4]. The calculation was done by solving three governing equation namely mass, momentum, and energy conservation equation. The general form of the mass, momentum, and energy conservation showed in equation 1, 2, 3 respectively [5]. The solver setting tabulated in table 2. The material used is water as coolant and stainless steel 316 (ss-316) as the pipe material with their properties is in table 3. A heat flux boundary condition is applied in the heater section while a constant temperature is applied in the cooler section. The outer pipe wall is modeled as adiabtic wall. The detailed boundary condition is displayed in figure 3.

$$\frac{\partial \rho}{\partial t} + \nabla \cdot (\rho \vec{u}) = 0 \tag{1}$$

$$\rho \frac{D\vec{u}}{Dt} = -\nabla P + \nabla \cdot \bar{\bar{\tau}} + \rho \vec{g} \tag{2}$$

$$\rho c_p \frac{DT}{Dt} = -\nabla \cdot q'' + q''' + \beta T \frac{DP}{Dt} + \psi \tag{3}$$

Table 2: Solver setting

Setting	Option chosen
Solver type	Pressure-Based
Time	Transient
Energy	On
Gravity	On
Viscous	Laminar
Pressure-velocity coupling	SIMPLE
Time-step (s)	2
Number of time step	1000

IOP Conf. Series: Materials Science and Engineering **1330** (2025) 012011 doi:10.1088/1757-899X/1330/1/012011

Table 3: Materials physical properties [6] [7]

Material	Density-Boussinesq (kg/m^3)	Heat capacity (J/kg-K)	Thermal conductivity (W/m-K)	Viscosity (Pa s)	Thermal expansion coefficient (K^{-1})
Air	994.08	4179.5	0.63	0.00064	0.000347
ss-316	7990	500	16.3	-	-

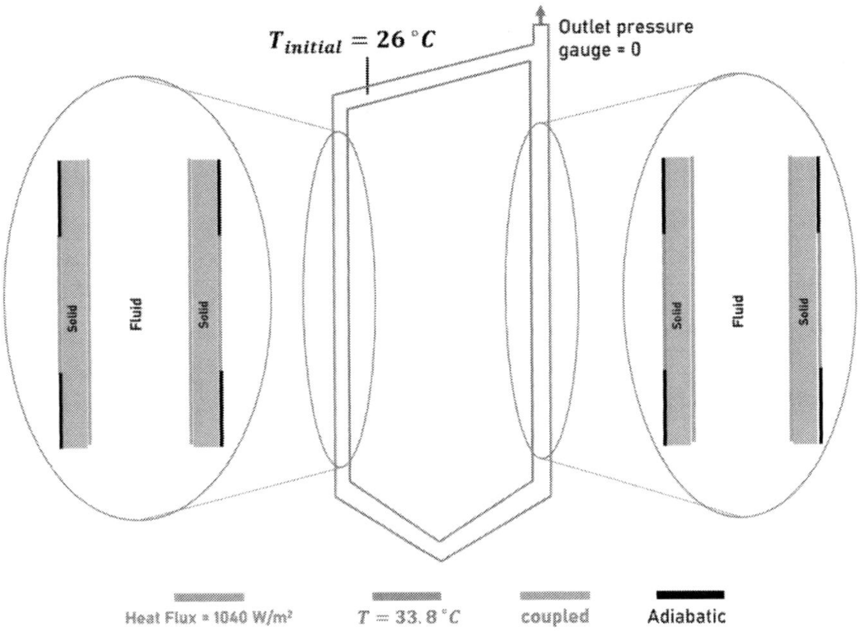

Figure 3: Boundary condition applied in the loop

Boussinesq approximation was used in this study to approximate the density change of water. The equation for boussinesq approximation showed in equation 4. The operating temperature (T_o) chosen is is 35 C. The flow Reynolds Number (Re) is also calculated in equation 5 using inner diameter of the pipe (D_{in}) as the characteristic length.

$$(\rho_o - \rho) = \rho\beta(T - T_o) \tag{4}$$

$$Re = \frac{\rho U D_{in}}{\mu} \tag{5}$$

3 Results and Discussion

3.1 Mesh Independence Test

Four mesh size was simulated to obtain its velocity profile in four places which is in heater, cooler, above heater, and above cooler. The velocity profile is displayed in figure 4. It can be seen that the 2 mm mesh have the most different result compared to the others, especially in above cooler position in figure 4(b). The mesh failed to capture the flow details because the elements size is too big. For the 1 mm mesh, the results is pretty good in above heater and above cooler position, but there is still a difference on heater and cooler position. Meanwhile, the 0.7 mm mesh results is already close enough to the 0.5 mm mesh. Therefore to reduce the computational cost, the 0.7 mm mesh is chosen.

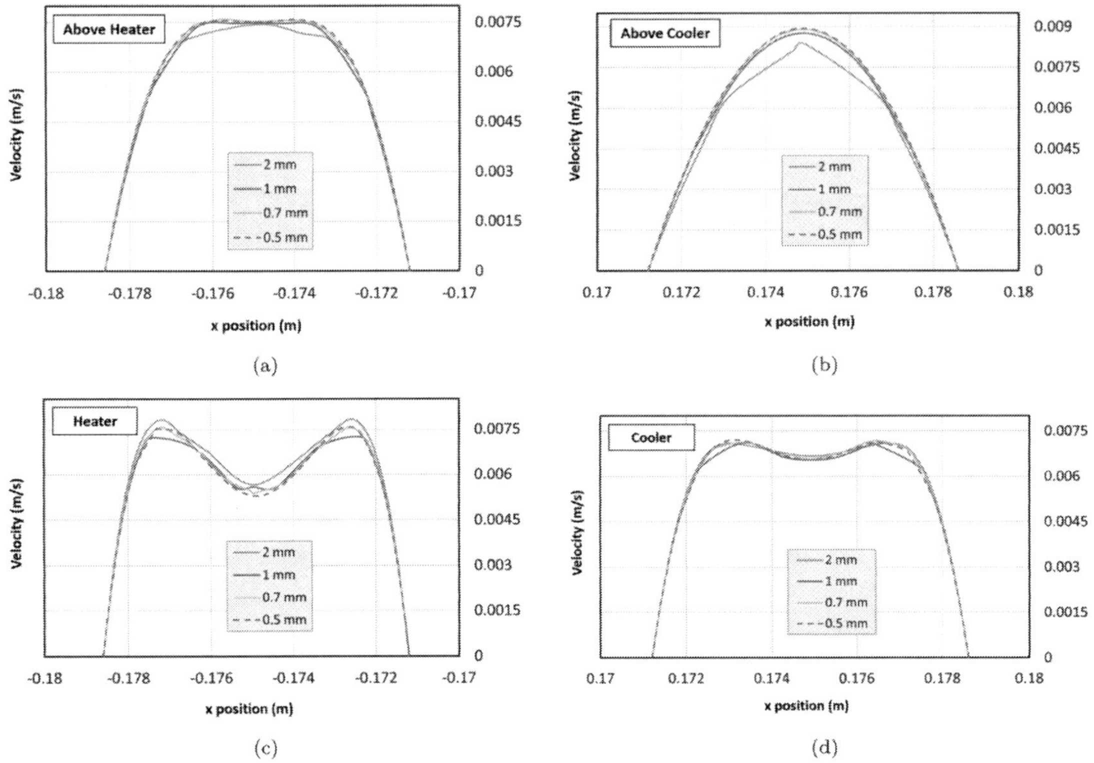

Figure 4: Mesh independence test result in (a) above heater, (b) above cooler, (c) heater, (d) cooler positions

3.2 Temperature Results

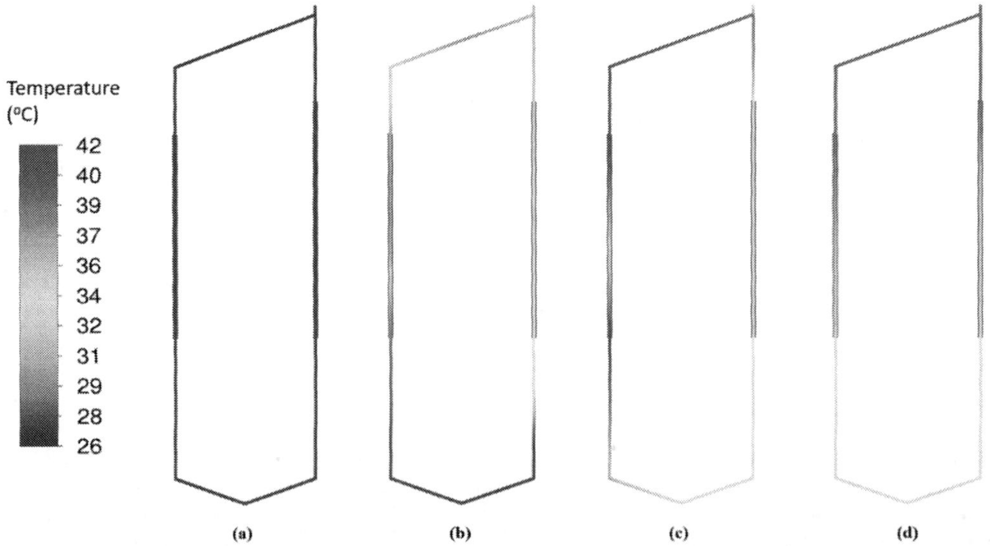

Figure 5: Temperature distribution at (a) $t = 0$, (b) $t = 300$, (c) $t = 600$, and (d) $t = 2000$ s

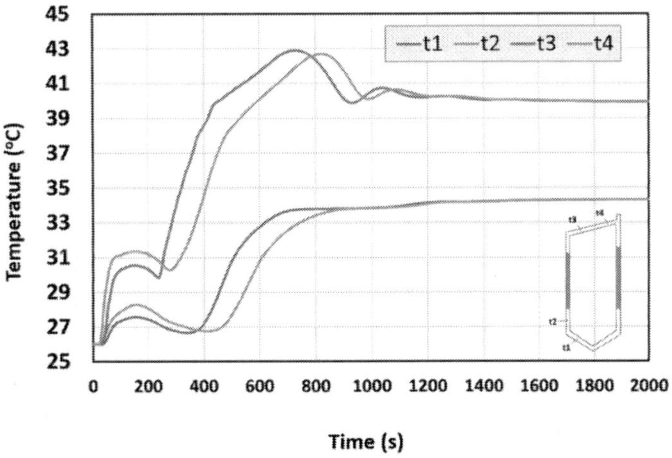

Figure 6: Measured temperature on four spots according to [3]

Figure 5 showed the temperature distribution for the loop. The heat can be seen move from the heater to the other parts of the loop due to water convection. The natural circulation direction happen in the clockwise direction. The detailed temperature change over time is displayed in figure 6. The temperature is measured in the cold legs part (t1 and t2) and hot legs part (t3 and t4) of the loop. Steady state temperature on the loop for t1 and t2 measured to be 34.29 C and for the t3 and t4 is 39.93 C. The time needed to reach this steady state condition is 1200 seconds. Before that time, the flow is still in transient condition with the temperature changing with respect to time. The upper parts of the loop (t3 and t4) experience a higher increase in temperature due to the bouyancy of water

Meanwhile from [3] study, the cold leg temperature and hot leg temperature are 33.8 and 41 C. The temperature change in figure 6 already have simmiliar trend compared to the [3] study. However there is a slight difference in the steady state temperature value with an error of 2.61 and 1.45 % for the hot legs and cold legs temperature.

If we compare the experiment results from [3] with the results in figure 6, there is a difference in the amount of time needed for the steady stated condition achieved. The experiment from [3] needed 3500 seconds to reach steady state condition, longer than the simulation results with just 1200 seconds needed to reach steady state. This difference in time is a result of ideal adiabatic condition applied in the simulation. In the real experiment, the heat from the loop can escape to the outside environment, resulting in a longer time to reach steady state condition.

The interaction between the loop and the environment can be captured in the simulation by also modelling the air surrounding the loop. If the air modelling was done, the steady-state time can be predicted more accurately. This is especially important when modelling a real reactor.

3.3 Reynolds Number Results

The flow velocity is measured in seven places named v1 - v7 using vertex average method. The results of the velocity change with respect to time is displayed in figure 7. In the steady state conditions, the seven velocity converge into three value. This three value is chosen to calculate the flow Reynolds Number (Re), displayed in figure 8. The time needed to reach steady state condition is around 1300 seconds, with the steady state Reynolds Number are 86.8, 101.13, and 132.6. Compared to the [3] study with a single Re value of 137, there is a difference in the steady state Reynolds number value with an error of 36.6, 26.18, and 3.2 %. The v6 spot have the smallest error and also have the most simmiliar trends compared to the reference.

This wide range of error happen because the possibility of different measuring spots between this study and reference study. In the [3] study, the author also use ANSYS Fluent. However, the velocity measuring spot and what the measuring method is not specified. Additionally, the physical properties of water and stainless steel 316 used as calculation input may also be different. Another possible cause is the difference of the operating temperature (T_o) used when applying Boussinesq approximation.

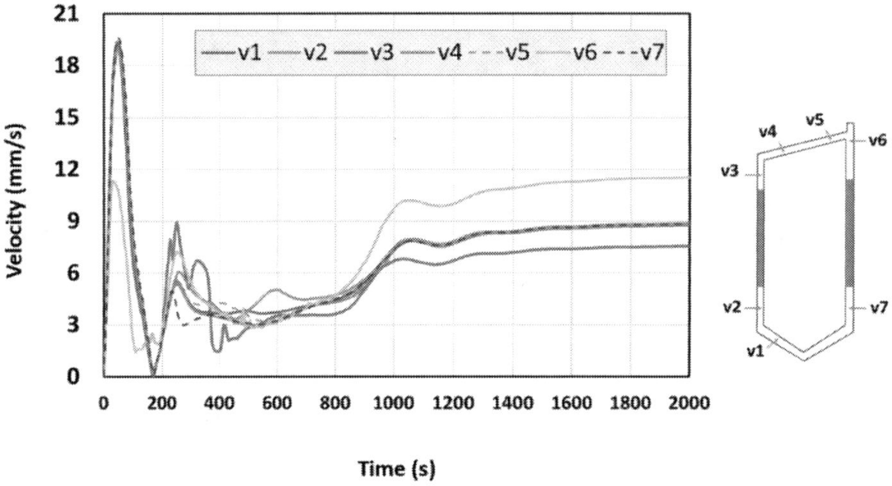

Figure 7: Velocity change with respect to time

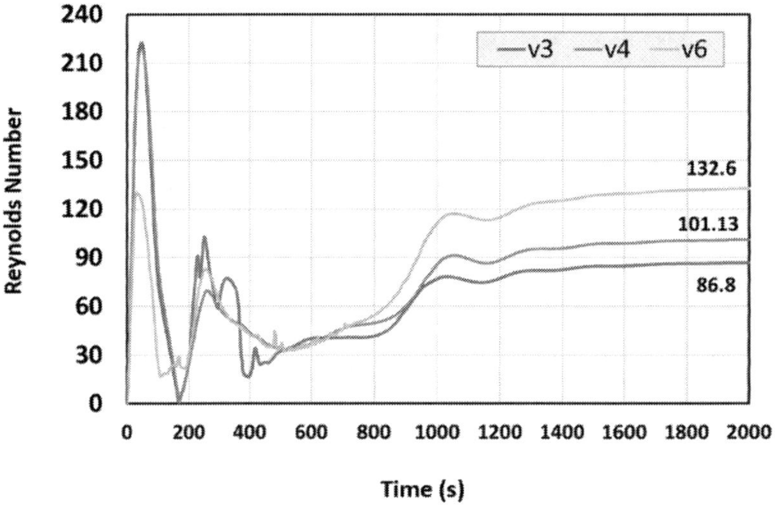

Figure 8: Reynolds number change with respect to time

4 Conclusion

A computationally efficient mesh size of 0.7 mm is obtained from the mesh independence test. 0.7 mm mesh size can capture the flow details similiar to the 0.5 mm size but with lower computational cost due to it's larger size. Then the flow parameters, namely the hot leg temperature, cold leg temperature, and Reynolds number are compared to the reference study with the errors are 2.61, 1.45, and 36.64-3.21 %. The flow Reynolds number have wide range of error due to the unknown measuring spot, materials properties, and operating temperature used in the reference. However, with the smallest error of the all parameters being under 3.5 %, this simulation model then can be used for the next study which is using molten salt as the coolant.

References

[1] Pioro I, Duffey R, Kirillov P and Panchal R 2016 1 - introduction: A survey of the status of electricity generation in the world *Handbook of Generation IV Nuclear Reactors*

Woodhead Publishing Series in Energy ed Pioro I L (Woodhead Publishing) pp 1–34 URL https://www.sciencedirect.com/science/article/pii/B978008100149300001X

[2] Khan A H, Hasan S and Sarkar M A R 2018 *International Journal of Nuclear and Quantum Engineering* **12** 72–77

[3] Hariyanto D, Permana S, Waris A, Mustari A P A, Kinoshita M, Aji I K, Maulana A and Setiadipura T 2024 *Nuclear Engineering and Design* **416** 112753 ISSN 0029-5493 URL https://www.sciencedirect.com/science/article/pii/S0029549323006027

[4] ANSYSInc 2011 *ANSYS Fluent Theory Guide*

[5] Bergman T 2011 *Fundamentals of Heat and Mass Transfer* (Wiley) ISBN 9780470501979

[6] Lemmon E W, Bell I H, Huber M L and McLinden M O 2023 *NIST Chemistry WebBook, NIST Standard Reference Database Number 69* (National Institute of Standards and Technology)

[7] Dincer I and Zamfirescu C 2015 *Appendix B Thermophysical Properties of Water* (John Wiley & Sons, Ltd) pp 457–459 ISBN 9781118534892 (*Preprint* https://onlinelibrary.wiley.com/doi/pdf/10.1002/9781118534892.app2) URL https://onlinelibrary.wiley.com/doi/abs/10.1002/9781118534892.app2

IOP Conf. Series: Materials Science and Engineering **1330** (2025) 012012 doi:10.1088/1757-899X/1330/1/012012

Parametric Analysis of Molten Salt Natural Circulation Loop using Computational Fluid Dynamics

Rahmad Dahana Saputra[1], Duwi Hariyanto[3], Syeilendra Pramuditya[1,2], Syaiful Bakhri[4]

[1]Nuclear Physics and Biophysics Research Division, Physics Department, Faculty of Mathematics and Natural Science, Bandung Institute of Technology, Jalan Ganesha 10 Bandung 40132, Indonesia
[2]Nuclear Science and Engineering Department, Faculty of Mathematics and Natural Science, Bandung Institute of Technology, Jalan Ganesha 10 Bandung 40132, Indonesia
[3]Industrial Technology Faculty, Institut Teknologi Sumatera, Jl. Terusan Ryacudu, Lampung, Indonesia
[4]Research Center for Nuclear Material and Radioactive Waste Technology (PRTBNLR), Research Organization for Nuclear Energy, National Research and Innovation Agency (BRIN), South Tangerang 15314, Indonesia

E-mail: rahmad.dahana@gmail.com

Abstract. Natural circulation is a form of passive safety in nuclear reactor. Natural circulation is important to prevent another accident like fukushima nuclear disaster that happen because the loss of power to the pump. An experiment on irregular pentagon natural circulation loop has done before using water as its coolant. In this study, the water coolant is replaced with molten salt to analyze its flow parameters. ANSYS Fluent which is a finite volume based computational fluid dynamics software is used to simulate the loop. First the loop heater and cooler temperature is varied to see the respons of the Reynolds number (Re), cold leg temperature (t1), and hot leg temperature (t3) for FLiBe, FLiNaK, and FNaB. For all molten salt, the Re is proportional to heater temperature when the cooler temperature is constant. For constant heater temperature condition, Re does not have a particular trends when the cooler temperature increase. Meanwhile, t1 and t3 always increase when the heater or cooler temperature increase. FNaBe is added for second analysis that focuses on comparing Re, t1, and t3 for constant heater condition and constant cooler condition. It is observed that FNaB have the highest Re due to its low viscosity. A correlation on the trends of t1 and t3 to the velocity trends is also observed.

1 Introduction

Generation IV International Forum (GIF) define eight technology goals for generation IV nuclear reactor [1]. One of the goals is related to the passive safety of the reactor. Passive safety means that no external power or human intervention needed when emergency situation happen. One of the example of passive safety in nuclear reactor is natural circulation. Natural circulation can help the heat transfer from the reactor core to the cooler parts of the reactor so that accident like core meltdown can be avoided.

Some research on natural circulation has been done before. [2] study a natural circulation loop using FLiBe to see the local and system level phenomena. [3] study the effect of heater and cooler position on a rectangular natural circulation loop using nitrate molten salt as the working fluid. In more recent study by [4], an irregular pentagon shaped natural circulation loop is studied using water as the working fluid. The irregular pentagon shape is chosen to minimize the salt solidification in the corners of the loop. This study is the development of the [4] study by replacing water with molten salt as the working fluid. But with the limited resource for accessing the molten salt material, a simulation study is chosen for analyzing the natural circulation loop.

2 Methodology

2.1 Geometry and Mesh Creation

The NCL geometry in this study is created based on [4] study. The geometry created in 2 dimension using ANSYS Discovery which is a Computer Aided Design (CAD) software. The loop has 7.4 mm inner pipe diameter and 9.8 mm outer pipe diameter. Other detailed dimension of the loop is showed in Figure 1(a). After that a mesh with the size of 0.7 mm is used to discritize the geometry resulting in 109404 number of elements. The magnified view of the mesh is displayed in Figure 1(b).

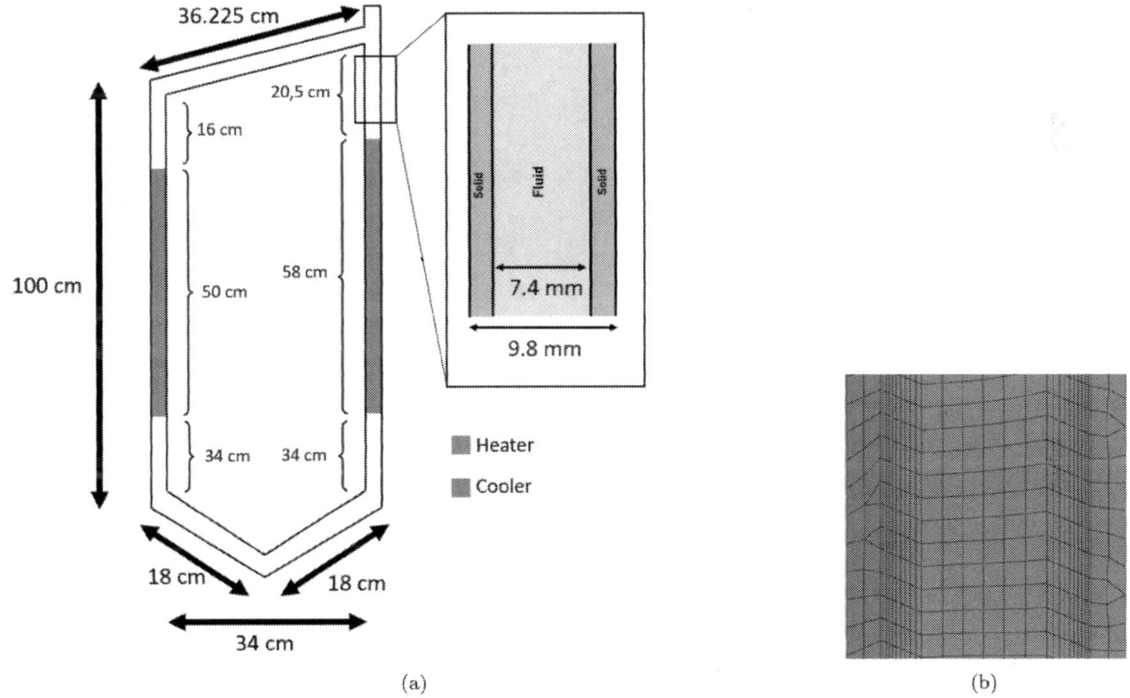

Figure 1: (a) Natural circulation loop dimensions, (b) Magnified view of the mesh

2.2 Parametric Analysis

ANSYS Fluent which is a computational fluid dynamics (CFD) software based on finite volume method is used to perform the simulation [5]. ANSYS Fluent solve three governing equation namely mass, momentum, and energy conservation showed in equation 1, 2, 3 in its general form [6]. The solver setting tabulated in table 1. The material used is FLiBe, FLiNaK, and FNaB as coolant and stainless steel 316 (ss-316) as the pipe material. The molten salt properties showed in table 2. The properties of FLiBe and FLiNaK is summarized by [7] while the properties of FNaB is summarize by [8]. A constant temperature boundary condition is applied in the heater and cooler section. The detailed boundary condition is displayed in figure 2. The flow Reynolds Number (Re) is calculated in equation 4 using inner diameter of the pipe (D_{in}) as the characteristic length.

IOP Conf. Series: Materials Science and Engineering **1330** (2025) 012012 doi:10.1088/1757-899X/1330/1/012012

$$\frac{\partial \rho}{\partial t} + \nabla \cdot (\rho \vec{u}) = 0 \tag{1}$$

$$\rho \frac{D\vec{u}}{Dt} = -\nabla P + \nabla \cdot \bar{\bar{\tau}} + \rho \vec{g} \tag{2}$$

$$\rho c_p \frac{DT}{Dt} = -\nabla \cdot q'' + q''' + \beta T \frac{DP}{Dt} + \psi \tag{3}$$

$$Re = \frac{\rho U D_{in}}{\mu} \tag{4}$$

Table 1: Solver setting

Setting	Option chosen
Solver type	Pressure-Based
Time	Steady
Energy	On
Gravity	On
Viscous	Laminar
Pressure-velocity coupling	SIMPLE

Table 2: Thermophysical properties of molten salt for parametric analysis

Molten Salt	Properties	Value	Reference
FLiBe (66% LiF-34% BeF_2)	Density (kg/m^3)	$2413 - 0.488 * T[K]$	[9]
	Viscosity (Pa-s)	$1.16 \cdot 10^{-4}\ exp(3755/T[K])$	[10]
	Heat Capacity (J/kg-K)	2386	[11]
	Heat Conductivity (W/m-K)	1.1	[12]
	Melting Point (^{o}C)	459	[7]
FLiNaK (46.5% LiF-11.5% NaF-42% KF)	Density (kg/m^3)	$2579 - 0.624 * T[K]$	[8]
	Viscosity (Pa-s)	$4 \cdot 10^{-5}\ exp(4170/T[K])$	[13]
	Heat Capacity (J/kg-K)	1884	[12]
	Heat Conductivity (W/m-K)	$0.36 + 0.00056 * T[K]$	[14]
	Melting Point (^{o}C)	462	[7]
FNaB (8% NaF-92% $NaBF_4$)	Density (kg/m^3)	$2446 - 0.711 * T[K]$	[8]
	Viscosity (Pa-s)	$8.77 \cdot 10^{-5}\ exp(2240/T[K])$	
	Heat Capacity (J/kg-K)	1500	
	Heat Conductivity (W/m-K)	$0.66 - 0.000237 * T[K]$	
	Melting Point (^{o}C)	384	

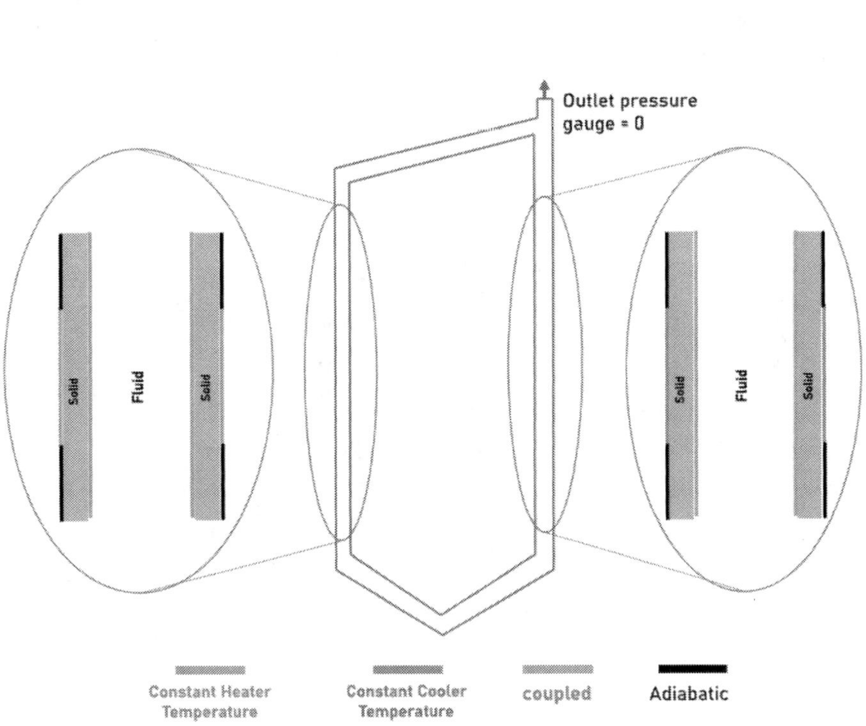

Figure 2: Boundary condition applied in the loop

2.3 Molten Salt Comparison
FNaBe is added for comparing its properties with the previous three molten salt. The properties of FNaBe is showed in Table 3. The comparison is analyzed in two conditions which is constant heater temperature of 625 C and constant cooler temperature of 550 C.

Table 3: FNaBe thermophysical properties [10]

Molten Salt	Properties	Value
FNaBe (57% NaF-43% BeF_2)	Density (kg/m^3)	$2270 - 0.37 * T[K]$
	Viscosity (Pa-s)	$3.4 \cdot 10^{-5}\ exp(5165/T[K])$
	Heat Capacity (J/kg-K)	2177
	Conductivity (W/m-K)	0.87

3 Results and Discussion

3.1 Parametric Analysis

3.1.1 FLiBe Flow velocity is obtained from simulation and then used to calculate the Reynolds number (Re), the results is showed in Figure 3. If we look at constant cooler conditions, the Re is proportional to the heater temperature. Re is also proportional to the heater temperature in same temperature difference (ΔT) condition. But if we look at constant heater condition, the Reynolds number doesn't have a particular trend.

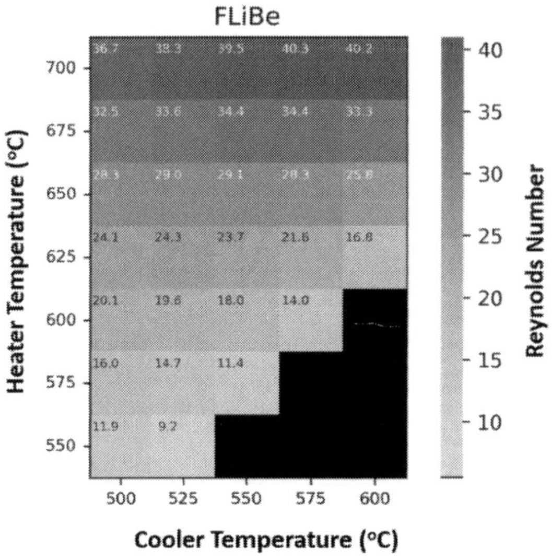

Figure 3: Reynolds number results for FLiBe

Cold leg temperature (t1) and hot leg temperature (t3) results is showed in figure 4. Both of t1 and t3 have the same trend which is proportional to heater or cooler temperature increase. This is because when some parts of the loop having a temperature increase, the overall loop temperature will rise also. One of the example of the loop temperature distribution is showed in Figure 5. A heater-cooler pair of 700-500 C is chosen

The difference that can be observed between t1 and t3 is at what time their temperature tends to increase more. t1 tend to increase more when the cooler temperature increase. This is because t1 is the lower parts of the loop which is when the flow just finished flowing from the cooler. On the contrary, t3 tend to increase more when the heater temperature increase. This is because t3 is the upper parts of the loop which is when the flow just finished flowing from the heater.

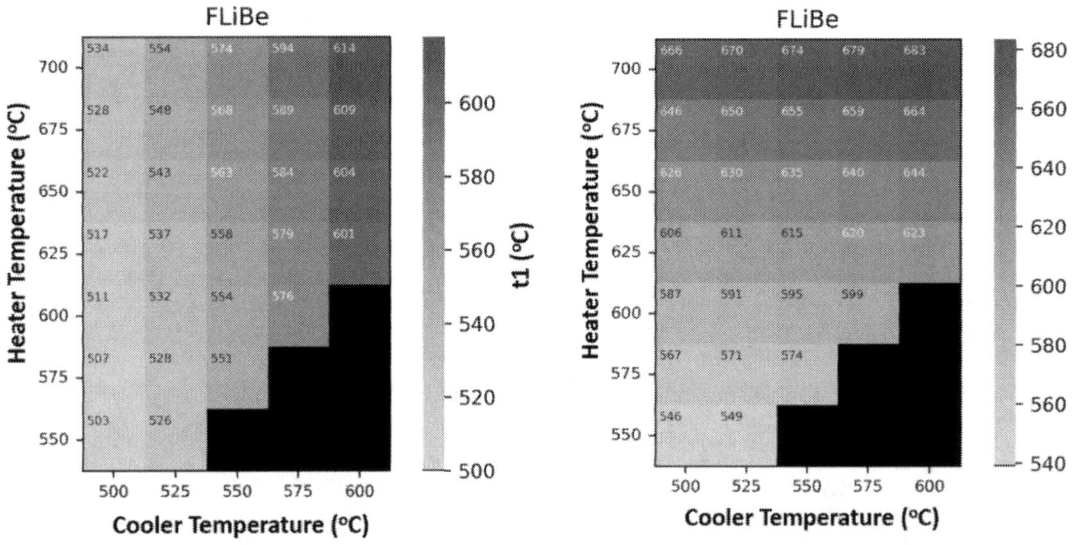

(a) (b)

Figure 4: FLiBe temperature results for (a) t1 and (b) t3

IOP Conf. Series: Materials Science and Engineering **1330** (2025) 012012 doi:10.1088/1757-899X/1330/1/012012

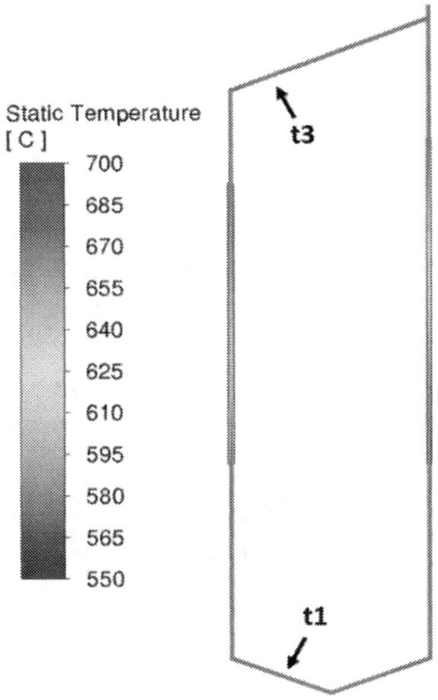

Figure 5: FLiBe temperature distribution at heater-cooler temperature of 700-550 °C

3.1.2 FLiNaK and FNaB In general, the result of parametric analysis obtained for FLiNaK and FNaB is the same as the results for FLiBe. Reynolds number (Re) is proportional to heater temperature when the cooler temperature is kept constant or when we look at the same ΔT. On constant heater condition, there is no particular trends for Re when cooler temperature increase. The result of Re for FLiNaK and FNaB is showed in Figure 6 and 7 respectively.

The results on t1 and t3 for FLiNaK and FNaB is showed in Figure 8 and 9. Both the t1 and t2 is proportional to the heater temperature and cooler temperature. Same as FLiBe, t1 tends to increase more when the cooler temperature increase, meanwhile t3 tends to increase more when heater temperature increase.

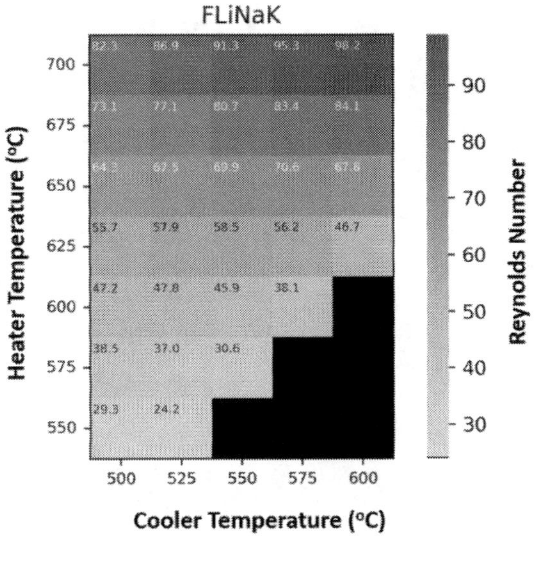

Figure 6: Reynolds number results for FLiNaK

IOP Conf. Series: Materials Science and Engineering **1330** (2025) 012012 doi:10.1088/1757-899X/1330/1/012012

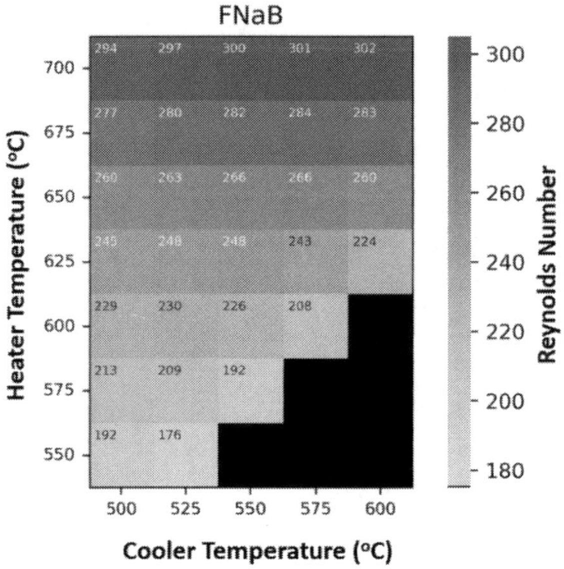

Figure 7: Reynolds number results for FNaB

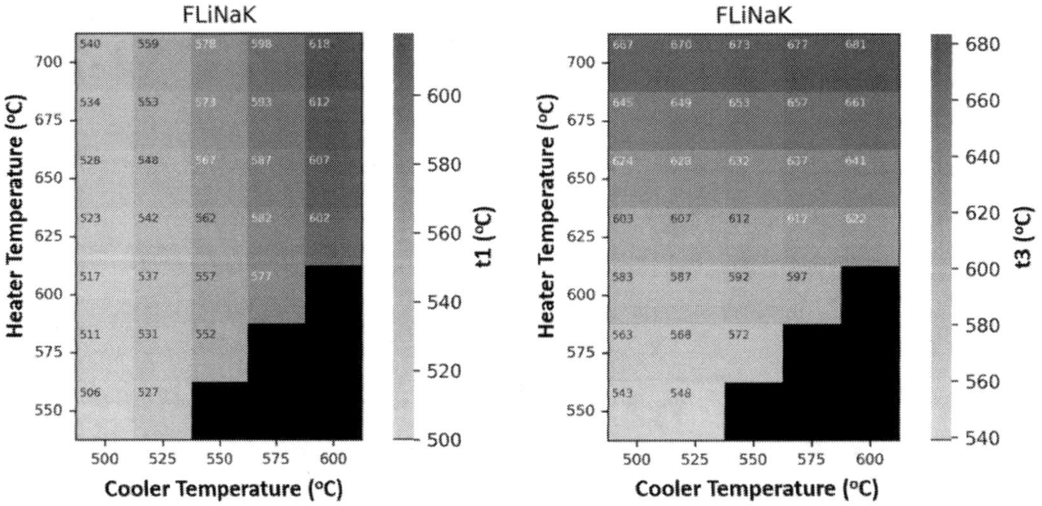

(a) (b)

Figure 8: FLiNaK temperature results for (a) t1 and (b) t3

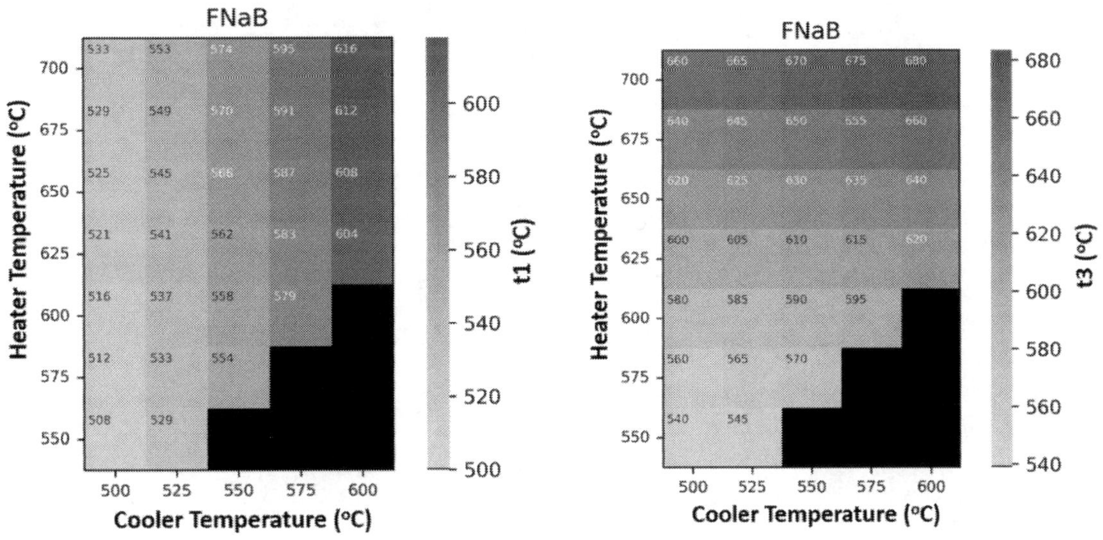

(a) (b)

Figure 9: FNaB temperature results for (a) t1 and (b) t3

3.2 Molten Salt Comparison

3.2.1 Constant Cooler Temperature The constant cooler conditon is done in 550 C temperature. FNaBe molten salt is added with the previous three molten salt to comapare its Re, t1, t3, and velocity. The comparison results is showed in Figure 10. FNaB have the highest Re showed in Figure 10(a). FNaB have the highest Re because of the very low viscosity compared to other molten salt. The viscosity comparison can be seen in Table 2 and 3. For t1 comparison, FNaB have the highest t1 initially. But as the heater temperature increase, FLiNaK have the highest t1 and FNaB is in second position. This t1 comparison have a trend correlation with the molten salt velocity showed in Figure 10(d). For t3 comparison, all molten salt have a simmiliar value in all heater temperature.

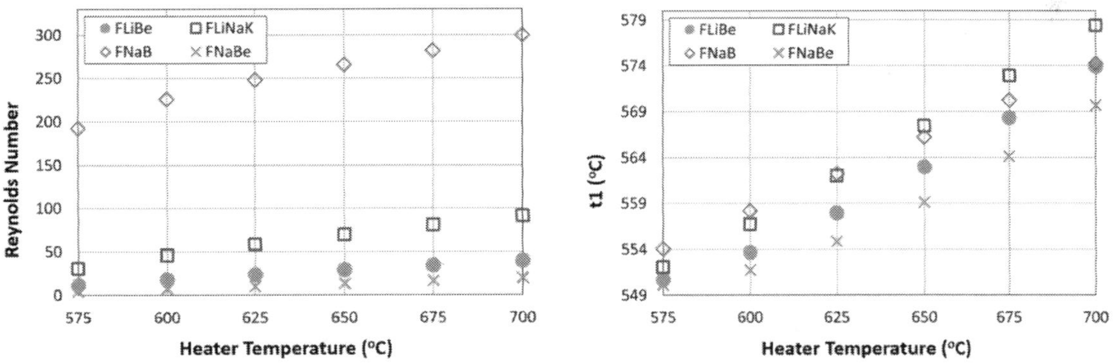

(a) (b)

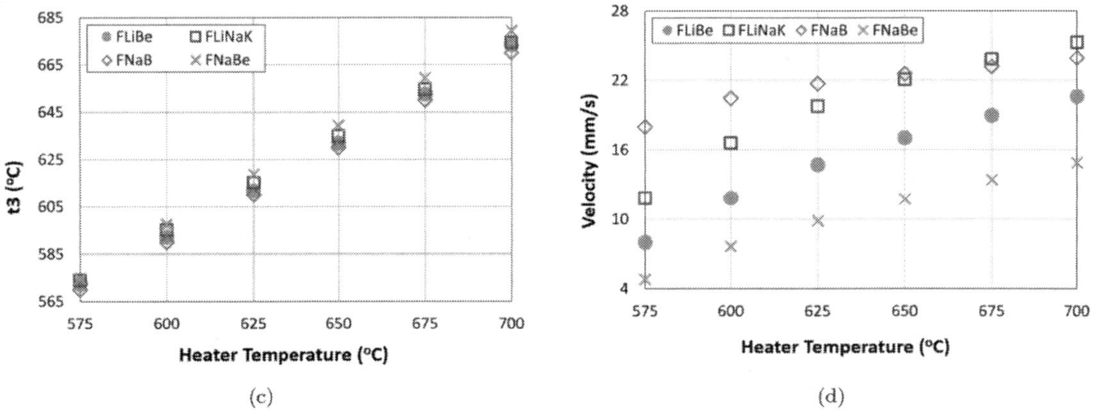

Figure 10: The comparison of (a) Re, (b) t1, (c) t3, (d) velocity at cooler temperature of 550 °C

3.2.2 Constant Heater Temperature A heater temperature of 625 C is chosen for the comparison. The results is showed in Figure 11. Same with constant cooler condition, FNaB have the highest Re due to its low viscosity. For t1 and t3 comparison, constant heater condition have an opposite trend compared to constant cooler condition. The t1 temperature is relatively the same for all molten salt. Meanwhile there is a trend correlation for t3 and velocity. But the trend between t3 and velocity now is inversly proportional.

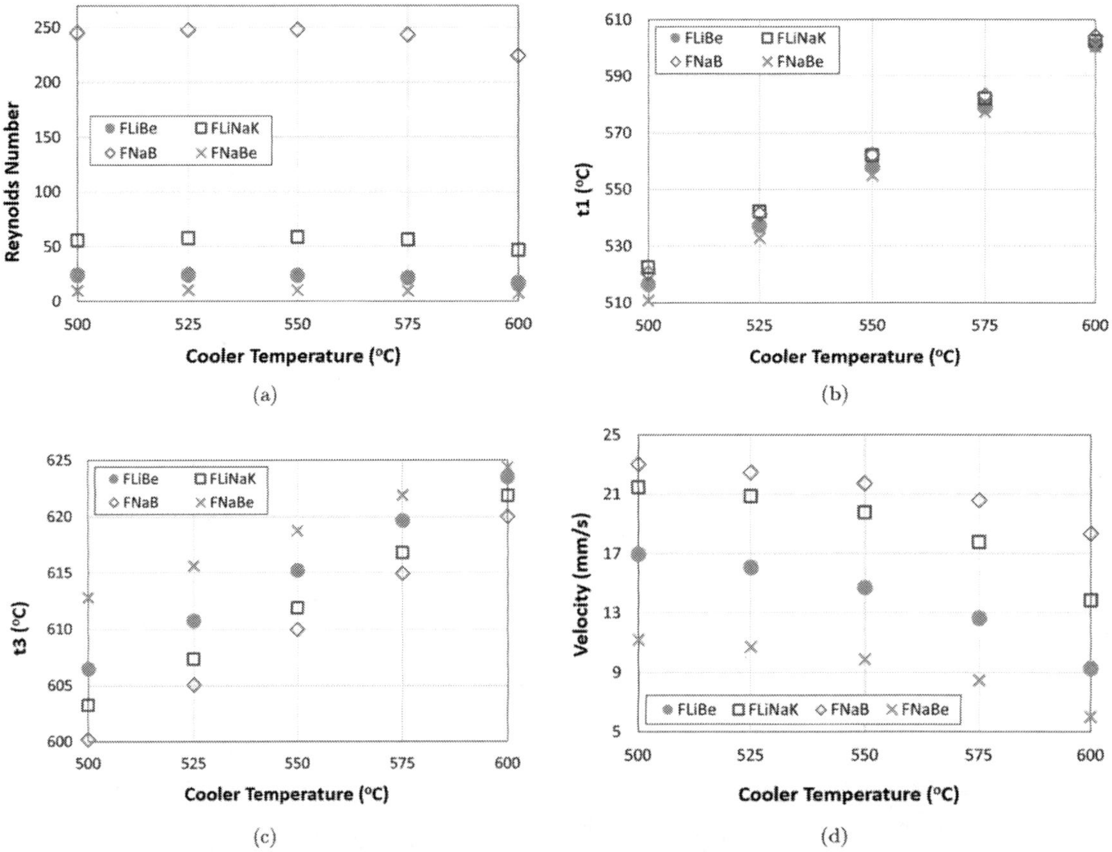

Figure 11: Comparison of (a) Re, (b) t1, (c) t3, and (d) velocity at heater temperature of 625 °C

3.3 Performance Comparison
From figure 10(a) and 11(a), it is observed that FNaB has the highest Reynolds number compared to the other molten salts. This potentially make FNaB the better coolant compared to the others. However, further analysis are needed, such as calculating heat transfer coefficient (h) for each molten salts for clearer heat transfer performance, consider how each salts can affect nuclear reactor neutronic aspects, and how the corrosion behaviour of each molten salts. Additionally, a correlation on non-dimensional parameters are needed to compare this NCL to other NCL. A correlation on Reynolds number and Grashof number can be done and then compared to other studies like [4] did.

4 Conclusion
The parametric analysis show that FLiBe, FLiNaK, and FNaB show an increase in Reynolds number when subjected to heater temperature increase. But an increase of cooler temperature can reduce the flow Reynolds number. For the cold leg (t1) and hot leg (t3) temperature, it is observed both are proportional to heater and cooler temperature increase. Next when FNaBe is added with three previous molten salt for comparison analysis, it is obtained that FNaB have the highest Reynolds number due to its low viscosity compared to other molten salt. A correlation on t1 with flow velocity in constant cooler condition and inverse correlation on t3 with flow velocity in constant heater condition is also observed.

Based on the Reynolds number results, FNaB potentially can be the better coolant compared to other salts. Further studies on other heat transfer and material aspects are required to better conclude the performance. Additionally, non-dimensional parameters correlation is needed so that this study can be compared to other studies.

References
[1] Pioro I 2016 2 - introduction: Generation iv international forum *Handbook of Generation IV Nuclear Reactors* Woodhead Publishing Series in Energy ed Pioro I L (Woodhead Publishing) pp 37–54 ISBN 978-0-08-100149-3 URL `https://www.sciencedirect.com/science/article/pii/B9780081001493000021`

[2] Britsch K, Anderson M, Brooks P and Sridharan K 2019 *International Journal of Heat and Mass Transfer* **134** 970–983 ISSN 0017-9310 URL `https://www.sciencedirect.com/science/article/pii/S0017931018346209`

[3] Srivastava A, Saikrishna N and Maheshwari N 2023 *Applied Thermal Engineering* **218** 119318 ISSN 1359-4311 URL `https://www.sciencedirect.com/science/article/pii/S1359431122012480`

[4] Hariyanto D, Permana S, Waris A, Mustari A P A, Kinoshita M, Aji I K, Maulana A and Setiadipura T 2024 *Nuclear Engineering and Design* **416** 112753 ISSN 0029-5493 URL `https://www.sciencedirect.com/science/article/pii/S0029549323006027`

[5] ANSYSInc 2011 *ANSYS Fluent Theory Guide*

[6] Bergman T 2011 *Fundamentals of Heat and Mass Transfer* (Wiley) ISBN 9780470501979

[7] Romatoski R and Hu L 2017 *Annals of Nuclear Energy* **109** 635–647 ISSN 0306-4549 URL `https://www.sciencedirect.com/science/article/pii/S0306454917301391`

[8] Beneš O and Konings R 2009 *Journal of Fluorine Chemistry* **130** 22–29 ISSN 0022-1139 fluorine & Nuclear Energy URL `https://www.sciencedirect.com/science/article/pii/S0022113908002030`

[9] Janz G J 1988 Thermodynamic and transport properties for molten salts : correlation equations for critically evaluated density, surface tension, electrical conductance, and viscosity data URL `https://api.semanticscholar.org/CorpusID:92860789`

[10] Williams D F 2006 URL `https://www.osti.gov/biblio/885975`

[11] Serrano-L'opez R, Fradera J S and Cuesta-L'opez S 2013 *Chemical Engineering and Processing* **73** 87–102 URL `https://api.semanticscholar.org/CorpusID:97305830`

[12] Davis C 2005 Implementation of molten salt properties into relap5-3d/athena URL `https://api.semanticscholar.org/CorpusID:101066703`

[13] Williams D F 2006 Assessment of candidate molten salt coolants for the ngnp/nhi heat-transfer loop URL https://api.semanticscholar.org/CorpusID:98525170

[14] Smirnov M V, Khokhlov V A and Filatov E S 1987 *Electrochimica Acta* **32** 1019–1026 URL https://api.semanticscholar.org/CorpusID:39641833

IOP Conf. Series: Materials Science and Engineering **1330** (2025) 012013 doi:10.1088/1757-899X/1330/1/012013

Neutronics Analysis of Heterogeneous Core of Small Modular Reactor Type GFR Thorium Nitride Fueled using OpenMC

Arroofi Candra Kartiko[1], Ratna Dewi Syarifah[1*], Nina Widiawati[2], Artoto Arkundato[1], and Fajri Prasetya[3]

[1] Department of Physics, Faculty of Mathematics and Natural Sciences, Jember University, Jember, Indonesia.
[2] Research Center for Nuclear Reactor Technology, BRIN, Indonesia
[3] Department of Nuclear Science and Engineering, Bandung Institute of Technology, Bandung, Indonesia

*E-mail: rdsyarifah.fmipa@unej.ac.id

Abstract. The use of mixed fuel in the form of thorium-uranium nitride (ThN-UN) offers the potential to improve thermal efficiency and optimize criticality, especially in fast reactors like Gas-cooled Fast Reactors (GFRs). To maintain its criticality, Th-232 absorbs neutrons and transforms into a new fissile material U-233. The inclusion of fertile material Th-232 necessitates a more precise reactor geometry design to ensure the reactor remains critical state until the end of the burn-up period. This study aims to compare variations in the percentage of U-233 enrichment in heterogeneous core geometries with five fuel variations (F1:F2:F3:F4:F5) using the OpenMC program at a power of 100 MWth. Benchmarking is performed by measuring the effective multiplication factor (k-eff) over 5 years of burn-up using reference data from previous studies. If the error value is less than 2%, further calculations will be conducted for both homogeneous and heterogeneous core configurations. The homogeneous calculations indicate that the U-233 enrichment percentage of 8.5% yields the most optimal results. This homogeneous data then is used for calculations in the heterogeneous core with 5 fuel variations. The heterogeneous core configuration is designed using five types different fuel percentages cases, each with variations in ring geometry. The comparison results show that the case 5 heterogeneous core geometry design, with a percentage distribution of 7%:7.5%:8.5%:9%:10.5%, achieves good optimization in terms of k-eff value, excess reactivity, neutron flux, and extended burn-up over a 10-years period.

1. Introduction

The GFR is one of the generation-IV reactors with the potential to be operational by 2030 [1]. GFR prioritizes the recycling of actinides to minimize the production of long-lived radioisotopes [2]. It operates using a fast neutron spectrum and cooled helium gas to achieve high thermal efficiency, with an outlet temperature reaching 850°C [3]. The use of high temperatures provides the opportunity to produce hydrogen gas as an alternative energy source and is suitable for several types of fuels, especially nitride-based [4]. The presence of nitride material (N-15) can enhance the conversion of fertile to fissile material [5]. Nitride materials can be combined with other nuclear fuels, particularly thorium (Th-232) and uranium (U-233). Thorium is much more abundant in nature compared to uranium, making it viable main fuel for GFRs. The use of thorium has the advantage in the long-term fuel breeding process, as Th-232 is converted into U-233. U-233 is a fissile material that can undergo fission directly when interacting with fast neutrons [6].

Research on thorium-uranium nitride fueled GFR has been conducted by previous researchers. Syarifah et al in 2018 [7] attempted to examine the use of UN-PuN fuel with ThN in GFRs using the SRAC program. The researchers used a heterogeneous core composition for UN-PuN with plutonium enrichment of 8%:10%:12% and ThN with U-233 enrichment of 7.8%:8%:8.8%. The calculation results show that the type of ThN fuel has a smaller maximum excess reactivity value than UN-PuN, which is 1.08%. The same thing was done in the study in 2022 [8], but the difference was in the reactor used, namely the PWR reactor. The heterogeneous core composition of UN-PuN fuel with U-235 enrichment of 5.5%:7%:8.5% had an excess reactivity value of 12.56%. Meanwhile, for ThN fuel with U-233 enrichment of 2%:4%:6% has an excess reactivity value of 7.67%. The geometric design of the GFR also affects the k-eff value and neutron flux [9]. Homogeneous geometry tends to produce quite high k-eff values due to the non-flattening distribution pattern of neutron flux. This causes a high peaking power in the center of the reactor. The use of heterogeneous core design in producing more flattening power. The use of heterogeneous core composition of 5 fuel variations with plutonium enrichment of 8%:9%:10%:11%:12% obtained an average neutron flux value of 1,107 × 10^{10} neutrons/cm^{2}s in begin of life (BoL) and 9,990 × 10^9 neutrons/cm^{2}s in end of life (EoL). The results of the comparison show that ThN fuel is more optimal in terms of criticality so that a high level of operational safety can be achieved.

Based on previous research, the researcher writing this paper is interested in exploring more deeper analysis of low-power GFR neutronics, namely SMR (Small Modular Reactor) type with several variations of heterogeneous core configurations using the OpenMC program and fueled by thorium-nitride (ThN-UN). The power used is 100 MWth, with considerations in the design of SMR, which prioritize scalability and modularity. The Heterogeneous core configuration uses 5 variations in fuel enrichment percentages (F1:F2:F3:F4:F5). The enrichment data was obtained from the optimization of k-eff values in homogeneous cores. Neutronic analysis includes analysis of k-eff values, excess reactivity, neutron flux, and extended burn-up. The calculation of extended burn-up is carried out to determine the extent to which the reactor can maintain its criticality until the specified burn-up time.

2. Methodology

2.1 Specifications and Design

The GFR design consists of several components, namely fuel, coolant, reflector, and absorber. The fuel used is a mixture of Th-232, U-233 enrichment, and nitride. The amount of U-233 enrichment ranges from 5 to 15% for homogeneous core design optimization. For heterogeneous core optimization, the average value of the U-233 enrichment across 5 variations in fuel composition was used. The reflector is made of Silicon Carbide (SiC) material. SiC materials have the advantage of maintaining mechanical properties from very high temperatures up to 1000°C and radiation resistance up to 50 dpa [10]. Boron Carbide (B4C) is used as an absorber material because of its neutron absorption capability and good irradiation performance for fast reactors [11]. For more detailed reactor specifications, please refer to Table 1.

Table 1. Parameter design of GFR

Parameters	Specifications
Fuel composition	ThN-UN (Th-232, U-233.,N15)
Enrichment U-233 (for homogeneous core)	5%-15%
Fuel/Gap/Cladding/Coolant volume fraction percentage	60%/0.5%/10%/29.5%
Lattice fuel	Hexagonal
Pin pitch	1.45 cm
Pitch assembly	17.04 cm
Active core diameter	250 cm
Outer reflector diameter	350 cm
Outer absorber diameter	390 cm
Total number of assembly	127

Homogeneous cores have a single variation in fuel composition, while heterogeneous cores use multiple variations in fuel composition [12] , as shown in Figure 1. The areas shown in specific colors have their respective functions : fuel, coolant, reflector, and absorber, as indicated in Figure 1. In a heterogeneous core, the arrangement of fuel composition variations involves a different number of rings for each fuel fraction (F_N), as shown in Table 2. The F1 fuel fraction consists of 3 rings and the F2, F3, F4, and F5 fuel fraction consists of 1 ring. The F1 area has more rings because the enrichment of U-233 used is lower when compared to the F2, F3, F4, and F5 areas. This arrangement aims to obtain a more uniform power distribution, especially in the inner area of the reactor.

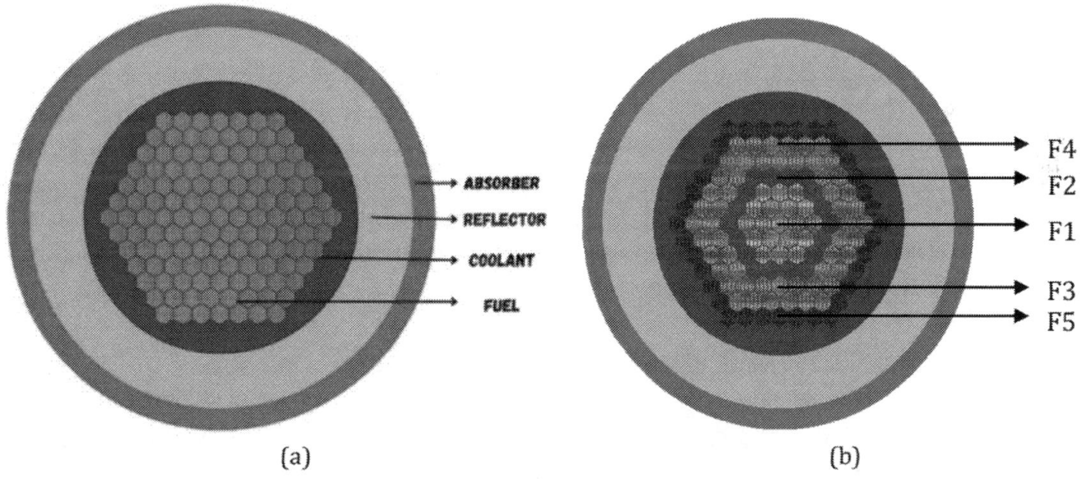

Figure 1. GFR geometry (a) Homogeneous core ; (b) Heterogeneous core

Table 2. Fuel fraction data for each ring for the heterogeneous core of the GFR

Fuel fraction	Number of rings	Number of fuel assembly
F1	3	19
F2	1	18
F3	1	24
F4	1	30
F5	1	36

2.2 Calculation method

The calculation of GFR optimization in this study used the Monte Carlo program OpenMC developed by MIT researchers since 2011 [13]. OpenMC uses stochastic calculations, where probabilities are involved and each simulation run yields a different outcome [14]. The OpenMC calculation used ENDFB-VIII.0 nuclear data library with fast reactor-type depletion chain files. To ensure the accuracy of the OpenMC code, validation againts other code is required. Validation parameters typically involve comparing the k-eff values obtained from the calculation with reference results. The reference k-eff value is derived from the data in paper [7], which used the SRAC calculation method. The code is considered validated if it has an error value ($\Delta k/k$) of less than 1% [15]. The error value can be calculated using the following equation:

$$\Delta k/k\ (\%) = \frac{k_{eff}\ OpenMC - k_{eff}\ SRAC}{k_{eff}\ SRAC} X\ 100\% \quad (1)$$

If the code has been validated, optimization calculations should then be performed for homogeneous core configurations with U-233 enrichment ranging from 5-15%. The results of these calculation should show k-eff values close to critical (k=1) and excess reactivity values that are minimal. The value of excess reacitivity (ρ) is calculated using the following equation:

$$\rho = \frac{k_{eff} - 1}{k_{eff}} \quad (2)$$

The case is then used for subsequent calculations in heterogeneous core. As supplementary data, additional information is provided in the form of neutron flux and extended burn-up. Extended burn-up aims to asses the reactor's ability to maintain a neutron life cycle within a specified burn-up range.

3. Results and Discussion

The calculation in the GFR, using k-eff parameters, provided benchmarking results between the OpenMC code and the SRAC code over five years of burn-up, as shown in Table 3. The benchmarking results indicate a maximum error ($\Delta k/k$) of 2.2%, with the increase in errors likely due to differences in nuclear data library used by OpenMC and SRAC. The OpenMC code will be used as the reference for calculating homogeneous core configurations with U-233 enrichment ranging from 5-15%.

Table 3. Benchmarking OpenMC and SRAC code results

Burn-up	k-eff		*Δk/k (%)*
	SRAC	OpenMC	
0	1.129	1.122	0.617
1	1.123	1.111	1.081
2	1.120	1.105	1.321
3	1.117	1.100	1.609
4	1.115	1.094	1.921
5	1.113	1.088	2.205

The results of the optimization calculation for the homogeneous core configuration are visualized in a burn-up time versus effective multiplication factor (k-eff) in Figure 2. The results of the optimization for the homogeneous core configuration with U-233 enrichment ranging from 5-15% indicated that the optimal enrichment value, which is closest to 1, was 8.5%. U-233 enrichment of 8.5% results in a very stable multiplication factor and is used as the basis for calculating the combined percentage of 5 types of fuels using a heterogeneous core configuration. These combined percentages use the results of a homogeneous core configuration with 8.5% U-233 enrichment as a reference, as shown in Table 4 for five different cases.

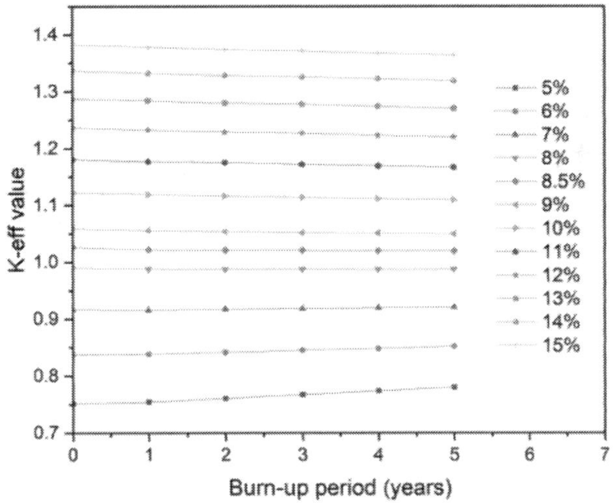

Figure 2. The k-eff value of U-233 homogeneous core enrichment optimization is 5%-15%

Table 4. Composition of U-233 enrichment percentage per fuel fraction (F_N)

Case calculation	Variation in U-233 enrichment percentage in heterogeneous core				
	F1	F2	F3	F4	F5
Case-1	6	7	8.5	9	12
Case-2	6	7	8.5	10	11
Case-3	6.5	7.5	8.5	9	11
Case-4	6.5	7	8.5	9.5	11
Case-5	7	7.5	8.5	9	10.5

The percentage of five varied cases was evaluated to identify the most optimal configuration, where the k-eff value was close to critical (i.e., closest to 1), as shown in Figure 3;a. The most optimal k-eff value was found in case-5, which has a fuel percentage distribution F1:F2:F3:F4:F5 = 7%:7.5%:8.5%:9%:10.5%. Case-5 achieved a maximum k-eff value of 1.028. This case was selected not only because of its k-eff value but also due to its smallest excess reactivity, as shown in Figure 3.b. According to Figure 3;b, Case-5 has the lowest excess reactivity value. Conditions related to the Beginning of Life (BOL) and End of Life (EOL) are detailed in Table 5, which includes both k-eff values and excess reactivity.

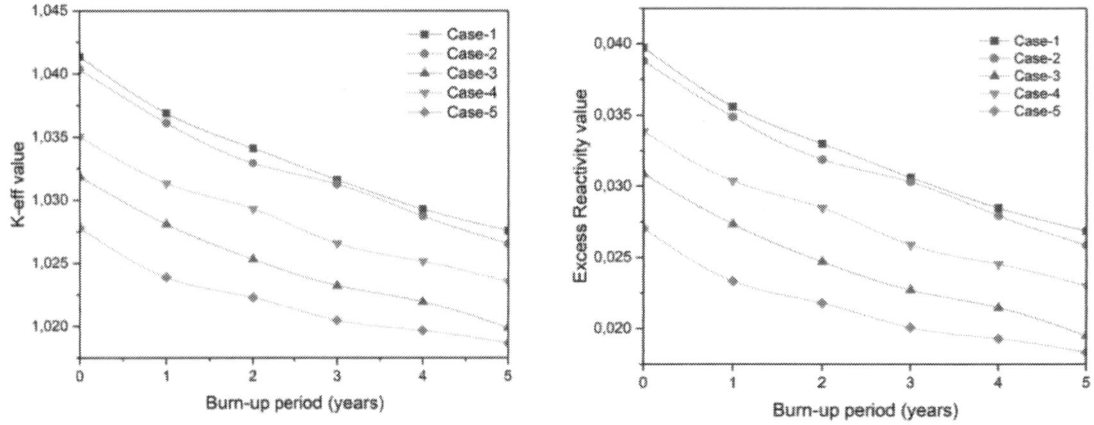

(a) (b)

Figure 3. Parameter data (a) k-eff value for each variation in heterogeneous core fuel composition; (b) Excess reactivity value for each variation in heterogeneous core fuel composition

Table 5. K-eff dan excess reactivity value for each variation in heterogeneous core fuel composition

Case calculation	K-eff in BOL	Excess reactivity in BOL	K-eff in EOL	Excess reactivity in EOL
Case-1	1.041	0.04	1.028	0.027
Case-2	1.040	0.039	1.027	0.026
Case-3	1.032	0.031	1.020	0.019
Case-4	1.035	0.034	1.024	0.023
Case-5	1.028	0.027	1.019	0.018

The flux distribution for each case of fuel composition in a heterogeneous core is shown in Figure 4. The calculation results show that the power flattening has been achieved, precisely in case-5. The flux distribution in case-5 tends to be flatter, compared to case 1 through 4, as the heterogeneous core design flattens the neutron flux distribution, minimizing flux peaks that can lead to excessive local heating and material degradation. A more uniform flux profile reduces the risk of power peaking and enhances overall fuel utilization. Consequently, the burn-up period is extended because the fuel is consumed more evenly. The power peaking in the F2-F4 area for case-1 to case-2 causes a very high flux value which is characterized by a deep red color in the area. In this context, the fission reaction by neutrons in the region has a high intensity and may lead to peaking power. The phenomenon of peaking power is highly avoided by reactors because it is prone to causing reactor accidents. Unlike case-5, the flux distribution in the F1-F5 area is more uniform, which can facilitate a more efficient nuclear fuel burn-up process. The average neutron flux values for case are 3.07 x 10^9 neutrons/cm^2s in BOL and 3.12 x 10^9 neutrons/cm^2s in EOL, respectively.

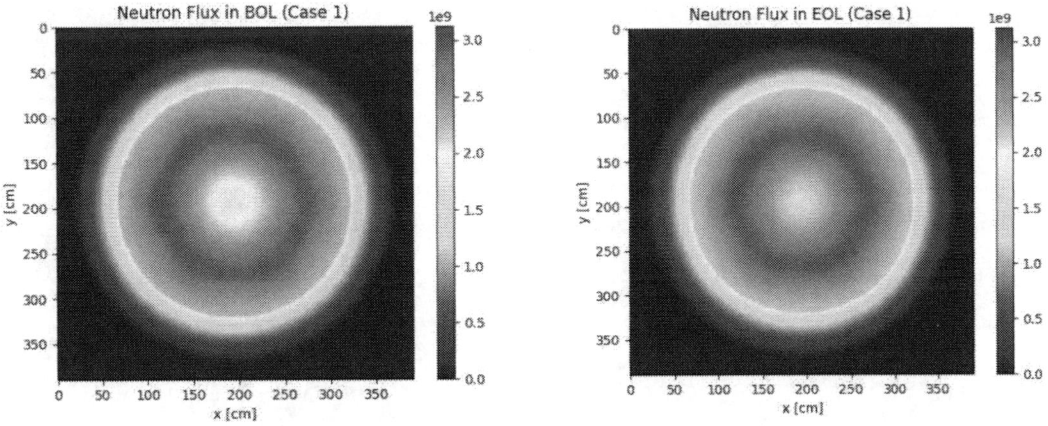
Neutron Flux in BOL (Case 1)
Neutron Flux in EOL (Case 1)
1e9
y [cm]
x [cm]

(a)

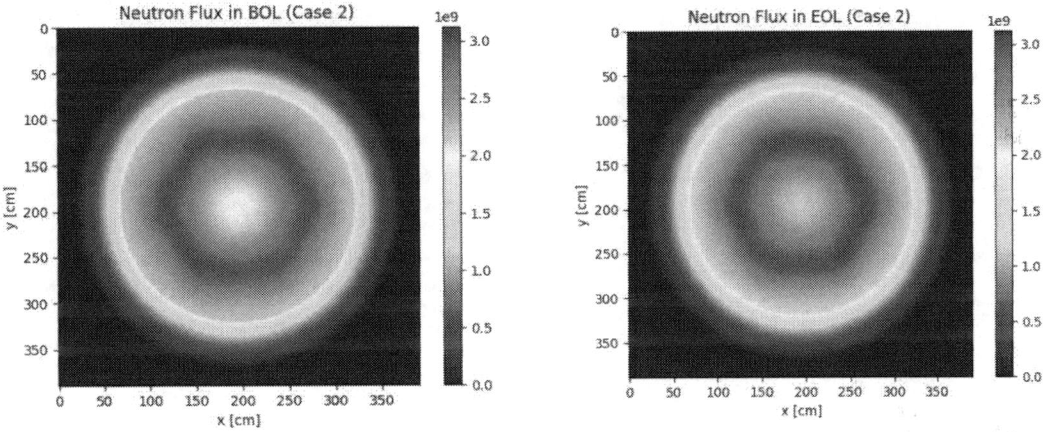
Neutron Flux in BOL (Case 2)
Neutron Flux in EOL (Case 2)
1e9
y [cm]
x [cm]

(b)

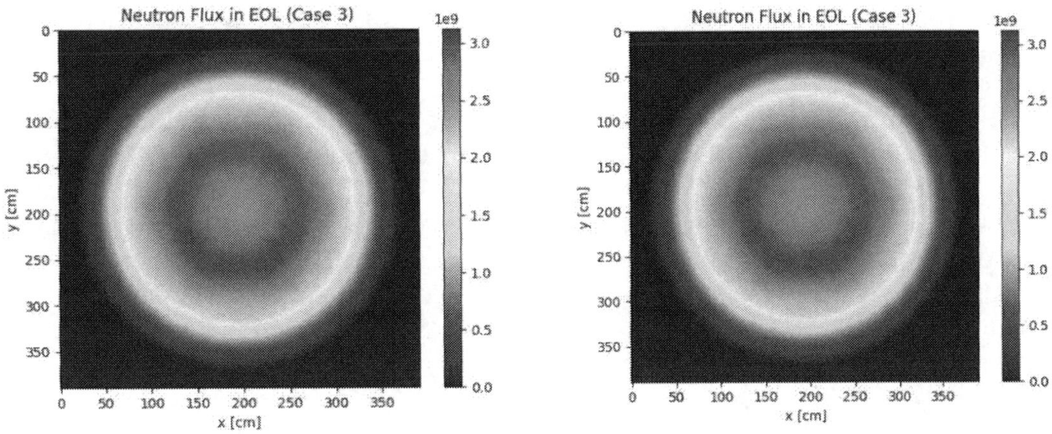
Neutron Flux in EOL (Case 3)
Neutron Flux in EOL (Case 3)
1e9
y [cm]
x [cm]

(c)

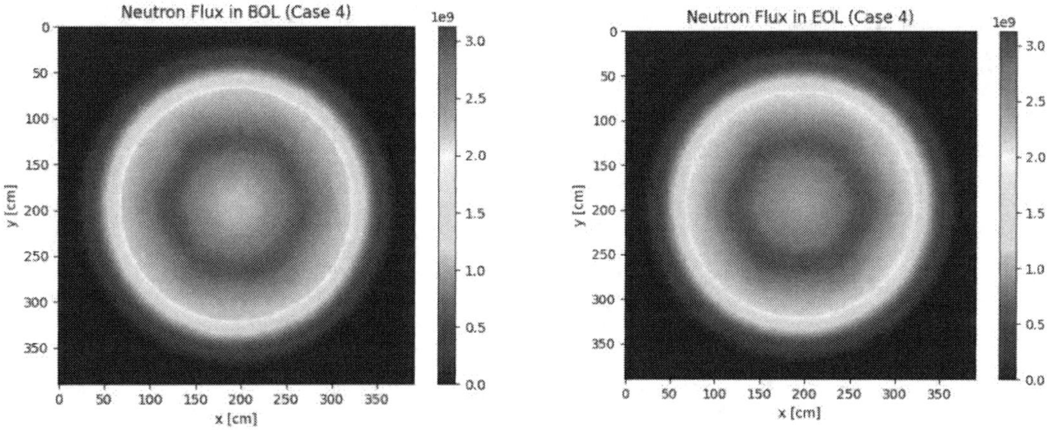

(d)

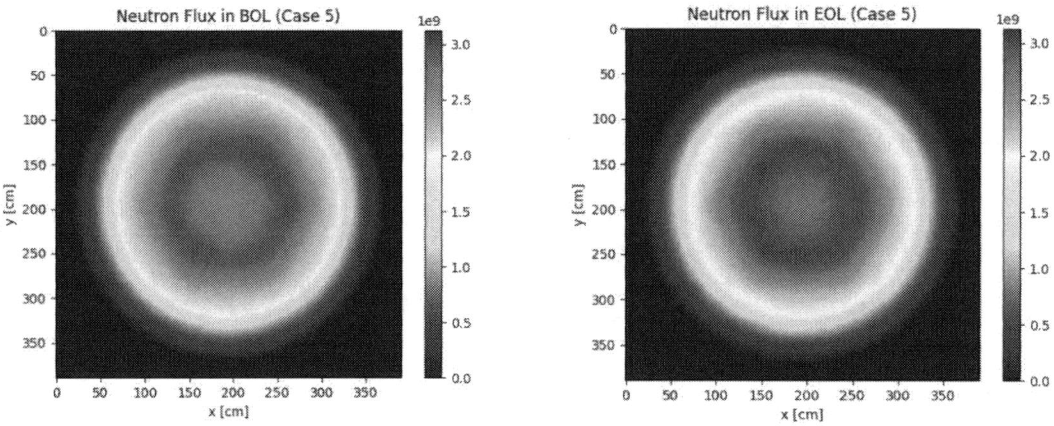

(e)

Figure 4. Heterogeneous nucleus flux neutron distribution under BOL and EOL conditions for case (a) case-1; (b) case-2; (c) case-3; (d) case-4; (e) case-5

The heterogeneous core in case-5 have been proven to have an optimal composition in terms of k-eff values, excess reactivity, neutron flux, and total product fission yield. To determine the ability of the fuel composition in case-5 to maintain the life cycle of neutrons during fission, extended burn-up is carried out over a predetermined time range. The results of the calculations show that the GFR is able to maintain its criticality for up to 10 years without becoming sub-critical, as shown in Figure 5. At the end of burn-up (10th year), the k-eff value is 1.00043. The main fuel, Th-232, breeds effectively, producing new fissile material in the form of U-233. The addition of U-233 from breeding can extend the life of neutrons.

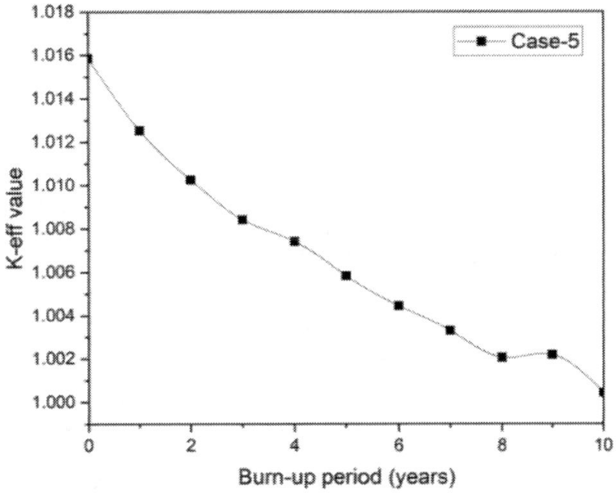

Figure 5. Analysis of the extend of the burn-up period for 10 years for case-5

Conclusion

A neutronic analysis of the heterogeneous core of GFR has been performed using OpenMC calculations. The results of the homogeneous core analysis indicated that a U-233 enrichment of 8.5% in case-5 demonstrated optimal criticality for 5 years of burn-up, with k-eff values of 1.026 at BOL and 1.019 at EOL. In further optimization, specifically in the heterogeneous core analysis, the most optimal configuration was identified in case-5, with a U-233 enrichment distribution of 7%:7.5%:8.5%:9%:10.5%. The k-eff and excess reactivity values at BOL were 1.0278 and 0.02705, respectively. Meanwhile, in EOL conditions are 1.01867 and 0.01833. The uniform distribution of flux contributes to an efficient active fuel burn-up process, resulting in minimal reactor operational waste. In the of extended burn-up calculations, case-5 was able to maintain criticality through the 10th year.

References

[1] GIF, "Technology Roadmap Update for Generation IV Nuclear Energy Systems," 2014.

[2] R. D. Syarifah, M. H. Aula, A. Arkundato, A. T. Nugroho, G. Jatisukamto, and S. Handayani, "DESIGN STUDY OF 300MWTH GFR WITH UN-PuN FUEL USING SRAC-COREBN CODE," vol. 18, no. 4, 2023, [Online]. Available: www.arpnjournals.com

[3] F. Prasetya, A. M. Mabruri, I. Karomah, R. D. Syarifah, I. K. Aji, and N. Trianti, "Validation of OpenMC Code Criticality Value Calculation for GFR with UN-PuN Fuel," in *Journal of Physics: Conference Series*, Institute of Physics, 2024. doi: 10.1088/1742-6596/2734/ 1/012065.

[4] GIF, "A Technology Roadmap for Generation IV Nuclear Energy Systems," 2002.

[5] System Design Report, "Development of Generation IV Advanced Gas-Cooled Reactors with Hardened/Fast Neutron Sectrum," Feb. 2005.

[6] S. Novalianda, P. Mahadi Sihombing, U. Al-Azhar, and P. Penerbangan Surabaya, "Studi Desain Teras Gas-Cooled Fast Reactor Berbahan Bakar Thorium Nitride Afiliation," *Jurnal Sains dan Teknologi Elektro*, vol. 14, pp. 2830–3512, 2024, doi: 10.47709/elektriese.v 14i01.4470.

[7] R. D. Syarifah, Z. Su'ud, K. Basar, D. Irwanto, S. C. Pattipawaej, and M. Ilham, "Comparison of uranium plutonium nitride (U-Pu-N) and thorium nitride (Th-N) fuel for 500 MWth gas-cooled fast reactor (GFR) long life without refueling," in *International Journal of Energy Research*, John Wiley and Sons Ltd, Jan. 2018, pp. 214–220. doi: 10.1002/er.3923.

[8] R. D. Syarifah, P. Investigator, M. H. Aula, A. Ardianingrum, L. N. Janah, and W. Maulina, "COMPARISON OF THORIUM NITRIDE AND URANIUM NITRIDE FUEL ON SMALL MODULAR PRESSURIZED WATER REACTOR IN NEUTRONIC ANALYSIS USING SRAC CODE," *Eastern-European Journal of Enterprise Technologies*, vol. 2, no. 8–116, pp. 21–28, 2022, doi: 10.15587/1729-4061.2022.255849.

[9] F. Prasetya, R. D. Syarifah, I. Karomah, I. K. Aji, and N. Trianti, "A COMPARATIVE ANALYSIS OF GAS-COOLED FAST REACTOR USING HETEROGENEOUS CORE CONFIGURATIONS WITH THREE AND FIVE FUEL VARIATIONS," *Eastern-European Journal of Enterprise Technologies*, vol. 1, no. 8(127), pp. 6–17, 2024, doi: 10.15587/1729-4061.2024.298202.

[10] S.J. Zinkle, "Nonfissile Ceramics for Future Nuclear Systems," *Proc. GLOBAL American Nuclear Society*, pp. 1066–1067, 2023.

[11] D. E. Mahagin, "FAST REACTOR NEUTRON ABSORBER MATERIALS," 1979.

[12] M. D. Afriani, R. Dewi Syarifah, F. Prasetya, and W. Maulina, "JIFP (Jurnal Ilmu Fisika dan Pembelajarannya) Efek Penambahan Protaktinium-231 (Pa-231) pada Nilai Faktor Multiplikasi Efektif Reaktor PWR Berbahan Bakar ThN-U233N Effect of Protactinium-231 (Pa-231) Addition on Effective Multiplication Factor Value of ThN-U233N Fueled PWR Reactor", [Online]. Available: http://jurnal.radenfatah.ac.id/index.php/jifp/index

[13] M. Ilham, H. Raflis, and Z. Suud, "Full Core Optimization of Small Modular Gas-Cooled Fast Reactors Using OpenMC Program Code," in *Journal of Physics: Conference Series*, Institute of Physics Publishing, May 2020. doi: 10.1088/1742-6596/1493/1/012007.

[14] P. K. Romano, N. E. Horelik, B. R. Herman, A. G. Nelson, F. Benoit, and K. Smith, "OpenMC: A state-of-the-art Monte Carlo code for research and development," *Ann Nucl Energy*, vol. 82, pp. 90–97, 2015.

[15] M. Ilham, H. Raflis, and Z. Suud, "Fuel Assembly Design Study for Modular Gas Cooled Fast Reactor using Monte Carlo Parallelization Method," in *Journal of Physics: Conference Series*, IOP Publishing Ltd, Feb. 2021. doi: 10.1088/1742-6596/1772/1/012025.

A Brief Comparison of Indonesian Prospective Nuclear Fuel Cycles

Dany Mulyana

Center of Nuclear Security Science and Policy Institute, Texas Engineering Experiment Station, Texas A&M University, College Station, TX, USA.
Research Center for Nuclear Reactor Technology, Badan Riset dan Inovasi Nasional, Jakarta, Indonesia.

E-mail: danymulyana@gmail.com

Abstract. A brief comparison study on fuel cycles of four prospective Indonesian power reactors has been completed. The study oversees 160 MWt NuScale Small Modular Reactor (SMR), 10 MWt Reactor Daya Eksperimental (RDE-10), 40 MWt Pembangkit Listrik dan Uap Panas Industri (PeLUIt-40) and Thorcon's TMSR-500. Both RDE-10 and PeLUIt-40 were fueled with 17wt% HALEU while the other two were fueled by 4.95wt% LEU, without thorium. The study was based on a set of data generated through Monte Carlo reactor physics simulation using OpenMC. The study found that NuScale SMR is the least ^{235}U consumer by 0.84 $g.MW^{-1}d^{-1}$ while the RDE-10 is the most ^{235}U consumer by 0.97 $g.MW^{-1}d^{-1}$. The simulation results show that TMSR-500 is the most uranium consumer by 1.41 $g.MW^{-1}d^{-1}$ while the uprating of RDE-10 to PeLUIt-40 makes PeLUIt-40 as the least uranium consumer by 1.14 $g.MW^{-1}d^{-1}$. As the least uranium consumer, both RDE-10 and PeLUIt-40 provide the highest attainable discharged fuel burnup of extractable energy of about 146 GWd/MTU. Assuming a one batch depletion scheme, NuScale SMR, RDE-10, PeLUIt-40 TMSR-500 has fuel consumption cycles in 2,210 days, 2140 days, 550 days, and 320 days, respectively.

1. Introduction

Indonesian government has been engaging a nuclear energy deployment since the early of 1970's. However, due to public resistance, it has been put on hold. With the global rise of zero carbon energy demand, the recent government is looking for another chapter to install the first nuclear power plant and have it connected to the grid in the near future. The expected electricity production using nuclear energy is about 9 GWe by 2060 [CNN Indonesia, 2024]. There are some industrial proposals for the basic nuclear reactor core to be deployed. At least three technologies are considered: light water-based reactor, pebble bed reactor, and molten salt reactor.

The light water reactor (LWR) is a proven technology deployed in almost in the world. Out of 442 installed nuclear reactors, about 96% of them are water-cooled [IAEA, 2024]. It typicaly uses uranium dioxide enriched up to 5% of ^{235}U as fuel and light water as neutron moderator, as cooling media, and as heat transfer media. Since LWR operates in a range of 275 to 315 °C, the water must be pressurized to about 15 MPa. NuScale Inc, sponsored by the United States of America government, is engaging the deployment of this type of reactor in Indonesia [US Embassy in Indonesia, 2023]. Differently, the pebble bed reactor (PBR) uses the concept of integrated High Assay Low Enriched Uranium (HALEU) fuel embedded in a 6-cm diameter sphere made from graphite which acts as neutron moderator. Instead of using water, PBR uses helium gas as its heat transfer medium that allow the reactor to operate in a range of 600 to 900 °C with a core pressure of up to 6.9 MPa. The development of PBR has been backed by the Indonesian government itself through its National Nuclear Energy Agency (BATAN) since 2014 and continued by the Research and Innovation Agency (BRIN) since 2021, resulting the *Reaktor Daya Eksperimental* (RDE) with a rated output thermal of 10 MW. It is also projected to produce 40 MWt with PeLUIt as a codename, to strengthen its industrial cogeneration functions. Meanwhile, the molten salt reactor (MSR) has a concept of liquid fuel flowing into a graphite structured reactor core. It is proposed to use a HALEU enriched up to 20% of ^{235}U, with an addition of thorium that is fertile to produce the fissionable ^{233}U as an option for a cost-competitive further fuel option. Thorcon Inc. has been actively proposing the deployment of MSR through its floating power plant concept using their reactor design of Thorium Molten Salt Reactor 500 (TMSR-500) delivering 557 MWt power [Devanney, et al., 2015]. Each of these reactors has its own mechanism to burn the fissile materials, as well as to breed the fertile materials. This creates a unique and distinguishable fuel cycle footprint which is started at the mining side, followed up by fuel enrichments and fabrications. Once the energy is extracted in a reactor, the spent fuel characteristics do not depend on the fresh fuel characteristics only, but also by the character of the reactors itself. The procurement of fresh fuel and the storage and/or reprocessing the spent fuel affect the national energy policy directly.

These chosen reactor designs for this study are currently being proposed by national and multi-national companies seeking to the early NPP deployments in Indonesia. These designs are classifiable as Small Modular Reactors with at least generation III+ reactor technologies, which are in line with the Act No. 59/2024 for the Indonesia's long term development plan. While any other technology providers are not explicitly ready to commercially launch their floating NPP design, TMSR-500's development is on its way to be in the floating NPP criteria. This work is limited to the small to medium reactor designs only.

This study describes the possibilities of fresh fuel consumption and leftover uranium quantities of the four candidate reactor types described previously. This study uses technical data generated through some reactor physics calculations. The outcome of this study is to compare the differences of possible nuclear fuel cycles of the respective reactors in providing a policy framework required for proper fuel management. The analysis in this study is limited to the quantities of uranium in the fresh and spent fuels only.

2. Methodology

The basic engineering data required for the policy analysis was generated through a reactor physics calculation involving a Monte Carlo (MC) particle transport method. MC was used due to its capability to perform a neutronic simulation on a complex geometry where the Boltzmann transport equation is not practically implemented deterministically. Instead, MC solves the equation stochastically based on nuclear data on the probabilities of neuron interaction with matters. It also follows the stochastics' nature of radiation mechanisms. However, compared to any deterministic method that is a rapid solver, MC is a resource exhaustive physics simulation procedure. This study used OpenMC code running on several 20-cores computers embedded in

IOP Conf. Series: Materials Science and Engineering **1330** (2025) 012014 doi:10.1088/1757-899X/1330/1/012014

a High-Performance Computer (HPC) system. Nevertheless, the process is straightforward and reliable for simulating material depletion for a reactor design.

This study did not perform a full core simulation to perform the reactor physics calculation. Instead, it performed an infinite lattice reactor technique that has a similar principle of calculation. The infinite lattice reactor technique performs the calculation by deploying reflective boundary conditions on all sides of a fuel bundle. This prohibits any neutron loss factors and enables fuel depletion to its maximum dischargeable burnup level. Hence, the four-factor formula applies, instead of the six-factor formula that applies to a full core calculation accounting the leakage terms. OpenMC can perform the depletion simulation using a first-order predictor-corrector algorithm to perform the neutron transport. This study compares the consumed uranium and U-235 by each reactor type normalized to a unit of mass per extracted energy, or g.MWd^{-1}. This study also deployed ENDF/B VII.1 nuclear data at each respective reactor's fuel temperature. This study used the CASL depletion chain [Romano et al., 2021]. The depletion simulations were to calculate the time required by each reactor system for fully consuming the fuel to its allowable energy extraction. Based on this maximum fuel discharged period, the fuel consumption rate and the waste generation rate could be calculated. This is indicated when the neutron multiplication factor approaches unity.

The LWR's fuel for simulation in this work used the NuScale's small modular reactor design that has a rated power output of 160 MWt which is equivalent to 45 MWe deploying 37 assemblies per core [NuScale, 2016]. This study used a ^{235}U enrichment of 4.4% and a density of 10.46 g.cm^{-3} for the fresh fuel. One fuel assembly of NuScale reactor has a height of 200 cm and sides of 21.617 cm (ref. Figure 1(a)) with a configuration of 17×17 fuel pin configuration. However, only 264 fuel pins are deployed by the fuels and the rest are water-filled channels due to their function for reactivity control channels. The pins have a radius of 0.392176 cm pitched by a 1.26 cm distance. The fuel is contained by a 0.05715 cm thick Zircaloy-4 cladding with a density of 6.55 g.cm^{-3}. Although the reactor operates at about 275 to 315 °C, the fuel can reach a higher temperature. Hence, the fuel model was set to 900 K. The water used in the model has a density of 0.7295 g.cm^{-3}. The fuel assembly was set to an equivalent thermal power of 4.32 MWt.

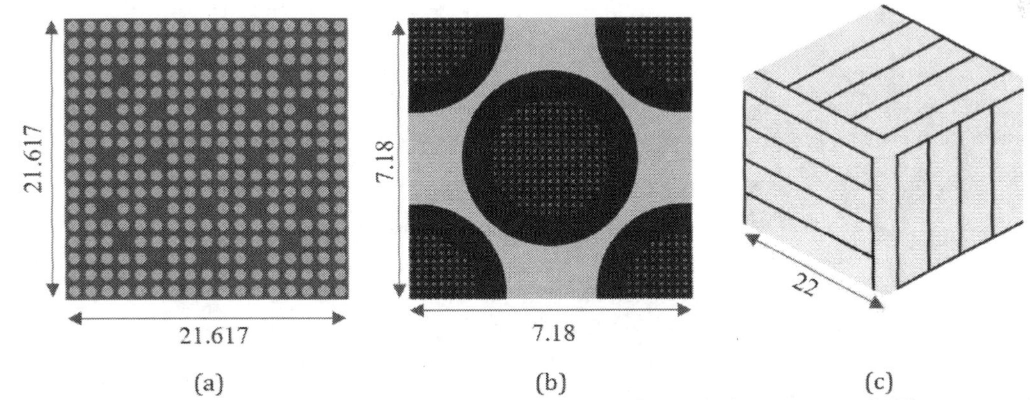

Figure 1. Fuel assembly lattice of (a) NuScale (bright grey = fuel pins, dark grey = water), (b) RDE-10 and PELUIT-40 (black = graphite, grey = helium, bright dots=TRISO), and (c) TMSR-500 (grey = graphite, black = fuel salt) designs. Their dimensions are not uniformly scaled for any comparison. All dimensions are in centimetres.

The PBR's fuel deployed in this work used the RDE design with a rated output power of 10 MWt. The fuel pebble has a diameter of 6 cm where its inner diameter of 5 cm is deployed by thousands of microsphere fuels and filled by graphite with a density of 1.73 g.cm^{-3}. The microsphere fuel, called TRISO, has a radius of 0.0455 cm with some PyC and SiC coatings, while the UO_2 (density of 10.4 g.cm^{-3}) fuel itself has a radius of only 0.025 cm and it is enriched to 17% of ^{235}U. A single pebble contains 5 g of uranium. A complete description of the fuel pebble, which is very similar to the one of HTR-10, can be found in the references [Mulyana and Chirayath, 2022a and 2022b; Wu et al, 2002]. In this work, the infinite lattice of the fuel has two fuel pebbles, without any graphite pebble, arranged in a Body Centered Cubic (BCC) lattice configuration (ref. Figure 1(b)). The fuel lattice has a side of 7.18 cm. Although having a higher output rated power of 40 MWt, the PeLUIt has the same core and pebble designs. The fuel lattice was set to have an equivalent thermal power of 687.14 W for RDE and 2.75 KW for PeLUIt. In terms of nuclear island design, RDE and PeLUIt designs are similar. However, due to the different target of rated power, their non-nuclear island's subsystems and their components are significantly different to adapt the faster flow of working fluid and the final energy conversion subsystems. In the nuclear island, the higher neutron flux in the PeLUIt's core demands substantially better materials with higher wear resistance, strength, structural stability, etc. to maintain the reactor's operation within its safety limits. Moreover, due to their different power, their operations' period will be different, leading to different the isotope production and loss mechanisms in the fuel pebbles.

The MSR's fuel assembly for this study used TMSR-500 specification having a rated power output of 557 MWt. This work deployed only LEU for the fuel with a ^{235}U enrichment of 4.95%. The LEU is mixed with F-Be-Na as a $NaF.BeF_2.UF_4$ composition with a mixed density of 3.2 g.cm^{-3}. The salt-to-graphite ratio within the assembly is about 11.85% while the graphite density is 1.7 g.cm^{-3}. Since TMSR-500 is designed to have 84 assemblies for 557 MWt, each assembly constitutes about 6.63 MWt. The assembly was set to 3.78 m which is the specified active core's height. The fuel temperature was set to 900 K while the graphite temperature was set to 600 K. The hexagonal side is about 22 cm (ref. Figure 1(c)).

The reactor designs considered in this study may have different uranium's consumption rate. Based on their U enrichments, their operation times will also be different. The idea of this study is to perform a fair comparison regardless to these differences, arising from their inherent technical specifications. The comparison of all reactors in this study is based on a normalization of total uranium and ^{235}U quantities in the core per 1 MW of generated power per day, as in the following equation:

$$mass_{norm} = \frac{mass_{initial} - mass_{discharge}}{Thermal\ Power \times Time_{discharged}}$$

Hence, the physical unit used to compare the U depletion in this study is g.MW^{-1} d^{-1}, which is the inversion of the fuel burnup measure itself. This way, the uranium consumption measure is normalized with regard to their different burnup levels.

3. Results and Discussion

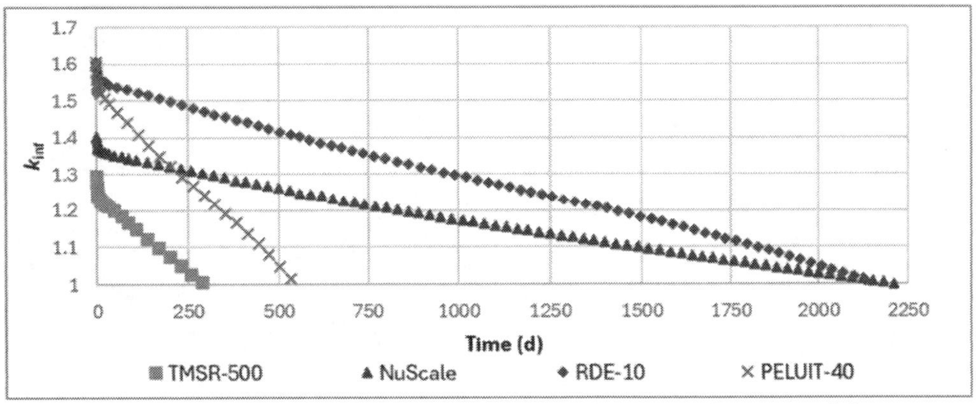

Figure 2. Infinite neutron multiplication factors of the simulated reactor models

The MC simulations calculated that after about 2,210 days (37.4 GWd/MTU), 2140 days (147.5 GWd/MTU), 550 days (146 GWd/MTU), and 320 days (34.1 GWd/MTU), the NuScale, RDE-10, PeLUIt-40 and TMSR-500 reactors approach subcriticality, respectively.

Per one core of NuScale reactor, there is 10,735.85 kg of fresh fuel containing 9,463.41 kg of uranium. With an enrichment of 4.4%, this means initially the reactor is loaded with 416.39 kg of ^{235}U. Once it approached its subcriticality, the reactor yields about 8,983 kg of leftover uranium. This also means that this reactor type consumes 1.36 g.MW^{-1} d^{-1} of uranium and spends about 5% uranium. At this point of subcriticality, the spent fuel has about 1.33% of ^{235}U, which is about 119.19 kg per core. Hence, the ^{235}U consumption is about 0.84 g. MW^{-1} d^{-1}.

Total uranium in the RDE-10 and PeLUIt-40 cores at the beginning of life are about 146.24 kg by assuming each reactor core contains 29,106 fuel pebbles. Once approached subcriticality, RDE-10 and PeLUIt-40 yielded 120.53 kg (82.42% of the initial quantity) and 124.16 kg (82.85% of the initial) of leftover uranium, respectively. The uranium leftover contains 3.43% and 3.83% of leftover ^{235}U for the RDE-10 and PeLUIt-40, respectively. These also mean 1.2 and 1.14 g.MW^{-1} d^{-1} of uranium consumptions, as well as 0.97 and 0.92 g.MW^{-1} d^{-1} of ^{235}U consumptions, for the RDE-10 and PeLUIt-40, respectively.

Meanwhile, total uranium in the TMSR-500 core at the beginning of life is about 5,226.83 kg. Once approached subcriticality, the TMSR-500 core yielded 4,974.63 kg (95.17% of the initial quantity) of leftover uranium, respectively. The uranium leftover contains 1.87% of leftover ^{235}U. These also mean 1.41 g.MW^{-1} d^{-1} of consumed uranium and 0.93 g.MW^{-1} d^{-1} of consumed ^{235}U.

At a common operation time of 320 days, the uranium consumptions of NuScale, RDE-10, PeLUIt-40 and TMSR-500 reactors are 1.14, 0.35, 0.72, and 1.41 g.MW^{-1} d^{-1}, respectively. Meanwhile, the U-235 consumptions are 0.67, 0.9, 0.63, and 0.93 g.MW^{-1} d^{-1} for NuScale, RDE-10, PeLUIt-40 and TMSR-500 reactors, respectively. The burnup levels at his common operation time are 5.4, 22.2, 88.6 and 34.1 GWd.MTU^{-1} for NuScale, RDE-10, PeLUIt-40 and TMSR-500 reactors, respectively.

The MC simulation results show that NuScale reactor is the least ^{235}U consumer while RDE-10 reactor is the most ^{235}U consumer although the Nuscale reactor consumes uranium more than the RDE-10 does. The PeLUIt-40 and TMSR-500 come in between where the later consumes more. RDE is a typical pebble bed reactor (PBR) while the NuScale is a typical pressurized water reactor (PWR) with small design. This aligns with a previous research results

[Mulyana and Chirayath, 2023] concluding that PBR has a higher rate of thermalization compared to PWR. Meanwhile, the RDE has higher U-235 enrichment compared to NuScale. Combination of these leads to a higher U-235 consumption since the Pu utilization in PBR is lower. The use of HALEU in PBR leads to a competitive consumption toward U-235, rather than to Pu.

Table 1. Consumed uranium and U-235 masses, as well as volume of waste predicted per 1 GWt power produced by NuScale, RDE-10, PeLUIt-40 and TMSR-500 reactor cores.

Time (d)	NuScale			RDE-10			PeLUIt-40			TMSR-500		
	U (t)	U-235 (t)	Waste (kL)	U (t)	U-235 (t)	Waste (kL)	U (t)	U-235 (t)	Waste (kL)	U (t)	U-235 (t)	Waste (kL)
200	0.27	0.17	1.96	0.24	0.19	50.34	0.23	0.18	48.97	0.28	0.19	1.83
400	0.54	0.34	3.91	0.48	0.39	100.69	0.46	0.37	97.94	0.57	0.37	3.67
600	0.82	0.51	5.87	0.72	0.58	151.03	0.68	0.55	146.91	0.85	0.56	5.50
800	1.09	0.67	7.82	0.96	0.77	201.37	0.91	0.74	195.88	1.13	0.74	7.33
1000	1.36	0.84	9.78	1.20	0.97	251.72	1.14	0.92	244.85	1.41	0.93	9.16
1200	1.63	1.01	11.74	1.44	1.16	302.06	1.37	1.10	293.82	1.70	1.12	11.00
1400	1.90	1.18	13.69	1.68	1.36	352.40	1.60	1.29	342.79	1.98	1.30	12.83
1600	2.17	1.35	15.65	1.92	1.55	402.75	1.82	1.47	391.76	2.26	1.49	14.66
1800	2.45	1.52	17.60	2.16	1.74	453.09	2.05	1.65	440.73	2.55	1.67	16.50
2000	2.72	1.68	19.56	2.40	1.94	503.43	2.28	1.84	489.70	2.83	1.86	18.33
2200	2.99	1.85	21.51	2.64	2.13	553.78	2.51	2.02	538.67	3.11	2.05	20.16
2400	3.26	2.02	23.47	2.88	2.32	604.12	2.74	2.21	587.64	3.40	2.23	21.99
2600	3.53	2.19	25.43	3.12	2.52	654.46	2.96	2.39	636.61	3.68	2.42	23.83
2800	3.80	2.36	27.38	3.36	2.71	704.81	3.19	2.57	685.58	3.96	2.60	25.66
3000	4.08	2.53	29.34	3.60	2.91	755.15	3.42	2.76	734.56	4.24	2.79	27.49
3200	4.35	2.70	31.29	3.84	3.10	805.49	3.65	2.94	783.53	4.53	2.97	29.32
3400	4.62	2.86	33.25	4.08	3.29	855.84	3.88	3.12	832.50	4.81	3.16	31.16
3600	4.89	3.03	35.21	4.32	3.49	906.18	4.10	3.31	881.47	5.09	3.35	32.99

4. Conclusions

The MC simulation results show that NuScale reactor is the least ^{235}U consumer while RDE-10 reactor is the most ^{235}U consumer although the Nuscale reactor consumes uranium more than the RDE-10 does. The uprating of RDE-10 to PeLUIt-40 slightly decreases both the uranium and ^{235}U consumptions. When fueled with LEU without thorium, the TMSR-500 is the most uranium consumer with a slightly higher consumption than the NuScale reactor does.

This study was focused only on the uranium consumption and leftover uranium quantities to provide a clear picture on the economical perspective of the reactors' fuel consumption. However, a further analysis on other actinides buildups is possible, especially for future non-proliferation and waste economics study.

IOP Conf. Series: Materials Science and Engineering **1330** (2025) 012014 doi:10.1088/1757-899X/1330/1/012014

References

[1] Romano, P.K., Josey, C.J., Johnson, A.E., and Liang, J., 2021. Depletion capabilities in the OpenMC Monte Carlo particle transport code. Annals of Nuclear Energy, Volume 152, Article Number 107989. https://doi.org/10.1016/j.anucene.2020.107989.

[2] Mulyana, D., and Chirayath, S.S., 2022a. Proliferation resistance assessment of a typical pebble bed reactor fuel. Annals of Nuclear Energy, 165, Article No. 108769, https://doi.org/10.1016/j.anucene.2021.108769.

[3] Mulyana, D., and Chirayath, S.S., 2022b. The impact of refueling schemes on the proliferation resistance of a pebble bed reactor. Annals of Nuclear Energy, 170, Article No. 109010. https://doi.org/10.1016/j.anucene.2022.109010.

[4] Mulyana, D., and Chirayath, S.S., Proliferation resistance of light water reactor versus pebble bed reactor. Annals of Nuclear Energy, 188, Article No. 109797. https://doi.org/10.1016/j.anucene.2023.109797

[5] Wu, Z., Lin, D., and Zhong, D., 2002. The design features of the HTR-10. Nuclear Engineering and Design, Volume 218, Issues 1–3, October 2002, Pages 25-32. https://doi.org/10.1016/S0029-5493(02)00182-6.

[6] IAEA, 2024. Water cooled reactors. International Atomic Energy Agency. https://www.iaea.org/topics/water-cooled-reactors, accessed on June 17, 2024.

[7] US Embassy in Indonesia, 2023. United States, Indonesia Announce Partnership on Small Modular Reactor Nuclear Clean Energy. https://id.usembassy.gov/united-states-indonesia-announce-partnership-on-small-modular-reactor-nuclear-clean-energy/, accessed on June 17, 2024.

[8] CNN Indonesia, 2023. Pembangkit Listrik Tenaga Nuklir Indonesia Dikomersilkan 2032. https://www.cnnindonesia.com/ekonomi/20231116070843-85-1024863/pembangkit-listrik-tenaga-nuklir-indonesia-dikomersilkan-2032. accessed on June 17, 2024.

[9] Devanney, J., Jorgensen, L., Livingston, J., Moir, R., Rodenburg, A.C., and Uhlik, C., 2015. ThorCon Executive Summary. Version 1.09. Martingale Inc.

[10] NuScale, 2016. The NuScale Design. PM-0616-49510-NP, Revision 0, nonproprietary version, NuScale Power, LLC.

Obtaining the Optimum Transuranic Layer Configuration for Transmutation in Molten Salt Reactor Fuel Channel

R. Andika Putra Dwijayanto[1*], Andang Widi Harto[2] and Azizul Khakim[1]

[1] Research Centre for Nuclear Reactor Technology, Research Organisation for Nuclear Energy, National Research and Innovation Agency, 80 KST BJ Habibie, South Tangerang, Indonesia 15314
[2] Department of Nuclear Engineering and Engineering Physics, Faculty of Engineering, Universitas Gadjah Mada, Jl. Grafika No. 2, Yogyakarta, Indonesia 55281

*E-mail: rand002@brin.go.id

Abstract. Incineration of transuranic (TRU) elements can be performed in various types of nuclear reactor. Thermal molten salt reactor (MSR) can be utilised for TRU incineration, typically by blending TRU elements with the carrier salt. This method has issues in low trifluoride solubility and lower TRU burnup than fast spectrum MSR. This study proposes an innovative method of TRU incineration in MSR as a heterogeneous configuration by inserting tubular TRU layer into MSR fuel channel. To obtain optimum configuration, TRU layers comprised of reactor grade plutonium (RGPu) and minor actinide (MA) were inserted in an MSR fuel channel in central, middle, and outer configurations. Neutronic analysis was performed in the MSR fuel channel using MCNP6.2 code, and decay analysis was done using ORIGEN2.1. The analysed parameters include infinite multiplication factor, temperature coefficient of reactivity (TCR), transmutation efficiency (TE), and post-irradiation decay activity. It was observed that TRU layer with a volume fraction of 5% in a narrow fuel channel and outer configuration exhibit the optimum TE and improved the TCR. After 1460 days of irradiation, the TE was found to be 30.98% and lower radioactivity than unirradiated TRU after 400 years of decay. This result is promising as an initial stage prior to whole core calculation.

1. Introduction

Radioactive waste is often considered as one of the obstacles of nuclear power expansion. The primary issue that is often raised is transuranic (TRU) elements, which consist of heavy nuclides with atomic number larger than uranium ($Z > 92$). TRU elements typically encompass plutonium and minor actinides (MAs), comprising of neptunium, americium, and curium. Berkelium and californium sometimes considered, although they exist mostly as trace elements. TRU waste have long half-life, requiring thousand until millions of years for them to decay into background radioactivity level. Such a long half-life is often touted as an argument against nuclear power deployment [1,2]. Although in fact, TRUs are alpha-emitter which requires them to be inhaled or ingested to give acute radiation poisoning. Both scenarios are extremely unlikely to happen under proper waste management [3].

Nonetheless, to prevent highly sensitive political issue regarding long-lived disposal of TRU elements, transmutation can be employed as an option to destroy the legacy TRU elements from operating nuclear power plants. TRU transmutation were widely studied in various fuel cycle and reactor types, both power [4–6] and non-power reactors [1,7]. TRU transmutation in light water reactors (LWRs) were studied both within the fuel rod [5,8,9] and as coating [6]. Similar methods were studied in fast reactors [10–12]. Using molten salt reactor (MSR) offers a different perspective in transmuting TRU, since the fuel is dissolved within the coolant in a liquid form. The TRU is mixed homogeneously with the fuel as opposed to be placed specifically within the reactor core in solid-fuelled reactors. Due to this fuel homogeneity, the average burnup of MSR fuel is generally lower than solid-fuelled reactors [13]. This may potentially affect the rate of TRU transmutation.

MSRs were studied as a TRU transmuter, both in thermal and fast spectra [14–17]. Using small molten salt fast reactor (SMSFR), a total of 5620 kg of MA with a thermal power of 500 MWt. Using identical core but FLiNaK salt, MA incineration can be increased to 8190 kg. Molten salt actinide recycler and transmuter (MOSART) concept with 2400 MWt power was shown to be able to transmute between 51-91% of TRU in a single fluid converter mode with Flibe carrier salt. On the other hand, thermal-spectrum MSR such as single fluid double zone thorium molten salt reactor (SD-TMSR) with 2250 MWt power can incinerate 40.96% of MA during 40 years of operation. By using reactor grade plutonium (RGPu) as the fissile driver, TRU incineration in passive compact molten salt reactor (PCMSR) with 570 MWt of power can reach 24.39% after only two years of operation. If only MA is considered, around 16.31% can be transmuted [18].

The above studies assumed that the TRU elements were dissolved within the fuel salt, i.e., homogeneous configuration. In a fast breeder MSR, this is not necessarily an issue since fast neutron offers excellent neutron economy in a TRU-based fuel. However, in a thermal breeder MSR using thorium fuel cycle, mixing TRU within the fuel salt can severely impact the already tight neutron economy. Consequently, breeding ratio (BR) can be reduced. On the other hand, dissolving too little TRU will diminish its transmutation rate. Thus, TRU transmutation in thermal breeder MSR by means of mixing the TRU elements within the fuel salt does not seem to be an attractive option.

This paper offers a novel innovation for TRU transmutation in a thermal breeder MSR is by separating the TRU from the fuel salt, in form of inserting solid TRU layer into MSR fuel channel. This concept is inspired by breed-and-burn molten salt reactor (BBMSR) [19] which uses depleted uranium blanket as a tubular layer within fuel salt tube to increase fuel cycle length. This concept is adopted here by replacing depleted uranium with TRU and inserted into thermal MSR fuel channel in place of fast MSR fuel tube.

The feasibility of this tubular TRU layer transmutation was analysed in the neutronic aspect. Initial assessment was previously done [20], and showed that such configuration has a potential to be used. However, it was not optimised, in term of TRU placement within the fuel salt and its volume fraction in which it replaces the fuel salt. This study expands from that previous assessment, in order to obtain the optimum TRU layer configuration. To maintain consistency with previous research, this analysis was performed in fuel channel level, using SD-TMSR fuel channels as the base model.

2. MSR Channel Model

SD-TMSR is an MSR concept developed in Shanghai Institute of Applied Physics (SINAP) [21]. It is a thermal breeder MSR working in thorium fuel cycle. Graphite is used as the neutron moderator.

IOP Conf. Series: Materials Science and Engineering **1330** (2025) 012015 doi:10.1088/1757-899X/1330/1/012015

Fuel channels are arranged as a hexagonal assembly, filled with molten salt fuel and surrounded by graphite moderator. The fuel enters the channels from the bottom and exits at the top. SD-TMSR uses single fluid stream, where fissile and fertile fuel are mixed in the same fuel salt. To improve thorium breeding, the fuel channels are divided into two zones; core zone with narrower fuel channel to optimise fission reaction and blanket zone with wider fuel channel to optimise radiative neutron capture by thorium.

This research used a single hexagonal fuel channel block model for neutronic calculations. Both core and blanket zones were modelled for TRU insertion. The radial cross-section of these fuel channels can be seen in Fig. 1. For full parameters, refer to Table 1.

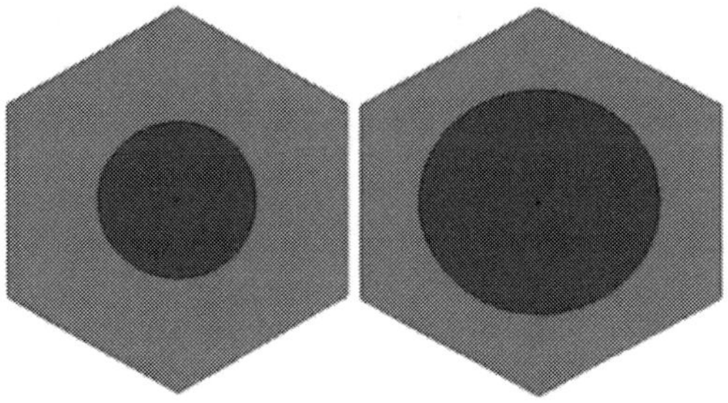

Figure 1. Radial cross section of SD-TMSR fuel channel model (a) core zone, (b) blanket zone

Table 1. SD-TMSR Core Parameters

Parameter	Value
Thermal power (MWt)	2250
Operational temperature (K)	900
Fuel density (g/cm^3)	3.3
Graphite density (g/cm^3)	2.3
Fuel channel height (cm)	460
Side length of hexagon (cm)	7.5
Core zone radius (cm)	3.5
Blanket zone radius (cm)	5
Fuel salt composition (%mol)	70LiF-17.5BeF$_2$-12.5(Th+^{233}U)F$_4$
^{7}Li enrichment (%mol)	99.995

In our previous study, the TRU was inserted in the fuel channel by replacing the fuel salt volume instead of moderator. This is to ensure that moderating condition remains identical with standard fuel channel without TRU. The same treatment is applied in this study. This time, the TRU layer placement within the fuel channel is simulated in three different positions; central, middle, and outer. The illustration of the placement is shown at Fig. 2. For the "middle" configuration, fuel volumes on the inside and outside of the TRU layer are set in a 1:1 ratio.

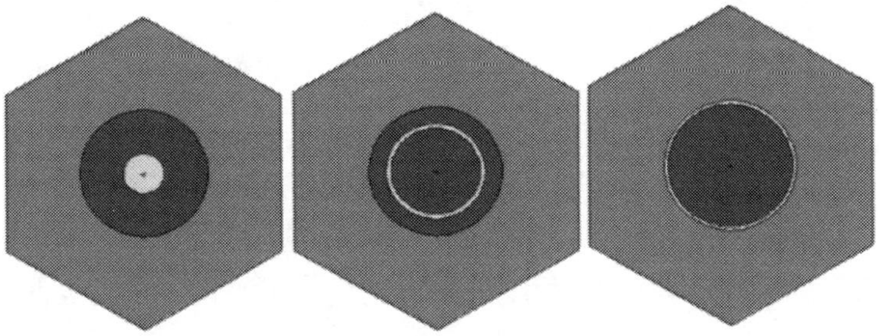

Figure 2. Configurations of TRU insertion within SD-TMSR fuel channel (a) central, (b) middle, (c) outer.

The TRU is an oxide compound ($[TRU]O_2$) with a density of 11.46 g/cm^3, consisting of RGPu and MA. The ratio of RGPu and MA was considered at 9:1. Typically, MA isotopes are considered as the primary burden among the TRU constituents. Nonetheless, RGPu was sometimes considered a burden as well, despite its potential use in the Generation IV reactors. For this study, both RGPu and MA were incinerated in the MSR. Study on transmuting only MA is reserved for future works.

The isotopic vector of the TRU elements is provided in Table 2 [22].

Table 2. TRU Vector Composition

Isotope	Mass Fraction	Isotope	Mass Fraction
Pu-238	1.58%	Np-237	42.25%
Pu-239	57.76%	Am-241	47.57%
Pu-240	26.57%	Am-243	8.50%
Pu-241	8.76%	Cm-242	0.32%
Pu-242	5.33%	Cm-243	0.01%
		Cm-244	1.26%
		Cm-245	0.07%
		Cm-246	0.01%

3. Methodology

As previously mentioned, there are three TRU layer placements simulated in this study; central, middle, and outer. Previous study modelled the TRU volume to replace 10% of the fuel salt volume. Here, TRU volume was varied from 5% to 20%, with a 2.5% increment. Thus, there are 48 possibilities on TRU incineration using the proposed heterogeneous configuration.

The analysed parameters were temperature coefficient of reactivity (TCR) and transmutation efficiency (TE). TCR depends on moderator and fuel, whilst the fuel component depends on the functions of salt density and Doppler effect. From there, TCR in MSR can be described as follows [23].

$$\frac{dk}{dT_{total}} = \frac{dk}{dT_{fuel\ density}} + \frac{dk}{dT_{Doppler}} + \frac{dk}{dT_{moderator}} \tag{1}$$

For TRU incineration as a tubular layer inserted into fuel channel, the temperature feedback from this component must be included in the equation, more specifically its Doppler effect. Therefore, Eq. 1 must be modified into Eq. 2 as follows.

$$\frac{dk}{dT_{total}} = \frac{dk}{dT_{fuel\ density}} + \frac{dk}{dT_{Salt\ Doppler}} + \frac{dk}{dT_{Tube\ Doppler}} + \frac{dk}{dT_{moderator}} \tag{2}$$

Calculating transmutation efficiency (TE) can be done by considering several factors. The simplest one is by only considering the TRU load prior to and after incineration, to evaluate the amount of transmuted TRU, as shown in Eq. 3 [1].

$$TE\ (\%) = 100\% - \frac{TRU_{end}}{TRU_{init}} \tag{3}$$

where TRU_{end} is the mass of TRU after incineration and TRU_{init} is the mass of TRU prior to incineration. This equation ignores the formation of new TR elements as a result of irradiation, only considering the increase and decrease in TRU mass. This approach may not be fully representative of the total change of mass, but used anyway for simplicity.

Neutronic analyses were performed using MCNP6.2 radiation transport code with ENDF/B-VII.0 neutron cross section library. MCNP6.2 is developed by Los Alamos National Laboratory (LANL) based on Monte Carlo method, allowing it to model a complex and repeated geometry without homogenisation. MCNP6.2 can calculate the reactor kinetic parameters simultaneously instead of two-step calculations [24], also supports burnup feature using CINDER90 module [25]. MCNP has been used to model several MSR designs such as MSBR [26], MSR-FUJI [27], and PCMSR [28], with sufficient accuracy.

Neutronic simulation for fuel channel level was done with 10.000 neutron history per iteration, 250 total iteration, and first 50 iterations discarded. This results in a statistical uncertainty of ± 42-44 pcm. For burnup calculation, the power for each channel is assumed at 2.23 MWt and the length of burnup at 200 days. MATMOD card was used to continuously remove gaseous fission products (H, He, Ar, Xe, Kr) and non-dissolved metals (Nb, Mo, Tc, Ru, Rh, Pd, Ag) from the fuel salt. All surface boundaries were treated as reflective.

4. Results and Discussion

The first evaluation is on the reactivity swing. The evolution of infinite multiplication factor (k_{inf}) is shown in Figs. 3 and 4.

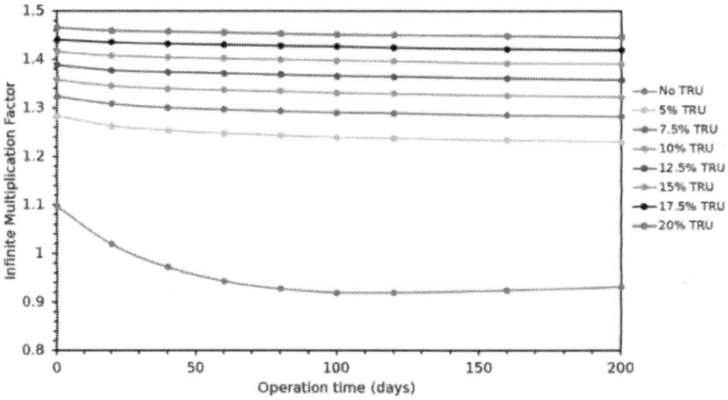

(a)

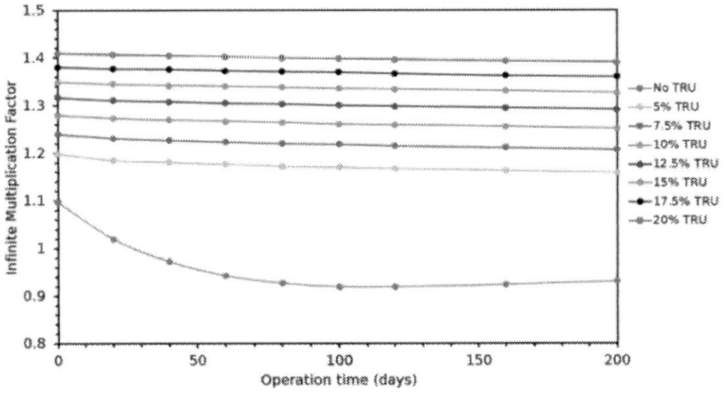

(b)

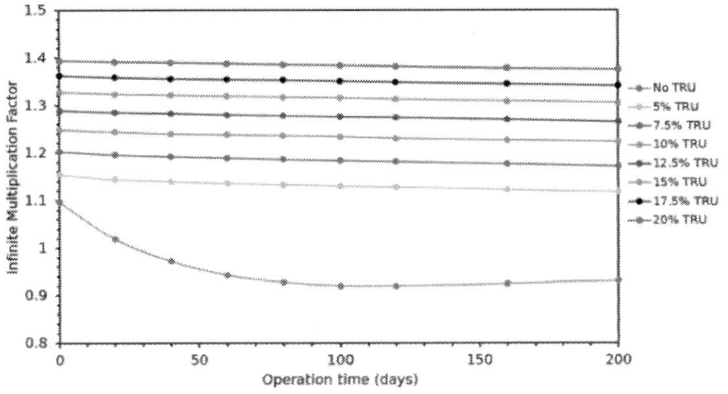

(c)

Figure 3. k_{inf} evolution in core zone (a) central, (b) middle, (c) outer

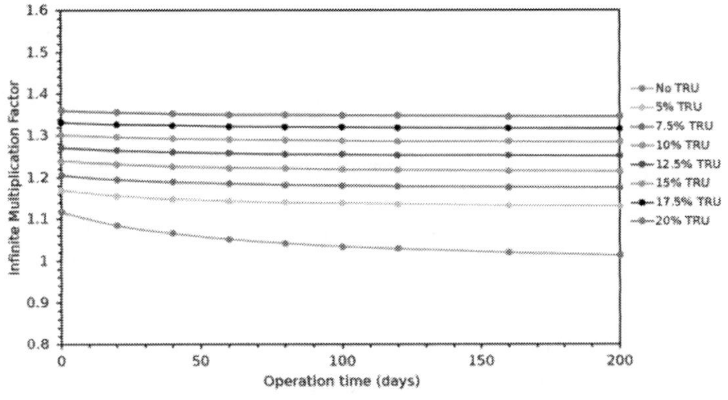

(a)

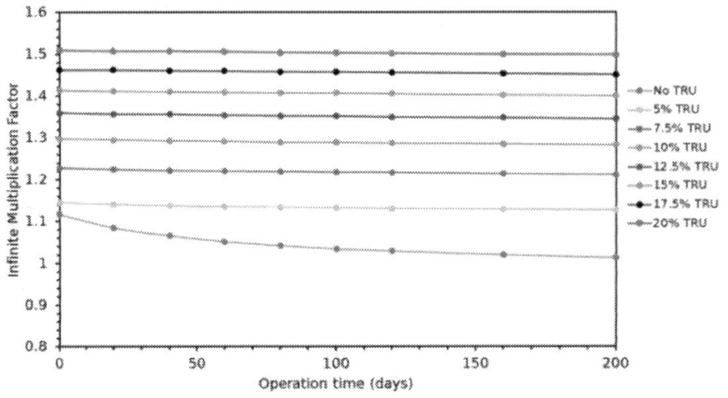

(b)

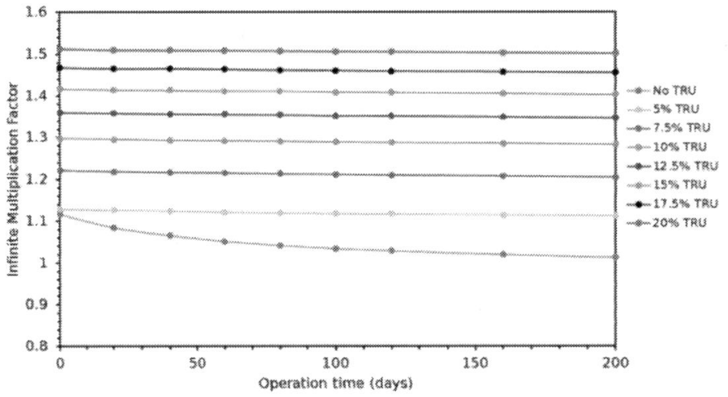

(c)

Figure 4. k_{inf} evolution in blanket zone (a) central, (b) middle, (c) outer

As observed from Fig. 3, central TRU position in core channel gives higher k_{inf} at the beginning of cycle (BOC) than middle and outer positions, which is understandable since the fission reaction is the highest at centre of the fuel channel, with the fission neutrons directly interacting with TRU cylinder which density is around three times larger than fuel salt, packing a larger fissile atom per volume unit. However, its reactivity swing is higher since U-233 formation from Th-232 is also lower, as a result of large self-consumption.

Since neutron flux is a function of area, and thermal power is a function of neutron flux, small surface area in central TRU position decreases the exposure of neutron flux in the TRU cylinder, so that its power fraction also decreases. Consequently, power fraction in the adjacent fuel salt increases. This slowed the U-233 breeding rate from Th-232, as more U-233 must be consumed to maintain criticality. Since fuel salt has lower density than TRU, fissile U-233 is depleted more rapidly, causing higher reactivity swing. On the contrary, despite lower k_{inf} at the BOC, outer TRU position showed a lower reactivity swing. Larger surface area in outer TRU position results in higher neutron flux and subsequently larger power fraction. Since fission dominantly occur in higher-density TRU layer, U-233 breeding in fuel salt can occur more efficiently. Thus, fissile mass change is lower, results in a smaller reactivity swing.

A different pattern was observed in blanket zone, where the highest k_{inf} is shown in outer position instead of the central. This is resulted from the smaller moderator volume, where in the central TRU position, the TRU receives smaller neutron flux as higher neutron populations were already absorbed in molten salt fuel. Consequently, the k_{eff} was suppressed. Meanwhile, the outer placement has its TRU layer directly in contact with the graphite moderator, which allows the thermalised neutron to be absorbed by the TRU. This results in higher k_{inf} at the BOC.

Next evaluated parameter is TCR. Table 3 summarised TCR values with various TRU layer, calculated at BOC.

Table 3. TCR values for SD-TMSR fuel channel with various TRU insertion configurations

TRU fraction	Central		Middle		Outer	
	Core	Blanket	Core	Blanket	Core	Blanket
0%	0.73	-2.93	0.73	-2.93	0.73	-2.93
5%	-2.77	-3.40	-5.89	-3.88	-5.22	-2.72
7.5%	-2.15	-2.45	-6.32	-2.45	-4.55	-1.13
10%	-3.07	-2.50	-3.88	-2.01	-3.94	-0.39
12.5%	-2.10	-3.05	-4.43	-1.31	-2.87	-1.04
15%	-2.52	-1.95	-2.92	-0.57	-2.30	-0.52
17.5%	-2.54	-1.67	-2.57	-0.29	-1.65	-0.54
20%	-1.80	-2.08	-2.47	-0.50	-2.39	0.22

From Table 3, generally, negative TCR is strengthened in the core zone and weakens in the blanket zone. Exceptions are seen in TRU fraction of 5% in blanket zone in Central and Middle placement. In Middle placement of blanket zone, the insertion of TRU improved the TCR at low TRU volume but worsens as the TRU volume increases. A similar pattern is shown in Central placement, but the decreasing trend is smaller. It implies that, in Central and Middle placement, small TRU volume (≤ 5%) strengthens the negative TCR.

The change of TCR values show no apparent pattern, likely due to high statistical uncertainty. However, in general, small TRU volume fractions (< 12.5%) in core channel are showing a tendency as the most positively impacted by TCR insertion. This is due to the strengthening of graphite moderator feedback which became negative, from originally has a strong positive value. Since the majority of fission reaction occurs in TRU layer, the spectral coincidence of graphite moderator with U-233 due to moderator heating, which increases fission reaction, does not occur, and moderator temperature feedback remains negative.

Despite the weakening in Doppler effect in fuel salt, whilst negative Doppler feedback from TRU is insufficient to compensate it, the total TCR value is significantly strengthening nonetheless, implying a positive benefit of TRU insertion. This is especially important as thermal breeder MSR often possesses weakly negative TCR, due to positive moderator feedback previously mentioned.

However, blanket channel exhibited a completely different result, showing a weakened of temperature feedback in all variations. This is due to the severe weakening of Doppler feedback from fuel salt as fission reaction occurs primarily in TRU layer instead of fuel salt. Meanwhile, improvement in moderator feedback is insufficient to compensate it due to smaller graphite volume. From these findings, the implication is that TRU insertion should be applied in core zone.

The last evaluated parameter is TE. Table 4 summarised the TE values for 200 days of burnup.

Table 4. Transmutation efficiency in various TRU layer configurations

TRU fraction	Central		Middle		Outer	
	Core	Blanket	Core	Blanket	Core	Blanket
5%	3.38%	1.12%	4.45%	2.04%	4.61%	2.12%
7.5%	2.52%	0.91%	3.07%	1.47%	3.18%	1.50%
10%	2.01%	0.75%	2.36%	1.13%	2.43%	1.15%
12.5%	1.69%	0.68%	1.91%	0.95%	1.96%	0.97%
15%	1.46%	0.60%	1.62%	0.78%	1.67%	0.80%
17.5%	1.27%	0.53%	1.41%	0.68%	1.42%	0.69%
20%	1.15%	0.49%	1.22%	0.62%	1.26%	0.63%

Outer TRU position is shown to have highest TE vis a vis against other positions. Smaller volume fraction has the highest TE, since small TRU content in small TRU fraction will be the fastest to be transmuted. Given identical thermal power, lower power fraction in central TRU position results in slower TRU transmutation, either through fission (fissile plutonium) or neutron capture (fertile plutonium and MA). Larger TRU loading in higher TRU volume fraction also slowed down transmutation rate since the thermal power remains unchanged. Therefore, the higher TRU volume fraction, the lower the TE value.

MSR fuel salt can be left indefinitely inside the reactor core owing to its highly radiation-resistant nature. However, the graphite moderator is subject to high exposure of neutron and gamma radiations, limiting its lifespan. Therefore, graphite moderator is periodically replaced. This gives little incentive to let the TRU layer stays inside SD-TMSR core for too long. The most efficient transmutation in the shortest time is expected. With this reasoning, the optimal configuration is chosen at outer TRU position in core zone, with 5% TRU volume fraction.

To observe a better understanding of the TRU transmutation performance on the optimal TRU position, a longer burnup time of 1460 days (4 years) was calculated. Graphite moderator in MSR is periodically replaced, with the replacement period depends on the power density, flux flattening, and volume ratio [23]. Considering the design parameters of SD-TMSR, to be conservative, 4 years was taken as the burnup time. The TE result is summarised in Table 5. Negative value means that the mass is increasing instead of decreasing.

IOP Conf. Series: Materials Science and Engineering **1330** (2025) 012015 doi:10.1088/1757-899X/1330/1/012015

Table 5. Transmutation result of optimum configuration

Isotope	TRU mass (g)		TE
	BOC	EOC	
Np-237	373.9	167.6	55.18%
Pu-238	126.4	439	-247.31%
Pu-239	4641	1763	62.01%
Pu-240	2144	1939	9.56%
Pu-241	709.7	822.4	-15.88%
Pu-242	433.6	489.4	-12.87%
Am-241	428.3	219.1	48.84%
Am-243	77.14	157.7	-104.43%
Cm-242	2.89	37.86	-1208.68%
Cm-243	0.091	3.33	-3567.92%
Cm-244	11.48	114.4	-896.52%
Cm-245	0.64	24.94	-3793.83%
Cm-246	0.092	1.79	-1843.84%
Total	8949.24	6179.52	30.95%

After 4 years-worth of incineration, total TRU mass decreases for 30.95%. This value is not directly comparable to several other studies such as [4,6,10], since most, if not all, of them discussed MA incineration instead of TRU as a whole. Also, this value can be an underestimation since it ignores the different power level in different places within MSR core and influx of neutron from adjacent fuel channel blocks. A whole-core study is required in the future studies to provide a better assessment of the TE values.

The mass evolution of plutonium and MA is provided in Figs. 5 and 6.

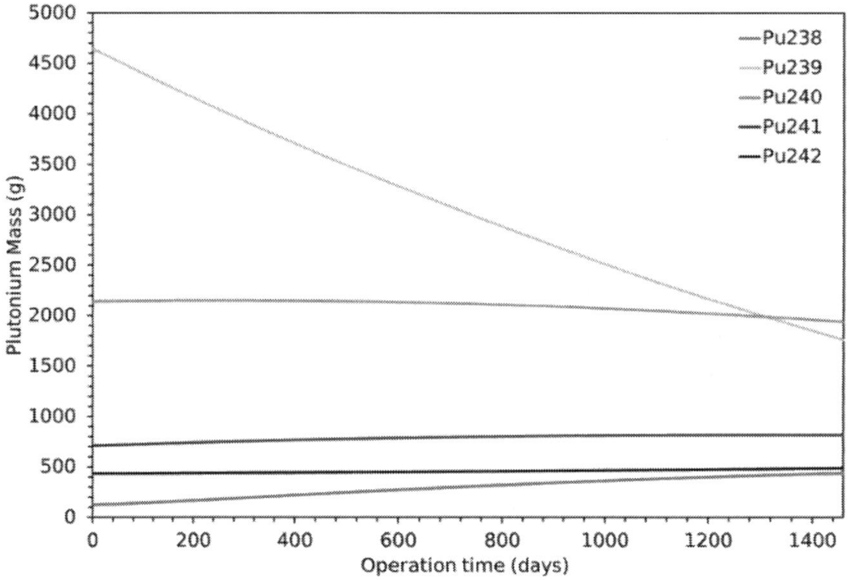

Figure 5. Mass evolution of plutonium isotopes in SD-TMSR fuel channel

For RGPu isotopes, mass change is predominantly occurred in Pu-239; a fissile isotope with the highest abundance. Since neutron is mostly absorbed by Pu-239, its mass diminished the fastest, whilst its replenishment rate is unable to compensate the loss due to small amount of Pu-238 and the absence of Np-239 precursor, which is a capture product of fertile U-238.

Another significant mass change occurs in Pu-238, which mass is increasing instead of decreasing. This is a result of Np-237 capture, a precursor to Pu-238. For other isotopes, their mass changes are comparably insignificant.

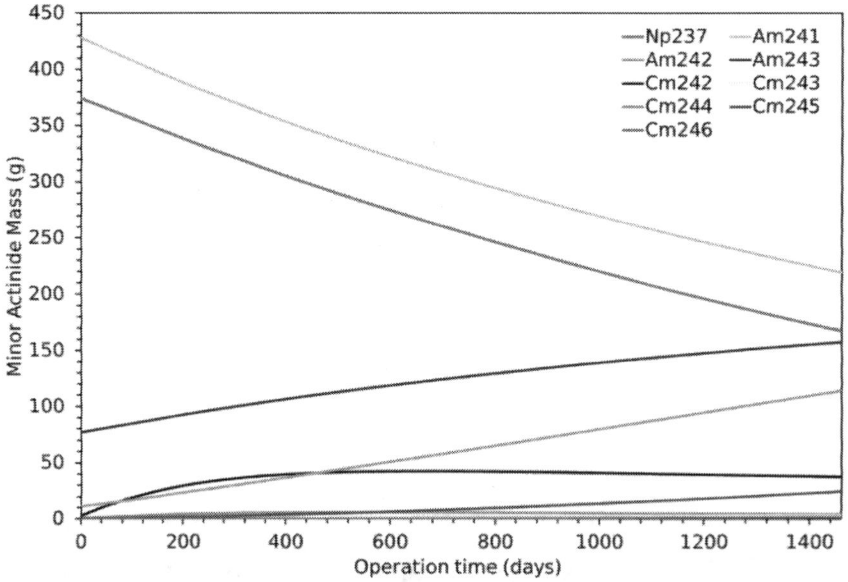

Figure 6. Mass evolution of MA isotopes in SD-TMSR fuel channel

For MA isotopes, the masses of Np-237 and Am-241 was transmuted to half their original values, with comparable transmutation rate. Meanwhile, Am-243 and all Cm isotopes had their mass increased significantly compared to the beginning, the most significant being Am-243 and Cm-244. Both have their replenishment rate faster than their destruction rate, owing to their smaller capture cross section than their precursors. For Cm isotopes, since basically they started with small mass to begin with, the mass buildup has not reached the equilibrium point, therefore continue to increase.

To observe the effect of TRU transmutation to the post-discharge radioactivity, decay calculation was performed until 1000 years after discharge. The time increment was set at 10 years for the first 100 years and 100 years for year 100 until year 1000 after discharge. Post-transmutation TRU radioactivity is shown in Fig. 7, compared to non-transmuted TRU as a reference value.

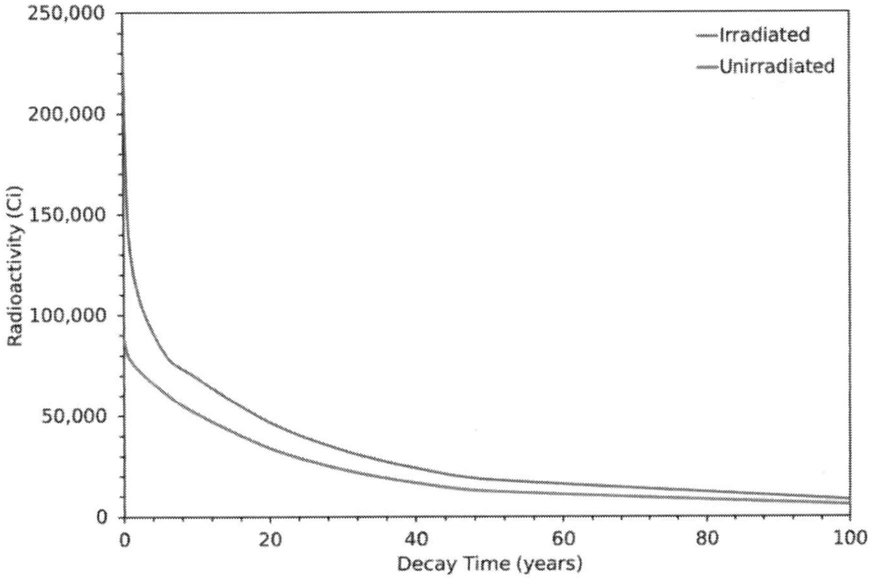

(a)

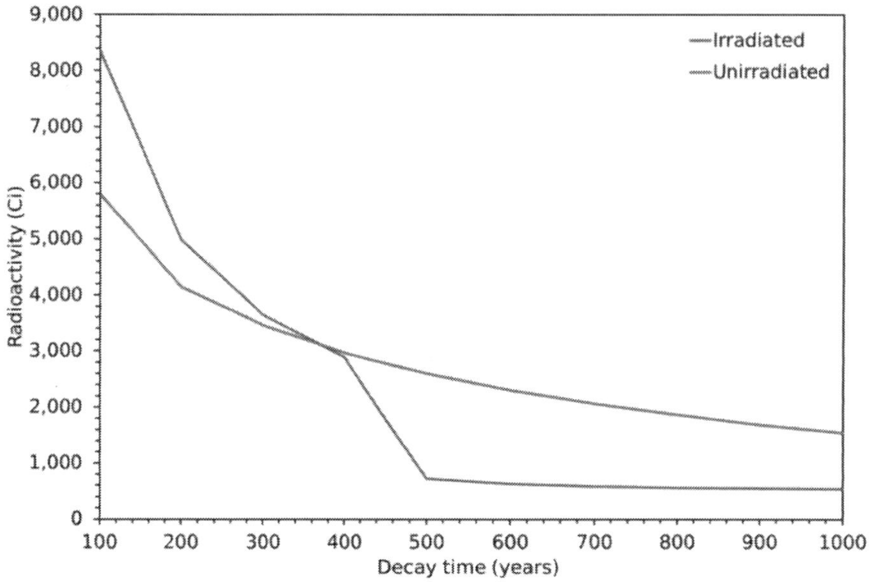

(b)

Figure 7. TRU radioactivity post-transmutation compared to non-transmuted TRU (a) first 100 years, (b) year 100 until year 1000

Moment after discharge, transmuted TRU has significantly higher radioactivity, almost three times as high as non-transmuted TRU. The finding is consistent with previous study [1]. The issue lies in short-lived Cm-242, which has high specific radioactivity combined with increased mass from the BOC. Both Pu-238 and Pu-241 also have their radioactivity increased. Since Cm-242 has short half-life, its radioactivity diminished quickly and became lower than Pu-241 radioactivity after a year. After 100 years, the radioactivity of transmuted TRU has not decreased below non-transmuted TRU, as the graph shows asymptotic pattern. Only after year 400 (Fig. 7b), transmuted TRU has its radioactivity lower than non-transmuted TRU.

Therefore, in long-term, TRU incineration can result in lower total radioactivity compared to unirradiated TRU, which benefits in lowering the burden of radioactivity. However, in short-term, the radioactivity will be higher. Therefore, short term TRU storage must be evaluated in accordance to the different radioactivity level.

5. Conclusions

From the findings of this study, the optimum configuration for TRU transmutation in SD-TMSR fuel channel was found to be the following: outer TRU position, core zone, 5% TRU volume fraction. At this configuration, the TCR was significantly improved and the TE is the highest (4.61% after 200 days of burnup and 30.95% after 1460 days of burnup). High power fraction of TRU layer compared to fuel salt contributes to the improvement of TCR and high TE. However, after transmutation, the TRU radioactivity remains higher than non-transmuted TRU for the first 400 years. This shows that heterogeneous TRU transmutation in MSR fuel channel is promising. Future research must be performed in whole MSR core, to assess the positioning of TRU-inserted channels within the core, the TRU load, and the impact of different transmutation rate in term of post-discharge radioactivity, as a justification of the need of TRU transmutation. MA-only transmutation must also be analysed.

IOP Conf. Series: Materials Science and Engineering **1330** (2025) 012015 doi:10.1088/1757-899X/1330/1/012015

References

[1] Dwijayanto R A P and Alfarisie M 2021 *GANENDRA Maj. IPTEK Nukl.* **24** pp 85-92
[2] Setiawan M B, Kuntjoro S, Husnayani I, Udiyani P M and Surbakti T 2019 *Malaysian J. Fundam. Appl. Sci.* **15** pp 577–9
[3] Cohen B L 1990 *The Nuclear Energy Option: An Alternative for the 90s* (New York: Springer New York)
[4] Liu B, Jia R, Han R, Lyu X, Han J and Li W 2018 *Ann. Nucl. Energy* **115** pp 116–25
[5] Washington J and King J 2017 *Nucl. Eng. Des.* **311** pp 199–212
[6] Tran V T, Tran H N, Nguyen H T, Hoang V K and Ha P N V 2019 *Sci. Technol. Nucl. Install.* **2019**
[7] Setiawan M B, Udiyani P M, Kuntjoro S, Husnayani I and Surbakti T 2020 *Nucl. Technol.* **206** pp 1945–50
[8] Hu W, Liu B, Ouyang X, Tu J, Liu F, Huang L, Fu J and Meng H 2015 *Ann. Nucl. Energy* **77** pp 74–82
[9] Hussain M and Sohail M 2017 *ICET 2016 - 2016 International Conference on Emerging Technologies* pp 16–9
[10] Liu B, Han J, Liu F, Sheng J and Li Z 2020 *Prog. Nucl. Energy* **119** 103148
[11] Tuček K, Carlsson J, Vidović D and Wider H 2008 *Prog. Nucl. Energy* **50** 382–8
[12] Kooyman T, Buiron L and Rimpault G 2017 *Nucl. Sci. Eng.* **185** 335–50
[13] International Atomic Energy Agency 2023 *Technical Report Series No. 489: Status of Molten Salt Reactor Technology* (Vienna)
[14] Ashraf O and Tikhomirov G V. 2020 *Ann. Nucl. Energy* **148** 107751
[15] Cui D Y, Zhang A, Li X X, Cai X Z, Chen J G and Zou Y 2023 *Ann. Nucl. Energy* **185** 109719
[16] Yu C, Li X, Cai X, Zou C, Ma Y, Wu J, Han J and Chen J 2017 *Ann. Nucl. Energy* **99** pp 335–44
[17] Ignatiev V, Feynberg O, Gnidoi I, Merzlyakov A, Surenkov A, Uglov V, Zagnitko A, Subbotin V, Sannikov I, Toropov A, Afonichkin V, Bovet A, Khokhlov V, Shishkin V, Kormilitsyn M, Lizin A and Osipenko A 2014 *Ann. Nucl. Energy* **64** pp 408–20
[18] Dwijayanto R A P, Sulistyo F Y, Zuhair Z and Harto A W 2024 *AIP Conference Proceedings* vol 2967 130002
[19] Cuoc E, Shwageraus E, Kasam A and Scott I 2021 *EPJ Web of Conferences* vol 247 p 01004
[20] Dwijayanto R A P, Harto A W and Khakim A 2023 *AIP Conference Proceedings* vol 2893 p 040006
[21] Li G C, Cong P, Yu C G, Zou Y, Sun J Y, Chen J G and Xu H J 2018 *Prog. Nucl. Energy* **108** pp 144–51
[22] Waris A, Aji I K, Pramuditya S, Novitrian, Permana S and Su'Ud Z 2015 *Energy Procedia* **71** pp 62–8
[23] Zou C Y, Cai X Z, Jiang D Z, Yu C G, Li X X, Ma Y W, Han J L and Chen J G 2015 *Nucl. Eng. Des.* **281** pp 114–20
[24] Zuhair, Suwoto, Setiadipura T and Su'ud Z 2020 *Acta Polytech.* **60** pp 175–84
[25] Pelowitz D B, Goorley J T, James M R, Booth T E, Brown F B, Bull J S, Cox L J, Durkee J W, Elson J S, Fensin M L, Forster R A, Hendricks J S, Hughes H G, Johns R C, Kiedrowski B C, Mashnik S G, McKinney G W, Pelowitz D B, Prael R E, Sweezy J E, Waters L S, Wilcox T A and Zukaitis A 2013 *MCNP6 User's Manual*
[26] Park J, Jeong Y, Lee H C and Lee D 2015 *Int. J. Energy Researc* **39** pp 1673–80
[27] Jaradat S Q and Alajo A B 2017 *Nucl. Eng. Des.* **314** pp 251–5
[28] Dwijayanto R A P, Oktavian M R, Putra M Y A and Harto A W 2021 *Atom Indones.* **47** pp 191–7

Computed radiography method for non-destructive testing to characterize cooling piping in research reactor G.A Siwabessy

Jepri Sutanto[1,2*], Baskan Hanurajie[2], Iwan Sumirat[2], and Asril P.A. Mustari[1]

[1] Nuclear Engineering, Bandung Institute of Technology (ITB), Bandung, 40132,Indonesia
[2] Research center for Safey, Metrology, and Nuclear Quality Technology, Reseach Organization for Nuclear Energy, National Research and Innovation Agency (BRIN), Tangerang Selatan, 15314, Indonesia

*E-mail: ananda052010@gmail.com

Abstract. A radiographic test is one of the NDT methods used to verify the material (non-destructive testing). It utilizes ionizing radiation, such as X-rays or gamma rays, to create detailed images that reveal hidden flaws, defects, discontinuities, or anomalies. A radiographic test method commonly checks welding connections, fabrication, forging, and casting. Radiography NDT technique is used across diverse sectors, including manufacturing, aerospace, infrastructure, energy, and even research reactors or nuclear power plants (NPP). This study will discuss the piping in the research reactor in South Tangerang, i.e., G.A. Siwabessy (RSG-GAS), in the context of maintenance and inspection. Industrial digital radiography substitutes conventional film radiography and is replaced by imaging plate (IP). Parameters such as SNR and CNR and basic spatial resolution (SRb) must be considered and understood. Digital radiography is currently more widely used in addition to the short process and interpretation of results with up-to-date software developments. The linear indication is attached to the result of the interpretation. The linear indication's orientation, size, and shape recommendations can provide valuable information to the nuclear energy research organization while replacing or substituting the piping. Image interpretation results were obtained to provide recommendations for improvements by standards and regulations. Other NDT methods are needed to compare the results.

Keywords: Non-destructive testing, industrial radiography, x-ray, material, computed radiography

1. Introduction

Nuclear technology has become an indispensable part of modern human life. Nuclear technology is widely used in all fields/sectors, such as agriculture, industry, medicine, energy, and information [1]. In the industrial sector, nuclear technology is pivotal in non-destructive testing techniques, petroleum drilling, quality control, quality assurance in paper mills, tracers, and material element testing. In the energy sector, there are currently 416 Nuclear Power Reactors in operation and 59 Nuclear Power Reactors under construction [2], [3]. Additionally, Indonesia has developed detailed designs of experimental power plants (RDE). The experimental power reactor program was created to introduce Indonesia to nuclear power plants (NPP) [4].

The G.A. Siwabessy Multi-Purpose Reactor, commonly called RSG-GAS, is the largest research reactor in Southeast Asia with a thermal power of 30 MW in South Tangerang - Banten. RSG-GAS is a pool reactor (pool type) with light water cooling and has been operating safely for over 30 years [5]. In addition to supporting aspects of operational safety and materials management burning in the core, non-destructive testing is needed for inspection and maintenance [6]. Maintenance is an organized preventive or repair activity, both administrative and technical, to continuously maintain structures, systems, and components to operate properly. Maintenance on non-power reactor (RSG-GAS) is divided into ruin maintenance (preventive maintenance and surveillance) and non-routine (repair maintenance and in-service inspection).

In the maintenance of RSG-GAS, several NDT techniques can be used, both surface, sub-surface, and internal. Radiography techniques are used to find hidden defects and geometry of the material. Ultrasonic testing is also used for volumetric materials other than cooling pipes on RSG-GAS. To check the surface of a component or material in the form of a pipe in RSG-GAS, qualified personnel can use the penetrant testing method, eddy current testing, leak testing, and magnetic particle testing. This study uses computed radiography (CR) with an imaging plate to replace conventional radiography. This CR test minimizes work time and facilitates image interpretation. This CR test is temporary; qualified personnel may further examine the component with conventional radiographic testing based on management policy or equipment availability.

Various methods for NDT have been developed and described in the technical literature. Based on the ISO 9712 standard, NDT methods or NDT techniques are acoustic emission testing, eddy current testing, leak testing, magnetic testing, penetrant testing, radiographic testing, strain gauge testing, thermographic testing, ultrasonic testing, and visual testing [7]. Radiography is one method that uses radiation, while other methods are not. Radiography is divided into two parts: with film (conventional) [8], [9], [10] and without film (digital). The radiographic method of NDT is based on the detection and analysis of ionizing radiation after its penetration into the test object [11]. This is one of the oldest and rather universal methods of NDT. This method did and continues to play an essential role in determining materials and product quality. Like the success story of digital photography, a significant upheaval has been seen in digital industrial radiography [12]. In the research reactors, CR can be used to characterize cooling piping and detect potential defects or damage. However, there are gaps in the current application of CR for NDT in research reactors that need to be addressed. There may be a lack of expertise in the application of CR for NDT in research reactors, which can lead to incorrect interpretation of results and potential safety risks. CR data analysis can be time-consuming and may require specialized software, which can limit the efficiency and effectiveness of the testing process.

Computed radiography, or cassette-based DR, is a complete system comprising a storage phosphor imaging plate (IP) and a corresponding read-out unit (scanner or reader), which converts the information from the IP into a digital image [13]. The X-Ray backscatter technique was used for more than 40 years to permit single-sided radiographic testing. Especially after the successful application in the security field, backscatter imaging returns to the focus of the NDT application. Image quality indicators are used to assess the radiographic quality of radiographs [14]. They are used to obtain evidence that the conditions for exposure and processing have been adequately chosen and fulfilled. They are not intended for use in judging the size of discontinuities nor for establishing acceptance limits for welding being radiographed.

Table 1. Factors considered in Computed Radiography[15]

Factors considered	Computed radiography (CR)
Imaging room	Retrofit traditional X-ray room or install detectors in the new room
Ease of use for technologists	Use a cassette with a phosphor plate; process in CR reader
Latent image Formation	X-rays strike the phosphor plate. X-ray energy deposited in the phosphor: energy is released from the phosphor when stimulated by light in the reader
Processing	Image processed by light; image processing takes place in a quality control station based on preset image algorithms
Exposure response	Linear; wide exposure latitude
Image contrast	kVp and grey value & SNR
Density	Image processing
Scatter radiation	Important for personal dose reduction and image processing; the phosphor can be more sensitive to low energy photons
Noise	Seen with inadequate mAs

In CR, X-ray or gamma-ray projection images are captured digitally using an imaging plate (IP). The digital images can be manipulated and enhanced using specialized software, resulting in higher accuracy and faster processing times. There are no routine costs of film and processing solutions (developer, stop batch, fixer, washing agent), shorter exposure time, and require less working time, thus increasing productivity. CR testing is preferred because it can more effectively inspect and evaluate the surface, subsurface, and inside of the material than other methods, such as ultrasonic testing, thermography testing, magnetic testing, and penetrant testing [12].

The X-ray radiography laboratory of the National Research and Innovation Agency of Indonesia (BRIN) has three functions, i.e., research, testing, and personnel certification laboratories. X-ray radiography checked the aging management [16] in the nuclear research reactor G.A. Siwabessy. Besides detecting the defects/flaws/discontinuities at an early stage, CR can also be used to control the quality of the material from time to time, which is why radiography tests become one of the methods in aging management in some materials.

2. Technique and measurement

2.1. Double wall double image (DWDI) or double wall double viewing (DWDV)

The DWDI or DWDV is a radiation technique with the source placed so that the radiation passes through two weld walls, whereas on the imaging plate (IP), only one weld wall is depicted, which is close to the film for interpretation. Figure 1 presents the direction and position of the object and the source. An elliptical DWDI technique may be used for materials and welds in components 3.5 in (89 mm) or less in nominal outside diameter [17]. Volumetric inspection [18], such as pipes, is adjusted to the diameter and thickness of the material; techniques other than ellipse can also be carried out, for example, with contact and panoramic techniques.

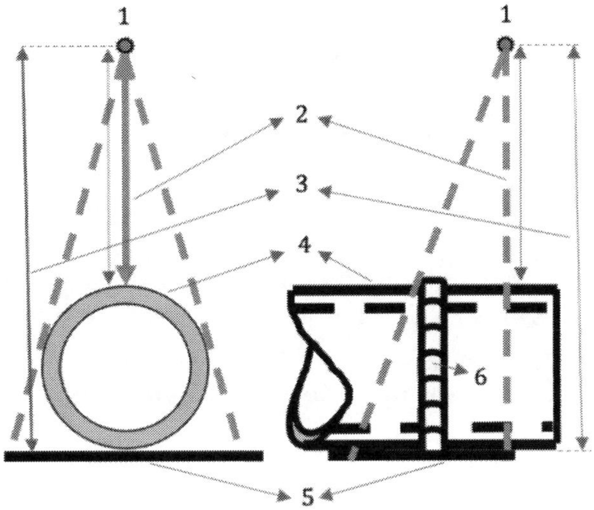

Keys

1	source (x-ray or gamma-ray)
2	source-to-object distance (SOD)
3	source-to-film distance (SFD) of source-to-detector distance (SDD)
4	material (in this study is pipe shape)
5	imaging plate (IP) and/or film
6	welding area

Figure 1. DWDI or DWDV technique

There are formulas to determine time and friction.

$$t = \left(\frac{SFD_{actual}}{SFD_{curve}}\right)^2 \frac{E}{i} \tag{1}$$

$$SFD_{ellips} = \sqrt{(SFD_{\perp} + P)^2} \tag{2}$$

$$P = \frac{1}{5}\, SFD + 2l \tag{3}$$

Keys

t	*exposure time (s)*
SFD	*Source to film distance (conventional) or source to detector distance (digital) (mm)*
P	*shift of source to from normal position (mm)*
E	*exposure value obtained from the exposure curve (kV)*
i	*current (mA)*
l	*width (mm)*

2.2. Signal-to-noise ratio (SNR)

The ratio of the mean value of the linearized grey values to the standard deviation of the linearized grey values (noise) in each region of interest in a digital image. Signal-to-noise ratio (SNR_N) by the basic spatial resolution of a digital image, SR_b^{image}, and calculated from the measured SNR by:

$$SNR_N = SNR.\frac{88,6}{SR_b^{image}} \quad (4)$$

Keys

SNR	*signal-to-noise ratio*
SNR_N	*normalized signal-to-noise ratio*
SR_b^{image}	*basic spatial resolution of a digital image (mm)*

2.3. Contrast-to-noise ratio (CNR)

The ratio of the difference of the mean signal levels between two image areas to the average standard deviation of the signal levels. The contrast-to-noise ratio is normalized by the basic spatial resolution of a digital image, SR_b^{image} as measured directly in the digital image with the duplex wire IQI on the object side and calculated from the measured CNR, i.e.

$$CNR_N = CNR.\frac{88,6}{SR_b^{image}} \quad (5)$$

$$\frac{CNR_N}{\Delta w} = c.\frac{\mu_{eff}\,.\,SNR}{SR_b^{image}} \quad (6)$$

Keys

CNR	*contrast-to-noise ratio*
CNR_N	*normalized contrast-to-noise ratio*
SR_b^{image}	*basic spatial resolution of a digital image (mm)*
c	*a constant (0,0886 mm)*
μ_{eff}	*the effective attenuation coefficient (mm^{-1})*

3. General preparations and requirements

3.1. Protection against ionizing radiation

Exposure to any part of the human body to an X-ray can be highly detrimental to health. Wherever X-ray equipment is used, appropriate health and safety requirements shall be applied based on local, national, and international regulations [19]. The use of a TLD or pendose is mandatory for personnel carrying out radiographic tests.

3.2. Surface preparation and stage of manufacture

In general, surface preparation is not necessary, but where surface imperfections or coatings can make detecting defects difficult, the surface shall be ground smooth, or the coatings shall be removed.

3.3. Location of the weld in the radiograph

Where the digital radiograph does not show the weld, high-density markers shall be placed on both sides of the weld area to evaluate (WAE).

3.4. Identification of radiographs

Symbols shall be affixed to each section of the object being digitally radiographed. Where possible, the images of these symbols shall appear in the digital radiograph outside the WAE to ensure unambiguous identification of the section.

The digital images shall be evaluated on a monitor. The monitor and the viewing conditions shall fulfill the minimum requirements of minimal brightness of 250 cd/m^2, display of at least 256 shades of grey, minimum displayable light intensity ratio of 1:250, display of at least 1 million pixels of a pixel size <0.3 mm.

3.5. Marking

Permanent markings on the object to be tested shall be made in order to accurately locate the position of each digital radiograph, e.g., zero-point, direction, identification, and measure. Where the nature of the material and its service conditions do not permit permanent marking, the location may be recorded by means of accurate sketches of photographs or from automated positioning systems.

3.6. Personnel

Personnel performing non-destructive testing by this study shall be certified in radiographic testing following ISO 9712 or an equivalent internationally or nationally accepted certification scheme to an appropriate level in the relevant industrial sector [7], [20].

The replacement of film radiography by digital techniques in NDT requires careful selection of suitable digital detectors and measurement procedures [21]. Standard practices were developed, e.g., ISO 17636-2 [22] for weld inspection and diverse ASTM standards for system characterization. The sensitivity of digital radiography systems depends on the achieved contrast-to-noise ratio (CNR) and SRb^{image}. For the measurement, the duplex wire shall be positioned directly on the detector surface of the cassette surface. The source-to-detector distance shall be (100 ± 5) cm. The mean grey value in the digital image shall exceed 50 % of the maximum grey value. The SNR shall exceed 100 for standard systems with pixel size $\geq 80\ \mu m$ or 70 for high-resolution systems with pixel size $< 80\ \mu m$ in the reference radiograph. The basic spatial resolution shall be read from the reference radiograph.

Three compensation principles (CP I, CP II, CP III) are applied in this study for radiography with digital detectors to achieve sufficient contrast sensitivity. Applying these principles requires a minimum contrast-to-noise ratio, CNR_N, normalized to the detector's basic spatial resolution per detectable material thickness difference Δw. Suppose the required normalized contrast-to-noise ratio (CNR_N per Δw) cannot be achieved due to an insufficient value of one of the compensation principle parameters. In that case, an increase in the SNR can compensate for this. Compensation principle I (CP I) for reduced contrast (e.g., by increased tube voltage) by increased SNR (e.g., by increased tube current or exposure time). Compensation principle II (CP II) for insufficient detector sharpness (the value of SR_b is higher than specified) by increased SNR (increase in the single IQI wire value for each missing duplex wire pair value). SR_b is $SR_b^{detector}$ for detector selection (IQI on the detector without object) or SR_b^{image} for quality evaluation of a production radiograph with the IQI on the source side of the object. Compensation principle III (CP III) for increased local interpolation unsharpness by increased SNR.

The digital data of the radiographic detector shall be evaluated with linearized grey value representation which is directly proportional to the radiation dose for the determination of SNR, SR_b and SNR_N. For optimal image display, contrast and brightness should be interactively adjustable. Optional filter functions, profile plots, and an SNR, SNR_N tool should be integrated into the software for image display and evaluation. For critical image analysis, the operator shall interpret the image with a zoom factor between 1:1 (meaning 1 pixel of the digital radiograph is

IOP Conf. Series: Materials Science and Engineering **1330** (2025) 012016 doi:10.1088/1757-899X/1330/1/012016

presented by one monitor pixel) and 1:2 (meaning 1 pixel of the digital radiograph is presented by four monitor pixels). The measuring range of radiation monitoring equipment must be suitable for industrial radiography. The minimum dose rate measurement range is between 20 $\mu Sv/h$ - 10 mSv/h.

4. Results and discussions

The radiography method should be used in a research reactor to evaluate the existence of flaw/defect/discontinuity, size location, characteristics, or unintended flaws. Pipe inspection was conducted with CR inside and outside research reactor areas [23]. CR is used to ensure the integrity of critical components in cooling pipelines and structural elements. CR technology offers advantages such as better image quality than conventional film, faster results and better data management. Maintenance or inspection of RSG-GAS uses several non-destructive testing techniques according to BAPETEN regulations for both routine and non-routine testing, for conventional radiography tests and digital radiography to test volumetric components, namely pipes, in this study checking the cooling pipe. Cooling pipes can also be done by ultrasonic testing methods. To check the surface of the components, both plates and pipes, can use the penetrant testing method, eddy current testing, and leakage testing, and magnetic particles testing.

There are 39 points for shooting with X-ray equipment, but this study only shows several pipes that indicate the parameters according to acceptance criteria. Voltage and current settings are done by trial and error. The following requirements shall contain for the application such as material type and thickness range, isotope or maximum X-ray voltage used, detector type, manufacturer, and model, minimum source-to-object distance (SOD), distance from source side of the object to the detector, focal size, image display parameters, storage media, radiation filters/masking, detector, pixel intensity or gray range, frame averaging. Commonly, single-wall radiographic techniques, and double-wall radiographic techniques are used to examine the pipe or tube welds. The kVp and mAs settings on the control panel are carried out to obtain the grey value and SNR based on Table 1.

Table 2. Data of specimen specification

Specification of the specimen [24]:	Relevant testing data:
Test object: pipe	Radiation source: X-ray
IQI: duplex wire and ASTM set A	Focal spot: 3 mm
Material: SS/Fe	Pre-filter at source port: 1 mm Al and 3 Al (Cu)
	Tube voltage, current: 170-200 kV; 3 mA
	Exposure time: 2 minutes – 3,3 minutes
	CR-plate-type: IP Blue and IP White
	Source-to-detector distance (SDD): 70-100 cm

Table 2 gives specimen pipe specifications using duplex or ASTM wire with SS/Fe material, an X-ray source with a tube voltage varying between 170-200 kV, a current of 3 mA, and a source-to-detector distance (SDD) of 70-100 cm. Geometrical magnification: factor M approaches the value "1" (the object is positioned directly on the detector). SR_b can be measured in different kinds and manners. In the standard committees, the duplex wire method was recommended due to its simplicity (EN 462-5, ISO 192325, ASTM E 2002). The region of interest (mm) can be measured at the welding area [25], [26]. SNR_N: 70 required (minimum). If the SNR_N is figured in the HAZ/parent material, these values shall be multiplied by 1.4. X-ray potentials >150 kV to 250 kV, the mean grey value in the digital image shall exceed 50% of the maximum grey value. The thickness of the metal filter is 0,5 mm to 2 mm by the penetrated material thickness (e.g., lead, tin, copper, or steel).

This study used several software tools for image interpretation, namely ISee (Fig.2), D-Tect (Fig.3), and Dinamix (Fig.4). Each software shows good image results for evaluating pixels, resolution, contrast and determining defects. Fig.2 shows the sketch of digital radiographs such as profiler, region of interest (ROI), IQI identification, and defect or flaw position. Fig.3 shows the identification area that measure the welding area with the DWDI technique. From Fig. 2, Fig. 3, and Fig.4, there is a profile of the pixel by drawing a line on the duplex wire adjacent to the welding area for the determination of the image unsharpness value.

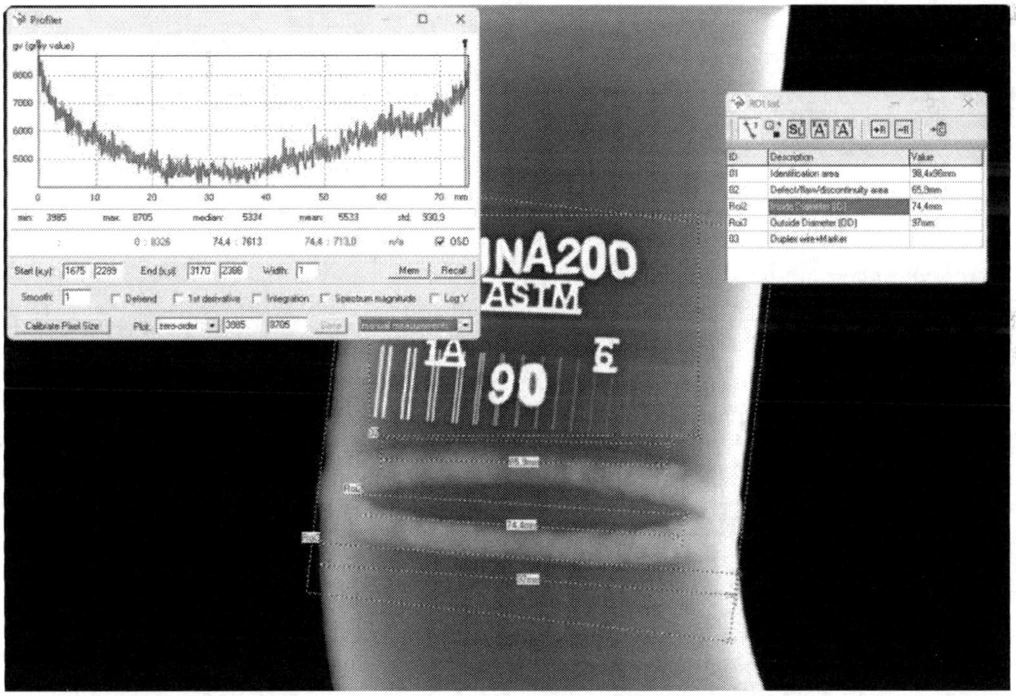

Figure 2. Sketch of digital radiograph

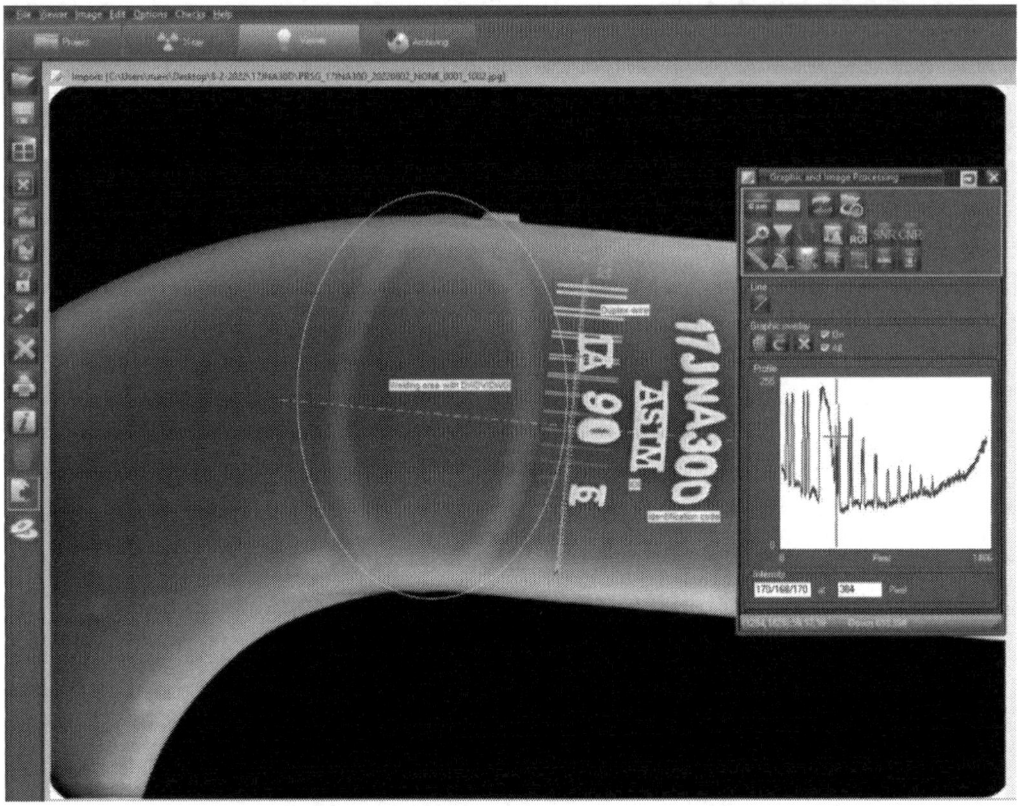

Figure 3. Identification area (measurement)

A duplex image quality indicator (IQI) ensures test procedure and radiograph quality. IQI is attached to the test object on the side of the radiation source. The duplex wire IQI (ISO 19232-5) shall be positioned directly on the digital detector. The basic spatial resolution or duplex wire value shall be determined to verify that the system hardware meets the requirements specified as a function of the penetrated material thickness. After the radiograph is obtained, the operator makes a judgment on the quality of the entire radiograph based on the developed IQI element [10]. According to Fig. 4, the function auto profiler automatically measures and evaluates duplex wire IQI by drawing a line on a duplex wire. The iSRb measurement results are also displayed. When measuring spatial resolution for a Duplex Wire, set "Height" to 21 or greater and "Sampling pixel" to 1.

From the results of radiographic tests using computed radiography (CR), there is a pipe that has relevant indications in the form of linear indications (LI), which are in ASME VIII UW-52 C.2 states that if the length of the indication exceeds $2/3t$ where t is the thickness material, then the defects that appear in the CR test results are declared rejected so that the pipe must undergo improvement. Furthermore, the results of this study provide considerations for installation/institution owners as a basis for the next steps in the maintenance of RSG-GAS. Detection of linear indications helps prevent catastrophic failure and ensures the safety, reliability, or lack thereof.

In industrial radiography, linear indication refers to a type of defect or anomaly that appears as a straight or nearly straight line on the radiographic image. These linear indications are often caused by various factors, including material defects, cracks, fractures, or foreign objects within the object being examined. Detecting and interpreting linear indications in radiography is essential for quality control and inspection purposes in RSG-GAS. Linear indication appears as dark lines or shadows on a radiographic film or digital image, contrasting with the lighter background. The orientation, size, and shape of the linear indication can provide important information about its nature and location within the object. Type of linear indications may consist of by propagation of cracks of fractures in the material; linear indications can also result from welding imperfections, such as incomplete fusion or lack of penetration in welds; linear indications may be caused by foreign objects or debris present in the object being radiographed, and variations in material composition or density can produce linear indications. The linear indication depends on the thickness. If the thickness is less than 0.75 inches, then any linear indication greater than ¼ inches is unacceptable.

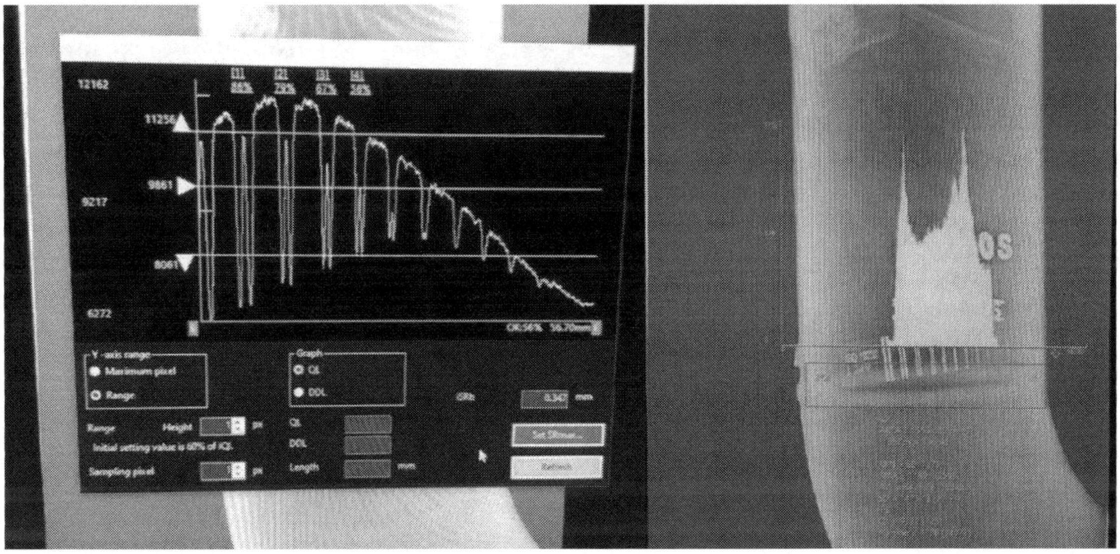

Figure 4. Profiler pixel values

Radiographic inspection plays a pivotal role in quality control, preventive maintenance, and failure analysis, helping organizations optimize production processes, reduce costly downtime, and ensure the reliability of their products and assets[27]. Like all research reactors, safety is a top priority at the RSG-GAS facility. It operates under strict safety regulations and guidelines to ensure the protection of personnel and the environment [28].

5. Conclusion

RSG GAS maintenance is an organized preventive or repair activity, both administrative and technical, to ensure that structures, systems, and components continue to operate properly. RSG-GAS maintenance and inspection uses several non-destructive testing techniques, including visual inspection, surface inspection, and volumetric inspection. The computed radiography testing method is used to determine the length and depth of defects and measure the thickness of an object, which is similar to ultrasonic testing. From the results of radiographic tests using computed radiography (CR), there is a pipe that has relevant indications in the form of linear indications (LI); of 39 measurement points with X-rays, there is 1 type of defect, i.e., linear indication, and 38 other points are still in good condition, and the results are no defects. Image interpretation results have been obtained to provide recommendations for improvements in accordance with regulations and standards. Every potential hazard has been identified and mapped at every stage of the process. As a result, risks that may occur can be managed, and worker safety is ensured, However, improvements still need to be made for every incident that has a high priority and high-risk level at nuclear facilities.

The study of CR can help improve image quality and help improve the efficiency of the NDT process, thereby reducing the time and cost involved. The practical and scientific implications of this study are increased safety and innovative solutions with the ability to detect defects or damage. Development of CR protocols, in addition to conventional radiography, as an alternative method, helps reduce noise and artifacts while improving image quality.

The computed radiography method is better than conventional radiography in evaluating flaws/defects/discontinuity, reducing radiation exposure time, and making it easier to interpret results using digitalization. CR is used to shorten work time and view images digitally. It can be managed digitally and is easy to store. Identifying and interpreting linear indications in radiography is crucial for quality assurance, safety, and compliance with industry standards and regulations. Highly trained radiographers (NDT personnel verifying the test specimen shall hold a certificate according to ISO 9712 or equivalent) play a significant role in this process, as their expertise is essential for accurately assessing the integrity of materials and structures.

6. Acknowledgement

The authors gratefully acknowledge the Research Organization for Nuclear Energy of the National Research and Innovation Agency, and Bandung Institute of Technology for funding the study.

References

[1] J. C. Imperial Trajano, "A policy analysis of nuclear safety culture and security culture in East Asia: Examining best practices and challenges," *Nuclear Engineering and Technology*, no. NET 792, pp. 1–26, 2019, doi: 10.1016/j.net.2019.04.014.

[2] IAEA, "REFERENCE DATA SERIES Nuclear Power Reactors in the World," Jul. 2024.

[3] IAEA, "Power Reactor Information System (PRIS), 2024." Accessed: Jan. 16, 2025. [Online]. Available: https://www.iaea.org/PRIS/home.aspx

[4] D. S. Wisnubroto, G. R. Sunaryo, Y. S. B. Susilo, S. Bakhri, and T. Setiadipura, "Indonesia's experimental power reactor program (RDE)," *Nuclear Engineering and Design*, vol. 404, Apr. 2023, doi: 10.1016/j.nucengdes.2023.112201.

[5] E. Hartini *et al.*, "Optimizing asset management for Structure System Components of RSG-GAS: A reliability-centric approach," *Nuclear Engineering and Technology*, 2024, doi: 10.1016/j.net.2024.06.055.

[6] M. I. Haith, P. Huthwaite, and M. J. S. Lowe, "NDT & E International Defect characterisation from limited view pipeline radiography," *NDT and E International*, vol. 86, no. May 2016, pp. 186–198, 2017, doi: 10.1016/j.ndteint.2016.12.007.

[7] ISO 9712:2021, *Non-destructive testing - Qualification and certification of NDT personnel*. 2021.

[8] Knud G. Boving, *NDE Handbook, Non-destructive examination methods for condition monitoring*, vol. 31, no. 5. 1989. doi: 10.1016/b978-0-408-04392-2.50039-4.

[9] J. R. Deepak, V. K. Bupesh Raja, D. Srikanth, H. Surendran, and M. M. Nickolas, "Non-destructive testing (NDT) techniques for low carbon steel welded joints: A review and experimental study," *Mater Today Proc*, vol. 44, pp. 3732–3737, 2021, doi: 10.1016/j.matpr.2020.11.578.

IOP Conf. Series: Materials Science and Engineering **1330** (2025) 012016 doi:10.1088/1757-899X/1330/1/012016

[10] S. Deivanai and M. Soni, "Non destructive testing and analysis of friction stir welded aluminium alloy 2024 pipes," *Mater Today Proc*, vol. 56, pp. 3721–3726, 2022, doi: 10.1016/j.matpr.2021.12.470.

[11] E. Nazemi, B. Rokrok, A. Movafeghi, M. H. C. Dastjerdi, and M. Dinca, "Obtaining optimum exposure conditions for digital X-ray radiography of fresh nuclear fuel rods," *Nucl Instrum Methods Phys Res A*, vol. 923, no. October 2018, pp. 88–96, 2019, doi: 10.1016/j.nima.2019.01.061.

[12] A. B. & P. V. Code, *ASME Boiler and Pressure Vessel Code (BPVC) Section V, Non-destructive examination (2021)*. 2021.

[13] ISO/TC 44, "ISO 17636-2, Non-destructive testing of welds - Radiographic testing - Part 2: X-and gamma-ray techniques with digital detectors," 2022.

[14] H. Alzahrani *et al.*, "Image quality determination of a novel digital detector for X-ray imaging and cone-beam computed tomography applications," *Nucl Instrum Methods Phys Res A*, vol. 968, no. February, p. 163914, 2020, doi: 10.1016/j.nima.2020.163914.

[15] R. Singh, *Applied Welding Engineering: Processes, Codes, and Standards*, 2nd ed. Butterworth-Heinemann, 2016.

[16] M. M. Zaky, S. Haggag, W. A. el Bar, and T. Mongy, "Ageing management of research reactors instrumentation using neutron radiography applications," *Alexandria Engineering Journal*, vol. 61, no. 4, pp. 3229–3235, 2022, doi: 10.1016/j.aej.2021.08.041.

[17] L. Yenumula *et al.*, "Radiographic evaluation of gas tungsten arc welded joints used in nuclear applications by X- and gamma-rays," *NDT and E International*, vol. 102, no. November 2018, pp. 144–152, 2019, doi: 10.1016/j.ndteint.2018.11.017.

[18] BAPETEN, *Regulation of The Head of The Nuclear Energy Regulatory Agency Number 9 of 2015 Concerning Provisions for Maintenance of Non-reactor Nuclear Installations*. Indonesia, 2015.

[19] IAEA Safety Guide No.RS-G-1.1, *Occupational Radiation Protection*, no. 5. 1999.

[20] Agustinus. B. P. S. D. P. S. S. Jepri Sutanto, "Study of the implementation of nuclear science and technology standards ISO 9712 NDT," *PPIS*, pp. 83–89, 2018.

[21] *ISO 17636-2, Non-destructive testing of welds - Radiographic testing - Part 2: X-and gamma-ray techniques with digital detectors*. 2022.

[22] E. Nazemi, B. Rokrok, A. Movafeghi, M. H. C. Dastjerdi, and M. Dinca, "Obtaining optimum exposure conditions for digital X-ray radiography of fresh nuclear fuel rods," *Nucl Instrum Methods Phys Res A*, vol. 923, no. October 2018, pp. 88–96, 2019, doi: 10.1016/j.nima.2019.01.061.

[23] R. Jarvis, P. Cawley, and P. B. Nagy, "Current deflection NDE for the inspection and monitoring of pipes," *NDT and E International*, vol. 81, pp. 46–59, 2016, doi: 10.1016/j.ndteint.2016.03.006.

[24] International Atomic Energy Agency, "Guidebook for the Fabrication of Non-Destructive Testing (NDT) Test Specimens," *Iaea-Tecdoc-Tcs-13*, no. 13, 2001.

[25] S. Tunde Azeez and P. Madindwa Mashinini, "Radiography examination of friction stir welds of dissimilar aluminum alloys," *Mater Today Proc*, vol. 62, pp. 3070–3075, 2022, doi: 10.1016/j.matpr.2022.03.225.

[26] H. P. Pydi, A. Pradeep, S. Vijayakumar, and R. Srinivasan, "Examination of various weld process parameters in MIG welding of carbon steel on weld quality using radiography & magnetic particle testing," *Mater Today Proc*, vol. 62, pp. 1909–1912, 2022, doi: 10.1016/j.matpr.2022.01.160.

[27] N. Meyendorf, N. Ida, R. Singh, and J. Vrana, "NDE 4.0: Progress, promise, and its role to industry 4.0," *NDT and E International*, vol. 140, Dec. 2023, doi: 10.1016/j.ndteint.2023.102957.

[28] IAEA Safety Standards No.GS-R-3, *The Management System for Facilities and Activities*. 2006.

Assessment of Material Attractiveness in Light Water-Based Reactors: Evaluating Plutonium Isotope Composition and Burnup Impact on Proliferation Risk

Fungky Iqlima Nasyidiah[1*] and Akfiny Hasdi Aimon[1,2,3]

[1]Department of Nuclear Science and Engineering, Faculty of Mathematics and Natural Sciences, Institut Teknologi Bandung, Indonesia
[2]Nuclear Physics & Biophysics Research Division, Department of Physics, Faculty of Mathematics and Natural Sciences, Institut Teknologi Bandung, Indonesia
[3]Physics and Technology of Advanced Materials Research Division, Department of Physics, Faculty of Mathematics and Natural Sciences, Institut Teknologi Bandung, Indonesia

*E-mail: fiqlimanasyidiah@gmail.com

Abstract. The assessment of nuclear safety and the potential risks associated with proliferation is a critical area of research in the context of nuclear energy utilization. This study focuses on evaluating the Material Attractiveness (ATTR) in several light water-based reactor designs, specifically NuScale, ESBWR, BWRX-300, and PWR, both during reactor operation and post-operation. The analysis employed Origen 2.2 for modeling radioactive decay and MCNP4C for assessing material attractiveness. The findings reveal an increase in the production of plutonium isotopes, with the exception of Pu-239, which decreases as a result of the fission. Initially, the Pu-240 composition across all four reactors is classified as Super-grade plutonium. However, with increased burnup, the Pu-240 composition transitions to a reactor-grade plutonium level. At the initial of irradiation, the ATTR values were determined to be 0.19 for the ESBWR, 0.20 for the PWR, 0.16 for the BWRX-300, and 0.21 for NuScale, categorizing them within the weapon-grade range. By the end of reactor operation, these values had significantly decreased, with ESBWR at 0.0125, PWR at 0.0148, BWRX-300 at 0.0111, and NuScale at 0.0133, placing them in the un-usable grade category. This marked reduction in ATTR values, correlated with increased burnup, indicates an effective decrease from weapon-grade to un-usable grade by the end of the reactor^s operational period. This trend is primarily influenced by the increased production of isotopes Pu-238, Pu-240, Pu-242, and the corresponding decrease in Pu-239, the primary fissile material.

Keywords: Non-proliferation- ATTR- Bare Critical Mass- Decay Heat- Spontaneous Fission Neutron- Neutron Prompt Life.

1. Introduction

The proliferation of nuclear materials poses a significant concern for global security, particularly in the context of civilian nuclear power generation. Light water reactors (LWRs), which constitute the most widely used type of commercial nuclear reactors, produce plutonium as a by product of their operation. This plutonium, depending on its isotopic composition, can vary in its attractiveness for potential misuse in nuclear weapons development. Understanding the factors that affect the proliferation risk associated with plutonium produced in LWRs is critical for assessing and mitigating security threats [1-2].

Plutonium generated in LWRs consists of several isotopes, including Pu-239, Pu-240, Pu-241, and Pu-242. The reactor's fuel type, burnup level, and operational conditions determine the relative abundance of these isotopes. Higher burnup levels, often associated with the more efficient utilization of nuclear fuel, tend to increase the proportion of less fissile and more neutron-absorbing isotopes, such as Pu-240 and Pu-242, thereby reducing the material's desirability for weapons use. However, the specific impact of burnup on the proliferation risk remains complex and depends on various technical factors, including reactor design and fuel cycle strategy [3-4].

This study aims to comprehensively assess the attractiveness of plutonium produced in LWRs by evaluating its isotopic composition under different burnup conditions. By analyzing the isotopic shifts in plutonium as a function of burnup, this research seeks to enhance the understanding of how fuel cycle choices in light water reactors affect the proliferation risk. The findings of this study are expected to contribute to policy decisions and technical guidelines that promote the secure use of nuclear energy while minimizing the risks associated with plutonium proliferation.

2. Research Methodology

This study uses a light water reactor-based reactor that uses UO_2 fuel, focusing on the fuel in the reactor terrace to review how the material attractiveness so that it is used as an initial discussion in research on non-proliferation aspects in terms of material attractiveness. This study performs calculations on only one fuel pin, which we assume is homogeneous for all pins in the reactor core and to facilitate comparison, we assume the same treatment for all selected reactors. The description of the fuel (one fuel pin) is as illustrated in Figure 1.

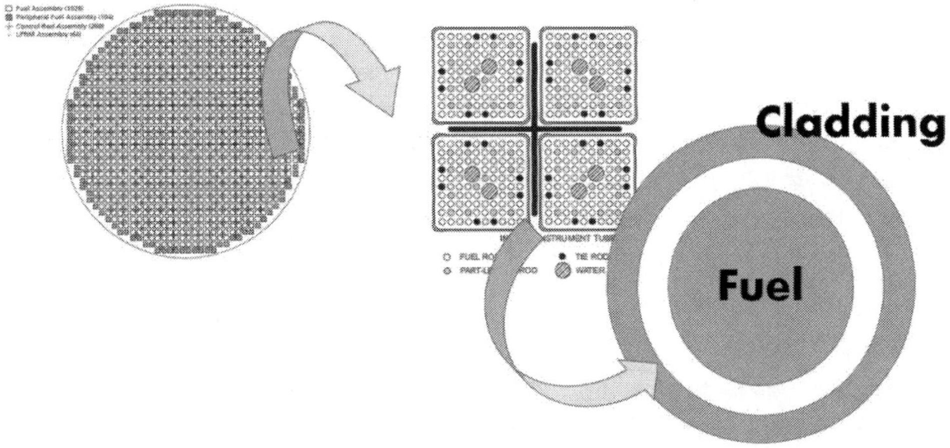

Figure 1. Fuel In The Reactor Core

The reactor specifications used in this study include four different types of LWR reactors, namely the Economic Simplified Boiling Water Reactor (ESBWR), Boiling Water Reactor X-300 (BWRX-300), Pressurized Water Reactor (PWR), and NuScale. Each reactor has different characteristics regarding power, fuel mass, U-235 enrichment percentage, and cladding material type as summarized in Table 1. [5-7]

Table 1. Reactor Specifications

Specification	ESBWR	BWRX-300	PWR	NuScale
Fuel	UO_2	UO_2	UO_2	UO_2
Power	4500 MW	870 MW	3358.82 MW	160 MW
Mass Fuel	163 kg Heavy Metal (U)	324 kg Heavy Metal (U)	305.88 kg Heavy Metal (U)	250 kg Heavy Metal (U)
Enriched Uranium	4,2% U-235	3,4% U-235	4,5% U-235	4,95% U-235
Cladding	Zircaloy-2	Zircaloy-2	Zircaloy-4	Zircaloy-4

2.1 ORIGEN Calculation

The ORIGEN2.2 code, developed by Oak Ridge National Laboratory, is used to calculate changes in fuel isotope composition due to neutron irradiation in the reactor. It can calculate the formation, decay, and processing of radioactive materials during and after irradiation. ORIGEN2.2 uses the "burnup" method or irradiation process with an average point approach to represent the entire reactor terrace, so the assumption of homogeneous fuel can be applied [8].

This method uses a first-order non-homogeneous ordinary differential equation (ODE) to calculate changes in atomic density of nuclides over time as a result of transmutation and radioactive decay processes. The parameters used in the calculation include the atomic density of the nuclide, the decay constant, the cross-section average spectrum for fission and neutron capture, the radioactive integration fraction, and the average neutron flux [8].

The use of ORIGEN2.2 in this study was modified to optimize the use of available computer memory and problem size, allowing the calculation of material composition both during reactor operation and after the reactor has stopped operating [9].

2.2 MCNP4C Calculation

The MCNP4C (Monte Carlo N-Particle) code is used to evaluate "material attractiveness" by simulating the transportation of neutrons, photons, and electrons in the reactor core. The simulation is supported by the JENDL3.3 data library to improve the accuracy of neutronic data [9]. The Monte Carlo method is applied to calculate the probabilistic distribution of particle interactions inside the fuel with higher statistical accuracy [10].

MCNP4C is used to evaluate the Rossi-alpha indices (n, p, and q), which include calculations related to the prompt neutron life inside the fuel. The results of these calculations provide information on the neutronic characteristics of the fuel relevant to the attractiveness of the material in a non-proliferation context [10].

IOP Conf. Series: Materials Science and Engineering **1330** (2025) 012017 doi:10.1088/1757-899X/1330/1/012017

2.3 Proliferation Resistance

To assess the proliferation risk and proliferation resistance of nuclear materials in this study, a proliferation resistance evaluation approach is used by considering aspects of material attractiveness as well as the influence of plutonium isotopes on material attractiveness for nuclear weapon use [11]. Assessment of proliferation resistance is carried out with reference to the safeguard measures that have been implemented by the International Atomic Energy Agency (IAEA). These safeguard measures are designed to ensure that nuclear material in nuclear facilities is not diverted from peaceful use and is not used for military purposes. Nuclear safeguards include activities necessary to control and supervise nuclear fuel potentially used in the production of nuclear weapons, as well as to prevent proliferation from civil nuclear activities including reprocessing and recycling of spent fuel [12].

The proliferation resistance measurement is performed by integrating the results from ATTR calculation, isotopic barrier analysis, and neutronic parameter evaluation obtained from simulations with ORIGEN2.2 and MCNP4C. The data obtained will be evaluated to determine the level of technical difficulty in developing nuclear weapons and the potential for nuclear materials in the civilian fuel cycle to be used for military purposes.

This research uses a quantitative approach to measure proliferation resistance with a focus on evaluating the plutonium produced during the reactor fuel cycle. Key parameters evaluated include the isotopic levels of plutonium, particularly Pu-240, which is considered an isotope with significant proliferation risk potential due to its properties that affect the explosive power and stability of nuclear weapons.

2.4 Evaluation of Isotopic Barrier and Material Attractiveness

This research evaluates proliferation resistance using the concept of isotopic material barrier, which emphasizes the technical difficulty in using certain materials for nuclear weapons manufacturing. This evaluation is based on the analysis of parameters such as decay heat, spontaneous fission neutron, and critical mass plutonium. These parameters are used to calculate the material attractiveness value (ATTR) based on the equation developed by Saito [13], namely:

$$ATTR = \frac{Explosive\ Energy}{Technological\ Diffulties} \quad (1)$$

$$ATTR = \frac{\left(\frac{\alpha_{SQ}^{i}}{\alpha_{\propto}^{9}}\right)^{n}}{\left[\frac{DH_{comp}}{DH^{8}}\right]^{p} + \left[\frac{SN_{comp}}{SN^{8}}\right]^{q}} \text{ ; index : n = p = q = 1} \quad (2)$$

The calculated parameters include Rossi-alpha to calculate potential explosive energy, decay heat (DH), and spontaneous fission neutrons (SFN). ATTR values were measured by normalizing against the ATTR value of pure plutonium (Pu-239) to determine the level of attractiveness relative to the use of nuclear weapons [14].

The evaluation method involves the use of the Rossi-alpha equation that measures the potential explosive energy and dynamic properties of atomic nuclei based on the relationship of supercriticality ($k_{eff\,-\,1}$) to fast neutron half-life (τ_0). The equation used to calculate Rossi-alpha is [15-16].

IOP Conf. Series: Materials Science and Engineering **1330** (2025) 012017 doi:10.1088/1757-899X/1330/1/012017

$$\alpha \equiv \frac{k_{eff}-1}{\tau_0} \tag{3}$$

The Rossi-alpha, DH, and SFN values are then used to determine the normalization and attractiveness of the material based on different ATTR scales (<0.002 exempt, 0.002<ATTR<0.02 unusable grade, 0.02<ATTR<0.1 usable grade, and >0.1 weapon grade). Using this ATTR scale, we can assess the potential proliferation risk associated with a particular nuclear material, and establish appropriate safeguards to prevent misuse of that material [17-18].

In analyzing the isotopic barrier, an evaluation is made of the relevant isotopic composition of plutonium, especially isotopes with even mass numbers (Pu-238, Pu-240, Pu-242). The properties of these isotopes, such as their high decay heat and spontaneous fission neutron (SFN) frequency, are used to measure the instrinsic resistance to proliferation. This evaluation is important because isotopes with high decay heat make handling plutonium technically more difficult and thus increase the barrier to its use for weapons purposes [19-21].

3. Results and Discussion

3.1 Plutonium Composition Calculation

The isotopic composition of plutonium obtained from ORIGEN2.2 simulations for each reactor type—ESBWR, PWR, BWRX-300, and NuScale—was compared both during irradiation and after reactor shutdown. The operation of each reactor was limited to a burnup of 33 GWd/t. The results, illustrated in Figure 2 reveal the distribution of plutonium isotopes in the fuel of each reactor. The reactors show varying operational times to reach the 33 GWd/t burnup: ESBWR takes 1315 days, PWR 975 days, BWRX-300 1421 days, and NuScale 2086 days. The isotopic composition vectors for all plutonium isotopes generally increase with burnup, except for Pu-239, which initially rises significantly before declining as burnup increases. This decrease is attributed to the fission reactions or neutron captures of Pu-239, causing its composition to decrease over time.

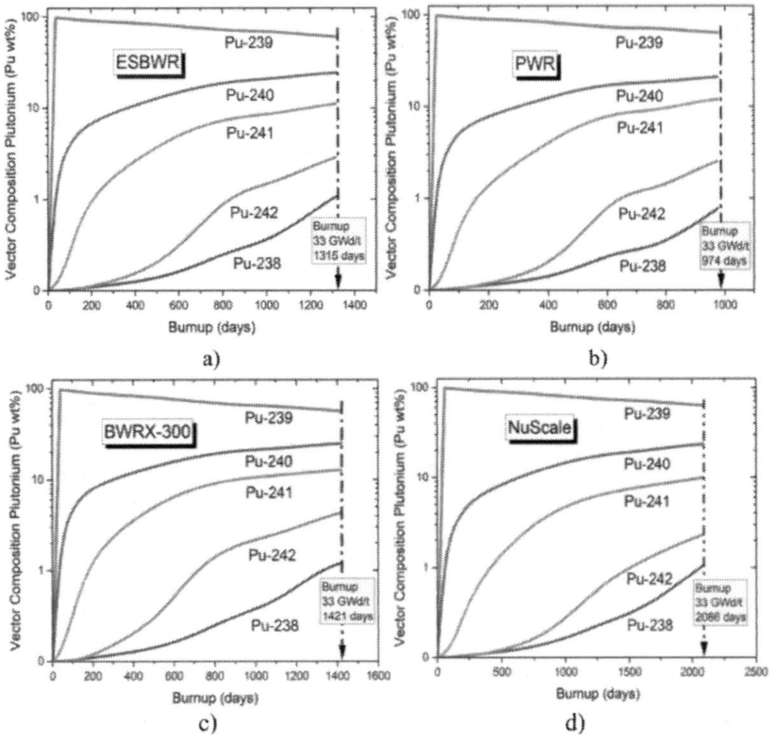

Figure 2. Vector Composition of Plutonium Isotope at Burnup Time, a) ESBWR, b) PWR, c) BWRX-300, d) NuScale

The production of plutonium isotopes is primarily driven by the conversion of U-238 to Pu-239 and other plutonium isotopes. The highest concentration is observed for Pu-239, as it is the first product in the neutron capture chain of U-238, followed by beta decays to Np-239 and further to Pu-239. Consequently, a higher U-238 content results in a higher plutonium isotope concentration. Among all isotopes, Pu-238 shows the lowest composition.

Figure 3 highlights that Pu-238 increases with burnup, with the BWRX-300 producing the highest Pu-238 (1.22%) from reactor start-up until reaching 33 GWd/t burnup. In contrast, PWR shows the lowest Pu-238 value (0.654%) after reaching 20 GWd/t, while ESBWR and NuScale produce Pu-238 values of 0.886% and 1.047%, respectively, at 33 GWd/t burnup. The Pu-239 concentrations for the four reactors intersect and initially increase significantly before declining after 1 GWd/t burnup. The highest Pu-241 concentration is produced by BWRX-300 and the lowest by NuScale at 33 GWd/t, with values of 12.836% and 9.895%, respectively. Lastly, Pu-242 concentrations are highest for BWRX-300 (4.329%) and lowest for NuScale (2.304%).

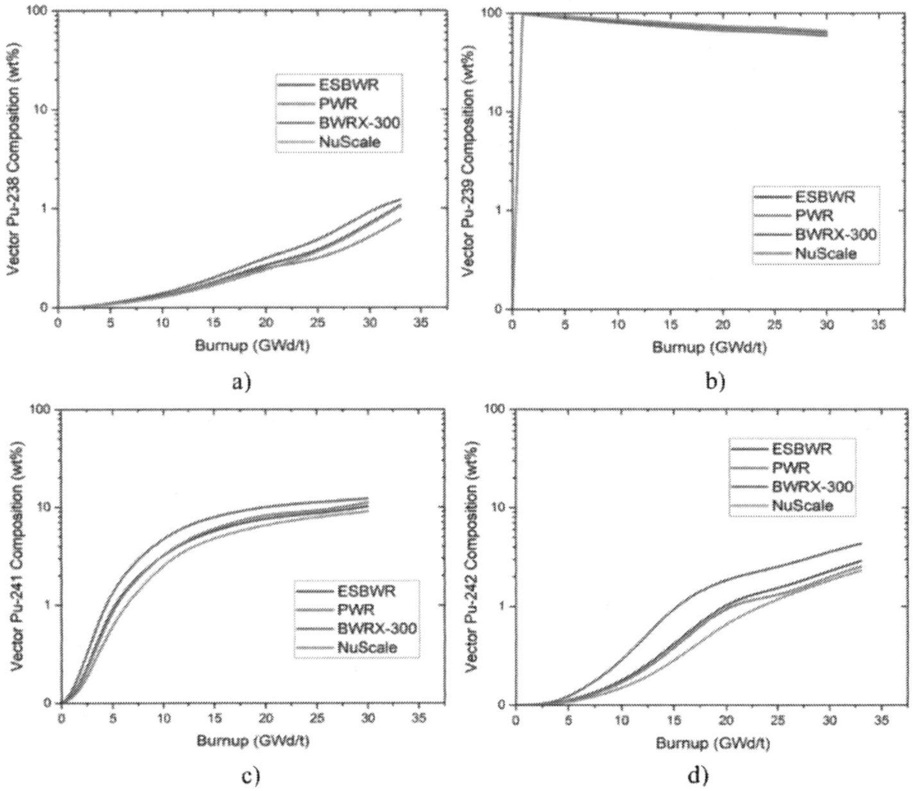

Figure 3. Vector Composition of Plutonium Isotope at burnup time, a) Pu-238, b) Pu-239, c) Pu-241, d) Pu-242

3.2 Analysis of Pu-240 Composition and Material Attractiveness

Special attention is given to Pu-240 due to its potential use in weapons. According to Pellaud's classification, Figure 4 shows that the Pu-240 produced in each reactor is initially classified as "super-grade" at 1 GWd/t burnup, indicating its high usability for weapon purposes. As burnup increases from 1-2.5 GWd/t, Pu-240 transitions to "super-grade," from 2.5-5 GWd/t to "weapon-grade," and from 5-20 GWd/t to "fuel-grade," which is not suitable for weapon programs. Finally, from 20-33 GWd/t, Pu-240 is categorized as "reactor-grade," indicating its safety for non-proliferation purposes as burnup effectively reduces Pu-240's usability for weapons.

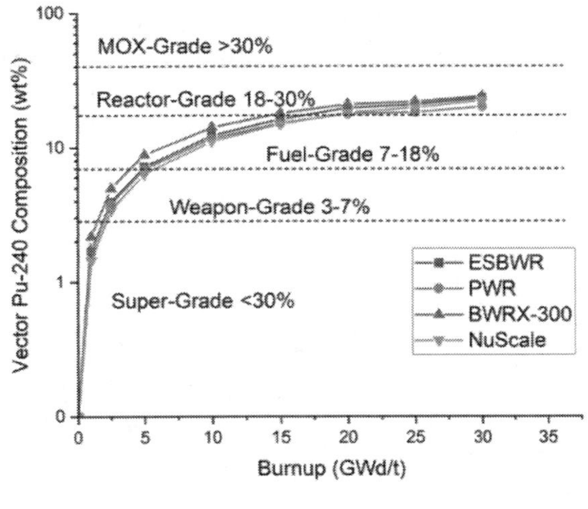

Figure 4. Pu-240 Isotope Technology Level Categorization at Burnup Time

Figure 5 shows the plutonium isotopic composition after reactor shutdown and a cooling period of 1,000,000 years for ESBWR, PWR, BWRX-300, and NuScale. The Pu-239 concentration significantly decreases due to its half-life of 24,110 years, leading to its eventual natural decay. Similarly, Pu-240 disappears from the chart after 6,561 years. Pu-238 and Pu-241 are not visible from the start due to their short half-lives (87.71 and 14.29 years, respectively). Interestingly, Pu-242 concentration increases, indicating that its decay characteristics differ from other isotopes.

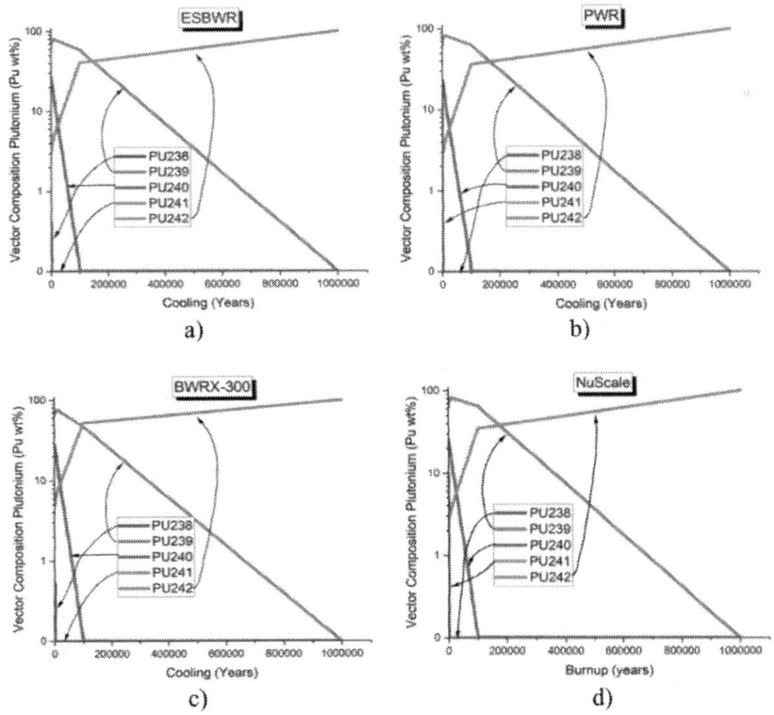

Figure 5. Vector Composition of Plutonium Isotope at Cooling Time, a) ESBWR, b) PWR, c) BWRX-300, d) NuScale

Figures 6 compare Pu-239 and Pu-240 values after reactor shutdown. BWRX-300 shows the lowest Pu-239 value, indicating a lower potential for weapon use compared to ESBWR, PWR, and NuScale. The Pu-240 concentrations decrease to "super-grade" levels within 100,000 years, with initial conditions classified as "reactor-grade," "fuel-grade," and "weapon-grade" transitioning to "super-grade" within about one year due to Pu-240's half-life. According to Pellaud, "super-grade" Pu-240 is highly usable with low radiation and heat generation, raising concerns for its potential misuse.

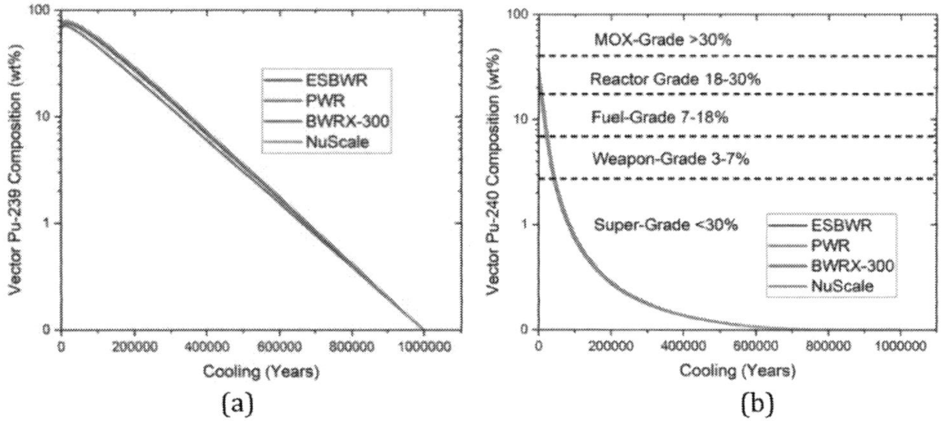

Figure 6. a) Vector Composition of Plutonium Isotope Pu-239, b)Pu-240 Isotope Technology Level Categorization at cooling time

3.3 Evaluation of Material Attractiveness of Fuel and Spent Fuel

The material attractiveness of plutonium isotopes is evaluated considering parameters like decay heat, spontaneous fission neutron (SFN), bare critical mass (BCM), neutron prompt life, and Rossi-alpha. Figure 7 shows the material attractiveness (ATTR) values for ESBWR, PWR, BWRX-300, and NuScale reactors, highlighting their categorization in nuclear weapon technology during initial irradiation phases (beginning of cycle/BOC). NuScale has the highest ATTR value (0.21) at 1 GWd/t burnup, while differences between large reactors (ESBWR and PWR) are minimal due to similar U-235 enrichment levels. However, significant differences are observed between small reactors (BWRX-300 and NuScale), mainly due to differing U-235 enrichment levels (3.4% for BWRX-300 and 4.95% for NuScale).

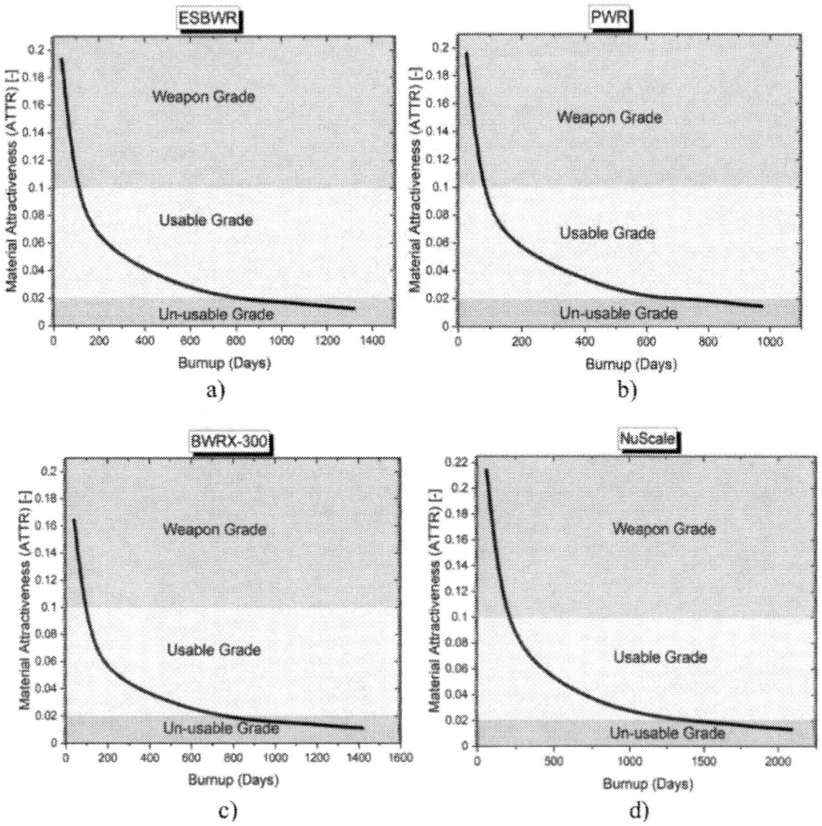

Figure 7. ATTR Value at burnup time, a) ESBWR, b) PWR, c) BWRX-300, d) NuScale

At the end of the cycle (EOC), the ATTR values of all reactors are categorized as "un-usable grade." ESBWR reaches EOC in 1319 days with an ATTR of 0.0125, PWR in 974 days with 0.0148, BWRX-300 in 1421 days with 0.0111, and NuScale in 2086 days with 0.0133. This demonstrates that higher burnup effectively reduces ATTR values. Figure 8 shows the value of material attractiveness at the ESBWR, PWR, BWRX-300, NuScale reactors, this value is very important in reviewing when the reactor is categorized as nuclear weapon technology when the reactor stops operating and this condition becomes a very important condition whether plutonium isotopes can be utilized (stolen) to become nuclear weapons or not. For ESBWR, PWR, and BWRX-300 reactors, when the reactor is shut down (no longer operating), the ATTR value is not categorized as weapon grade or usable grade but is categorized as un-usable grade, in contrast to NuScale, when the reactor stops operating, the ATTR value is still categorized as Usable Garde, indicating that there is still an opportunity to be misused.

IOP Conf. Series: Materials Science and Engineering **1330** (2025) 012017 doi:10.1088/1757-899X/1330/1/012017

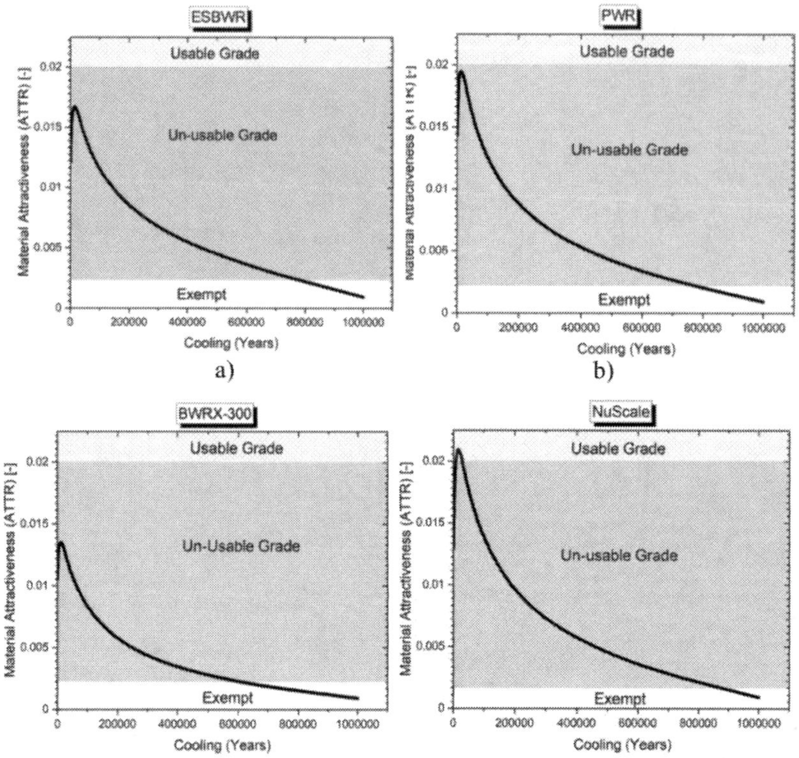

Figure 8. ATTR Value at cooling time, a) ESBWR, b) PWR, c) BWRX-300, d) NuScale

Figure 9 evaluates the impact of compression on the NuScale reactor at the beginning (BOC), middle (MOC), and end of the cycle (EOC). At BOC and MOC, the ATTR value decreases slightly with compression, indicating that high isotopic compositions of plutonium maintain weapon-grade classification. When EOC, the material is categorized as "un-usable grade," and compression has little impact on ATTR values, suggesting minimal influence from compression on non-proliferation resistance. The compression of the nuclear material causes its surface area to decrease, reducing the number of neutrons lost of leakage. When the number of neutrons lost is equal to the number produced by fission, the system is critical. As the surface area continues to decrease, the system becomes supercritical [22]. More neutrons are produced by fission than are lost to leakage and the number of nuetrons increases exponentially, resulting in an nuclear explosion [22].

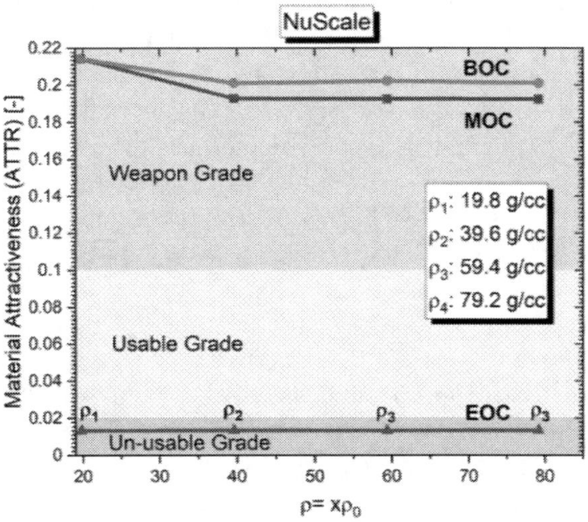

Figure 9. Effect Of Compression On ATTR For Nuscale Reactors

Overall, the analysis underscores the dynamic behavior of plutonium isotopic compositions and their material attractiveness over time, emphasizing the importance of reactor design and fuel management strategies in enhancing nuclear non-proliferation efforts.

4. Conclusions

It can be concluded based on the non-proliferation assessment aspects that have been presented. Based on plutonium isotope production, PWR and BWRX-300 tend to be better at non-proliferation than ESBWR and NuScale, as they produce more Pu-238. The presence of Pu-238 and other minor actinide components may provide technical constraints in the purification of plutonium for weapons. On the other hand, ESBWR and NuScale produce more Pu-239 and Pu-241, which are fissile isotopes and may add proliferation risks in case of fuel reprocessing. However, this does not necessarily make the PWR or BWRX-300 completely safe, just more favorable from a non-proliferation perspective. In general, small reactors such as NuScale and BWRX-300 tend to produce less Pu-239 and more other isotopes, which are harder to use for nuclear weapons, so they are better in the aspect of non-proliferation. ESBWRs and PWRs, with higher power, can produce more Pu-239 at lower burn rates, which can be a non-proliferation issue if not closely monitored. Reactors with high burn rates and greater efficiency tend to produce more Pu-242 and Pu-238, which is good for reducing the risk of nuclear proliferation.

The gap between decay heat and spontaneous fission neutrons reflects the fuel design, enrichment and combustion characteristics of each reactor, with a low proliferation risk thanks to the moderate enrichment level of U-235 and the slow accumulation of fission isotopes that contribute to neutron production.

From a nuclear non-proliferation point of view, based on the relationship between ATTR values and burnup values, it shows the importance of keeping the fuel at a high burnup to reduce the potential use of plutonium in nuclear weapons. Although at the beginning of the reactor screw is at the weapon level, the greater the burnup value indicating the longer operating time (irradiation) will reduce the activity of replacing fuel too frequently. So as to avoid the theft and formation of weapon-grade plutonium. Weapons-grade plutonium at low burnup can be of international concern, as it can be used in nuclear weapons if accessed by irresponsible parties. Therefore, a fuel cycle designed for high burnup helps reduce the risk of proliferation.

And the effect of fuel compression can reduce the ATTR value. This can be seen by applying compression three times to the NuScale reactor. So that the decrease in the ATTR value also affects the proliferation capability of the fuel and is able to increase the level of technology towards non-nuclear weapons.

References

[1] Permana, S., et al. (2015). Analysis on isotopic plutonium barrier based on spent nuclear fuel of LWR. *Annals of Nuclear Energy, 75,* 116–122. https://doi.org/10.1016/j.anucene.2014.07.036

[2] Permana, S., et al. (2013). Study on material attractiveness aspect of spent nuclear fuel of LWR and FBR cycles based on isotopic plutonium production. *Energy Conversion andManagement, 72,* 19–26. https://doi.org/10.1016/j.enconman.2013.01.038

[3] Kessler, G. (2007). Plutonium Denaturing by Pu-238. Nucl. Sci. Eng. 155, 53-73.

[4] Pellaud, B. (2002). Du Combustible Nucléaire Aux Déchets : Recherches Actuelles From Nuclear Fuels To Waste: Current Research Proliferation aspects of plutonium recycling. In C. R. Physique (Vol. 3).

[5] Status report 100-Economic Simplified Boiling Water Reactor (ESBWR). (t.t.)

[6] Status Report-BWRX-300 (GE Hitachi and Hitachi GE Nuclear Energy). (2019).

[7] NuScale Power Inc., 2011. NuScale Power Modular and Scalable Reactor (NuScale), Status report 106.

[8] ORNL. (1973). ORIGEN – The ORNL Isotope Generation and Depletion Code.

[9] Permana, S., & Suzuki, M. (2011). Basic evaluation on material attractiveness of isotopic plutonium barrier. *Progress in Nuclear Energy, 53*(7), 958–963. https://doi.org/10.1016/j.pnucene.2011.04.014

[10] Permana, S., Suzuki, M., & Saito, M. (2011). Basic Analysis on Isotopic Barrier of Material Attractiveness Based on Plutonium Composition of FBR. *Journal of Nuclear Science and Technology, 48*(5), 724–733. https://doi.org/10.1080/18811248.2011.9711755

[11] Mladineo, S. V. et al. (2003). *Guidelines for the Performance of Nonproliferation Assessments.*

[12] IAEA. *Basic Principles Objectives IAEA Nuclear Energy Series Technical Features to Enhance Proliferation Resistance of Nuclear Energy Systems.* (1464). http://www.iaea.org/Publications/index.html

[13] Saito, M. et al., (2008). Development of methodology for plutonium categorization (ii) – improvement of evaluation function "attractiveness". Trans. Am. Nucl, Soc. 98 (1). 669.

[14] Kimura, Y., Saito, M., & Sagara, H. (2012). Improvement of evaluation methodology of plutonium for intrinsic feature of proliferation resistance based on its isotopic barrier. Annals of Nuclear Energy, 40(1), 130–140. https://doi.org/10.1016/j.anucene.2011.10.008

[15] Permana, S. (2020). Analysis on different material attractiveness concept based on plutonium composition of FBR blanket system. *Progress in Nuclear Energy, 127.*

https://doi.org/10.1016/j.pnucene.2020.103404
[16] King, W. E. et al. (2010). The application of a figure of merit for nuclear explosive utility as a metric for material attractiveness in a nuclear material theft scenario. *Journal of Nuclear Engineering and Design*, *240*(11), 3699–3707. https://doi.org/10.1016/j.nucengdes.2010.09.015
[17] Saito, M. et al., (2005). Development of methodology for plutonium categorization – the challenge of Attractiveness. Trans. Am. Nucl. Soc. 96, 763–764.
[18] Saito, M. et al., (2005). Advanced core concepts with enhanced proliferation resistance by transmutation of minor actinides
[19] DeVolpi, A., 1982. Denaturing Fissile Materials. Prog. Nucl. Energy 10 (2), 161-220
[20] Kessler, G., (2007). Plutonium Denaturing by Pu-238. Nucl. Sci. Eng. 155, 53-73.
[21] Kessler, G., et al., (2008). A new scientific solution for preventing the misuse of reactor-grade plutonium as nuclear explosive. Nucl. Eng. Des. 238, 3429-3444.
[22] Jones, G.S., (2019). Reactor-grade plutonium and nuclear weapons: ending the debate. *Nonproliferation Rev., 26, 61* . https://doi.org/10.1080/10736700.2019.1603497

IOP Conf. Series: Materials Science and Engineering **1330** (2025) 012018 doi:10.1088/1757-899X/1330/1/012018

Burnup Performance of modified CANDLE shuffling in axial direction on gas cooled fast reactors with UN-Th fuels

Jean Pierre Ndayiragije, Zaki Su'ud, Abdul Waris

[1]Physics Department, Institut Teknologi Bandung, Indonesia.

[2]Nuclear Science and Engineering Department, Institut Teknologi Bandung, Indonesia.

ndayirajeanpierre@gmail.com

Abstract. The modified CANDLE (Constant Axial shape of Neutron flux, nuclide densities, and power shape During Life of Energy production) strategy shuffling in an axial direction has been successfully applied to gas cooled fast reactor. This paper investigated the utilization of natural uranium, enriched nitride and Thorium ($^{238}U^{15}N$ and ^{232}Th) as fuel on 3000MWt reactor power and refuelling every 10 years of burnup. The reactor core is partitioned into ten regions with the same volume in the axial direction. Initially, a fuel input ($^{238}U^{15}N$ and ^{232}Th) is placed in region 1. After ten years of burnup, a fuel in region 1 was moved to region 2, then to region 3, and so on until the fuel in region 9 was moved to region 10. The fuel in 10th region was removed in the core. The neutronic computations were performed in two different ways by using SRAC 2006 code and JENDL4.0 as a nuclear data library. Firstly, PIJ was used for fuel cell calculations and secondly CITATION for a reactor core calculation. The results show that the effective multiplication factor is greater than one, at the beginning of life, its value is approximately 1.0028 and at the end of life is approximately 1.0279. This indicates that the reactor is capable to operate through a burn-up period by employing $^{238}U^{15}N$ and ^{232}Th (10%) as fuel cycle input. At the beginning of life, the integral conversion ratio is approximately 9.9455 and at the end of life is approximately 1.2906. The burnup level at the end life is about 30.29% HM.

1. Introduction

The utilization of uranium nitride in modified CANDLE as fuel has been widely employed by several researchers. However, the combination of UN and ^{232}Th can be an alternative option to minimize the amount of UN loads in reactor. The benefits of the thorium fuel cycle have recently been addressed in discussions about nuclear power resources and impacts on the environment, because of its low radio-toxicity waste and its natural abundance. Natural thorium is 3 to 4 times more abundant than natural uranium, predominantly in the form of the fertile ^{232}Th isotope.[1]. However, natural thorium resources do not contain any fissile isotopes, which could explain why thorium is only considered a potential resource. Thorium can be used to reduce fuel cycle costs, make reactors safe and reliable to operate due to lower excess reactivity required[2]. ^{232}Th generates the fissile isotope U-233, which has superior fissile properties in epithermal and thermal energy ranges than U-235, helps to reduce reactivity swings during the irradiation and reducing the demand for control materials[3],[4].

The utilization of combination of uranium nitride and thorium will minimize the amount of natural uranium employed in modified CANDLE burnup scheme.

This work aims to perform the neutronic analysis of modified CANDLE shuffling in axial direction applied on gas cooled fast reactors with uranium nitride combined with natural thorium (10% of Th-232) fuels. As depicted in Figure. 1, in the CANDLE, the core is divided into three regions: fresh fuel, burning fuel and spent fuel[5],[6]. Fission reactions occur in the burning region and produce fast neutrons[7]. The CANDLE burn-up technique gets changed by the addition of different regions. The reactor cores are divided into numerous regions with the same volume in either the axial or radial directions, hence the term modified CANDLE[8],[9],[10],[11]. Figure 2 displays the Modified CANDLE strategy shuffling in an axial direction. The core is subdivided into 10 regions with equivalent volume directed axially. A fuel (UN and Th-232) is firstly placed in region 1, and after ten years it is moved to region 2 and region 1 is charged with fresh UN and Th-232 fuel. This concept is basically done to entire regions and the fuel in region 10 is moved out of the reactor core. Regions 1-5 are primarily breeding zones where plutonium is accumulated in fuels, but regions 5-10 have enough plutonium accumulated to contribute considerably to power production. Region 5-10 is known as the burning zone.

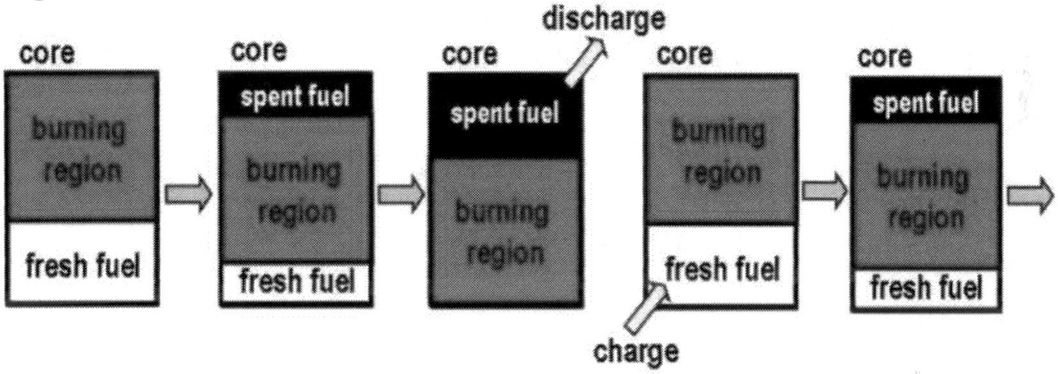

Figure 1. CANDLE concept and management of fuel[12].

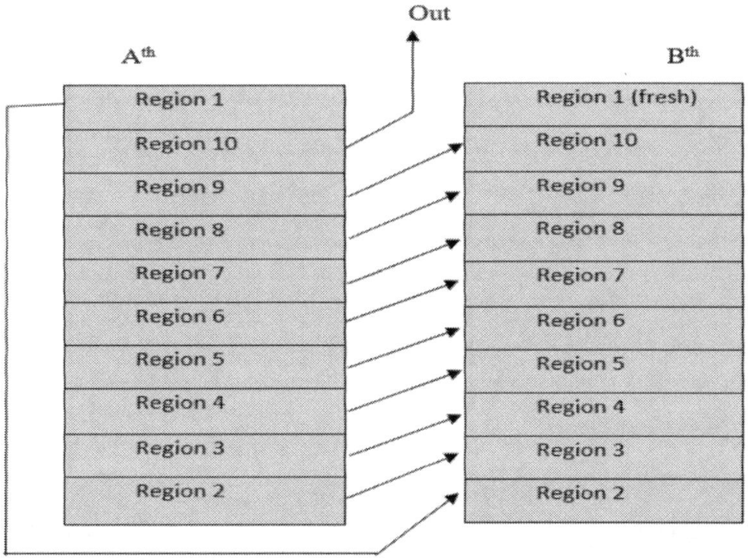

Figure 2. Modified CANDLE shuffling in axial direction.

IOP Conf. Series: Materials Science and Engineering **1330** (2025) 012018 doi:10.1088/1757-899X/1330/1/012018

2. Methodology

2.1. Reactor design

The parameters employed in this paper are presented in Table 1. Uranium nitride (UN) and 10% thorium-232 are employed as fuel. The thermal power output is 3000MWt and helium was employed as a coolant. HT-9 and Lead (Pb) were used as cladding and reflector materials respectively. The selection of HT-9 as a cladding was based on its high conductivity and its high resistance to irradiation, however, the selection of Lead as a reflector material was based on its high reflection coefficient, oxidation resistance, a low absorption cross-section and limited swelling under neutron irradiation[13],[14]. Helium gas was employed as a coolant because of its less radioactive and its less neutron absorption[15]. Figure 3 shows the cell geometry design while Figure 4 shows the reactor core design. Uranium Nitride with enriched ^{15}N isotope is used as fuel due to its smaller neutron capture cross-section than ^{14}N[16]. Using the ^{15}N isotope aimed to boost the neutron economy[16]. 10% of thorium was added to UN as fuel to minimize the uranium needed because if thorium is inexpensive, more secure and minimize the production of toxic waste[17].

Table 1. Design parameters

Parameters	Values
Power (MWt)	3000
Number of regions	10
Refuelling time (years)	10
Fuel type	$^{238}U^{15}N$, ^{232}Th
Fuel quantity (%)	65
Cladding quantity (%)	10
Coolant quantity (%)	25
Pin Pitch	1.4cm
Coolant type	Helium
Core radius	180cm
Core height	250cm
Width of reflector	70cm
Material of reflector	Lead (Pb)
Cladding material	HT-9
Geometry of pin cell	Cylindrical
Geometry of the reactor core	Cylinder in 2D

IOP Conf. Series: Materials Science and Engineering **1330** (2025) 012018 doi:10.1088/1757-899X/1330/1/012018

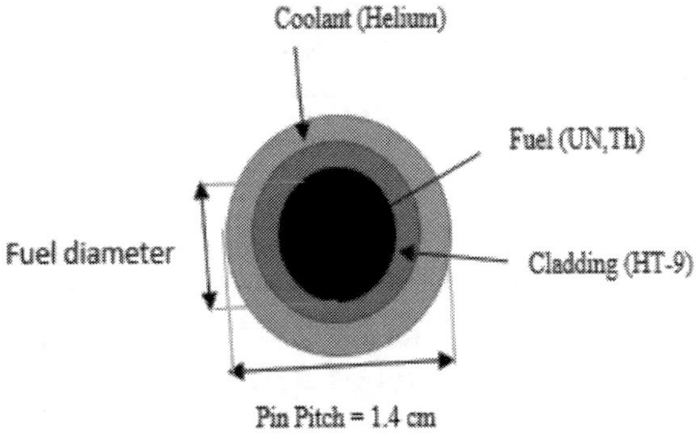

Figure 3. Cell geometry design.

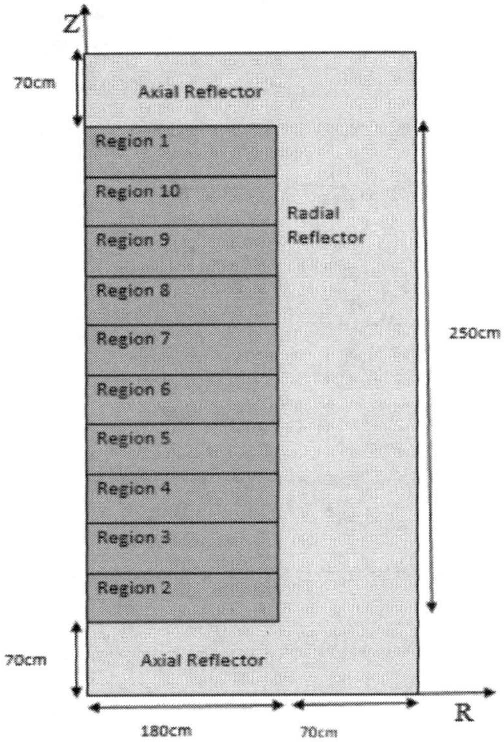

Figure 4. The reactor core design (1/2 core).

2.2. Numerical analysis method

The SRAC2006 as well as the JENDL 4.0 nuclear data library are used in the calculation method. SRAC (Standard Thermal Reactor Analysis code) is a Japanese Atomic Energy Agency (JAEA)-developed code system for neutronic analysis of various types of reactors. In the first step, the fuel pin cell is calculated using the collision probability method (PIJ). The neutron flux, macroscopic, and microscopic are the results of the PIJ. The results of PIJ are used in the second step, which is the reactor core calculation. The core calculation is conducted employing the CITATION. Figure 5 depicts the SRAC Calculation Scheme used in this study.

IOP Conf. Series: Materials Science and Engineering **1330** (2025) 012018 doi:10.1088/1757-899X/1330/1/012018

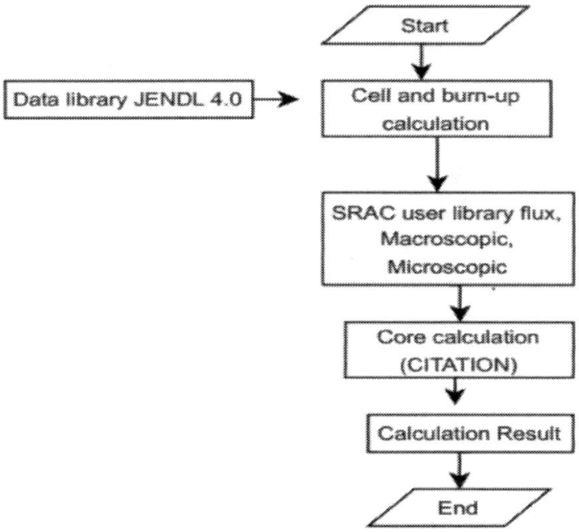

Figure 5. SRAC Calculation Scheme

The SRAC code employs a diffusion equation technique. One group neutron diffusion equations, two group neutron diffusion equations, and a multigroup neutron diffusion equation are all part of the neutron equilibrium diffusion equation. According to the Multigroup Neutron Diffusion Equation, neutrons on the reactor core are spread throughout a wide energy spectrum. By dividing neutron energy into energy groups, multigroup neutron diffusion equations are derived. Consider neutron energy group scheme shown in Figure 6.

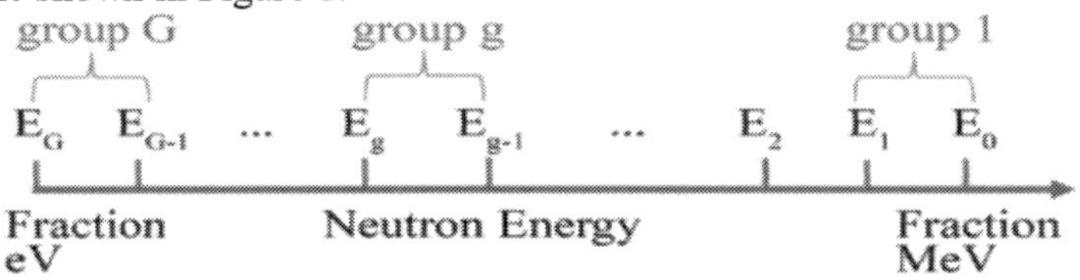

Figure 6. Neutron energy group scheme.

The balance of neutrons will be:

$$\begin{bmatrix}\text{Time rate}\\ \text{of change}\\ \text{of neutrons}\\ \text{in group g}\end{bmatrix} = - \begin{bmatrix}\text{change}\\ \text{to}\\ \text{leakage}\end{bmatrix} - \begin{bmatrix}\text{absorption}\\ \text{in}\\ \text{group}\end{bmatrix} + \begin{bmatrix}\text{source}\\ \text{neutrons}\\ \text{appearing}\\ \text{in group g}\end{bmatrix} - \begin{bmatrix}\text{neutrons}\\ \text{scattering}\\ \text{out of}\\ \text{group g}\end{bmatrix} + \begin{bmatrix}\text{neutrons}\\ \text{scattering}\\ \text{into}\\ \text{group g}\end{bmatrix} \quad (1)$$

The mathematical representation of the equation (1) is

$$\frac{1}{v_g}\frac{\partial \phi_g}{\partial t} = \nabla \cdot D_g \nabla \phi - \Sigma_{ag}\phi_g + \frac{\chi_g}{k_{eff}} S_g - \Sigma_{sg}\phi_g \sum_{g'=1}^{G} \Sigma_{sg'g}\phi_{g'}\ , \quad g = 1,2,\ldots.G \quad (2)$$

The source (S_g) due to fissions can be written

$$S_g = \chi_g \sum_{g'=1}^{G} v_{g'}\, \Sigma_{fg'}\phi_{g'} + S_g^{ext}, \quad (3)$$

$$\frac{1}{v_g}\frac{\partial \phi_g(r,t)}{\partial t} - \nabla \cdot D_g(r)\nabla\phi_g(r,t) - \Sigma_{tg}(r)\phi_g(r,t) = \sum_{g'=1}^{G} \Sigma_{sg'g}(r)\phi_g(r,t) + \frac{\chi_g}{k_{eff}} \sum_{g'=1}^{G} v_{g'}\Sigma_{fg'}(r)\phi_g(r,t) \quad (4)$$

The equation (4) is called the multigroup diffusion equations.

Where

V_g = The change of neutrons in a group of energy g

ϕ_g = neutron flux for group g

χ_g = The Probability of fission neutron with energy will be born in group g

D_g = Coefficient of diffusion for the g energy group

Σ_{tg} = Total cross section for group g

$\Sigma_{sg'g}$ = group-transfer cross section
k_{eff} = effective multiplication factor
$V_{g'}$ = Change of energy in a group of energy g'
$\Sigma_{fg'}$ = the fission cross section for group g'

3. Results and discussions

Figure 7 presents the burnup level during operation time. The burnup value increases slowly during the first 40 years of burn up (breeding zone) owing to lack of the fissile material and increases rapidly after 40 years of burnup (burning zone) because of fuel was entering in the burning region. At the end of life, the burnup level is approximately 302976MWd/ton or approximately 30.29%HM. Figure 8 presents an infinite multiplication factor (k_{inf}) during operation time. During the beginning of life, k_{inf} is lower than unity owing to the fresh fuel and the absence of Pu-239 atomic density. On the other hand, after 60 years of burning, the k_{inf} increased considerably over one because of fuel was entering in the most active region, and then reduced slightly after 78 years till the end of life owing to the reduction of U-238 atomic density. The quantity of fissile material in the core is still small in the beginning of the burnup period, but it rapidly grows and causes the burnup rate to accelerate. Because fertile material releases fissile material, k_{inf} grows until the end of life.

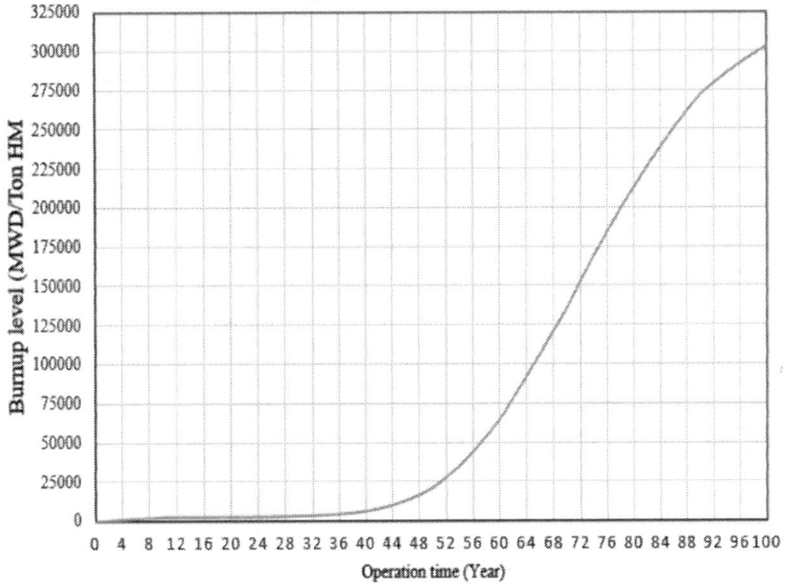

Figure 7. Burnup level

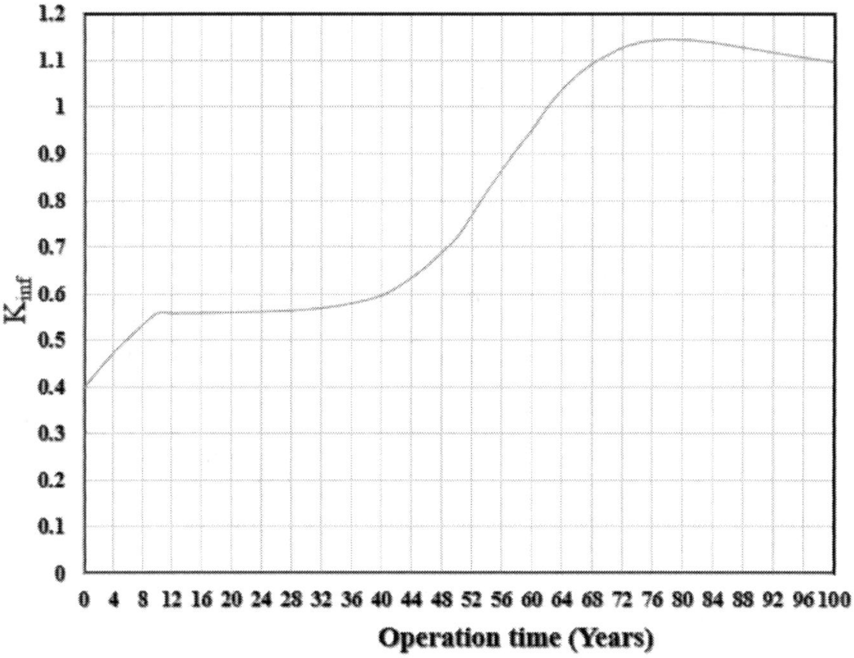

Figure 8. K_{inf} during operation time

The integral conversion ratio (Integral C.R) versus time of burnup is presented in Figure 9. Integral conversion ratio decreased from around 10 to approximately 1 during the operation time. The integral conversion ratio diminished significantly during the first burnup period, then slowly from 10 years to near half-life of the reactor operation time, and finally more slowly to the end of life because of positioning of first region nearby tenth region. Once the burnup process progresses, the density of the U-238 isotope drops, the amount of plutonium-239 generated grows, and the atomic ratio of the fertile to fissile material diminishes. As a result, the integral conversion ratio decreases. Figure 10 presents the total fission product during operation time. From the beginning until about 40 years of burnup the fission product (FP) accumulation is slow owing to the fuel was in the breeding zone. However, from 40 years to 100 years of burnup, the accumulation of fission product (FP) increases considerably because of the fuel reached to most active regions of the core (burning zone).

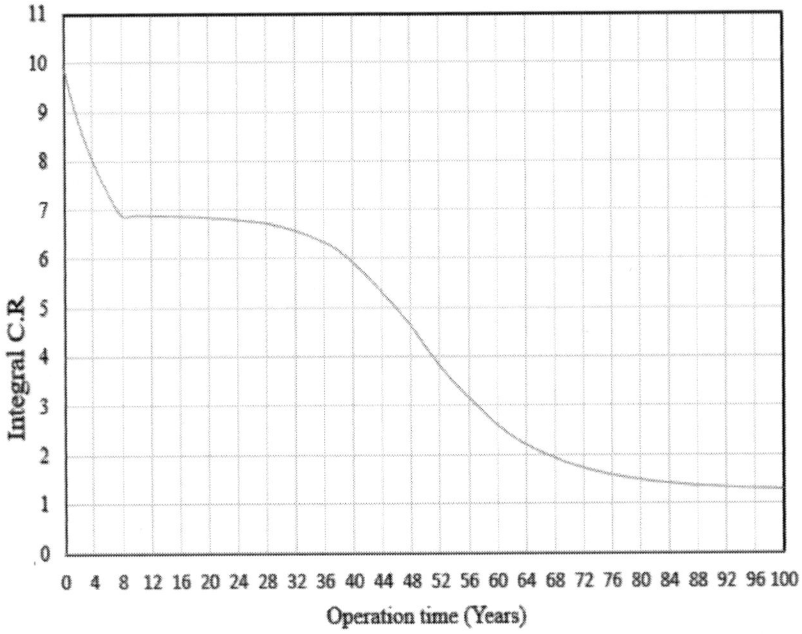

Figure 9. Integral conversion ratio during operation time.

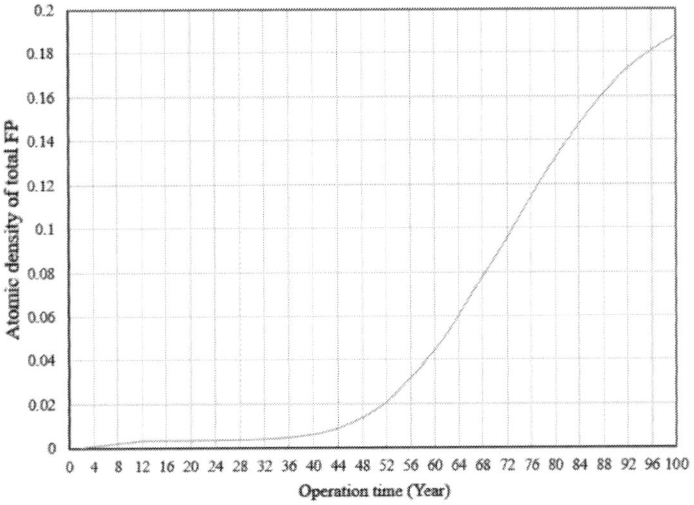

Figure 10. Fission product during operation time.

Figure 11 illustrates the effective multiplication factor (K_{eff}). During the reactor operation, the K_{eff} value is greater than one (supercritical condition), indicating that the reactor can operate using UN and 10% of Th-232 as fuel cycle input. Figure 12 presents U-238 atomic density versus time while Figure 13 shows Pu-239 atomic density versus time of burnup. According to the fission process in the core, U-238 atomic density diminished steadily throughout reactor operation. In the beginning of operation, the Pu-239 atomic density is zero due to that it has not yet been created. After about 34 to 72 years, Pu-239 augmented quickly because of fertile materials transform into fissile materials. After 72 years of operation time, Pu-239 atomic density reduces till the end of life because of reduction of fuel.

IOP Conf. Series: Materials Science and Engineering **1330** (2025) 012018 doi:10.1088/1757-899X/1330/1/012018

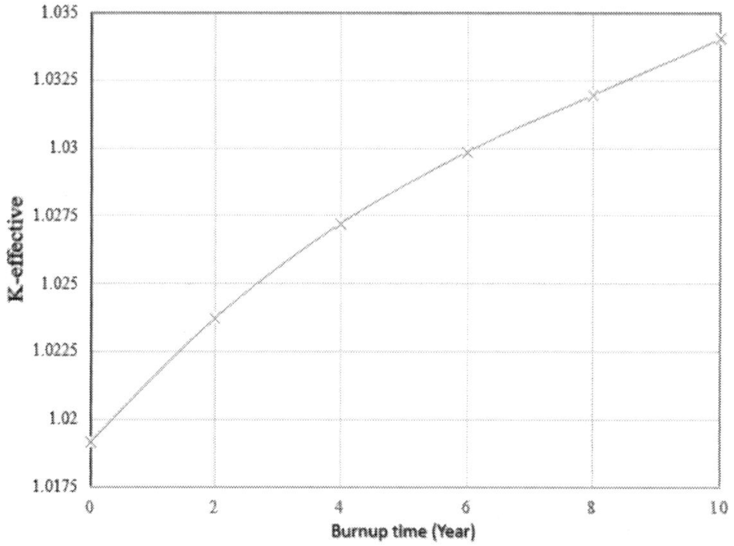

Figure 11. Effective multiplication factor.

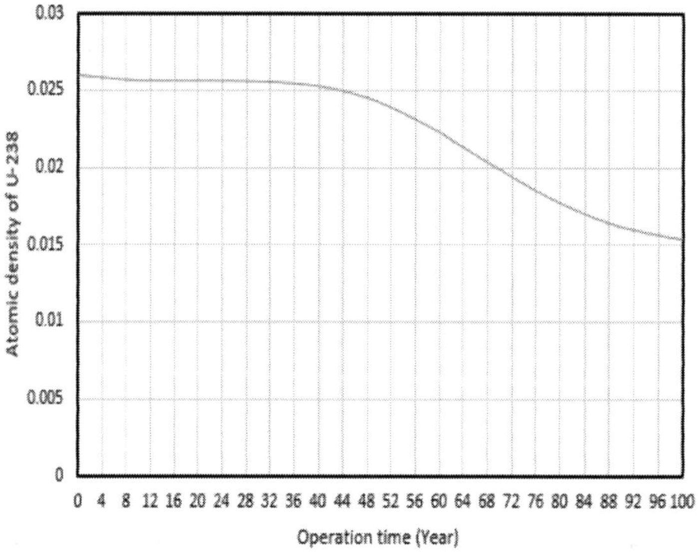

Figure 12. U-238 atomic density.

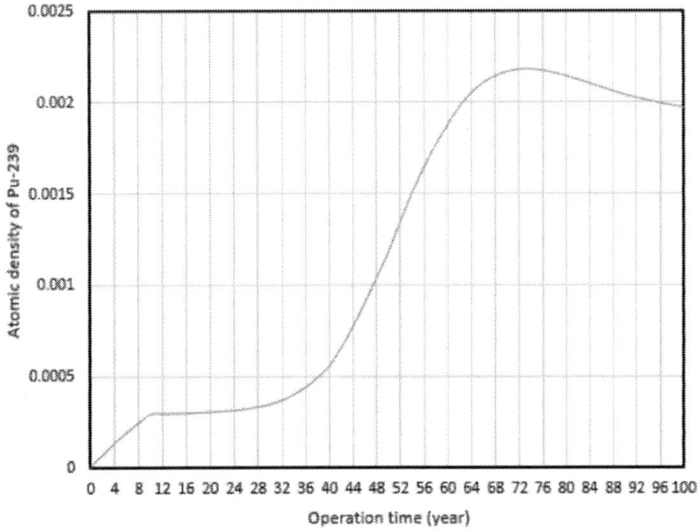

Figure 13. Pu-239 atomic density.

Figure 14 presents U-235 atomic density versus time. During first ten years, U-235 is burned fast owing to the placement of the first region near the mainly active regions. However, after ten years, U-235 burns slowly until roughly fifty years of burnup. After fifty years, the U-235 burning rate increased again till approximately 76 years and then, the U-235 burning process decreased. Figure 15 presents the atomic density of Americium-241 (AM-241) during operation time. In the beginning, the atomic density of AM-241 is approximately zero until about half-life (50 years of burnup) of the reactor owing to the placement of the first region near the more active regions. After 50 years AM-241 increased till the end of reactor operation owing to the conversion of fertile into fissile.

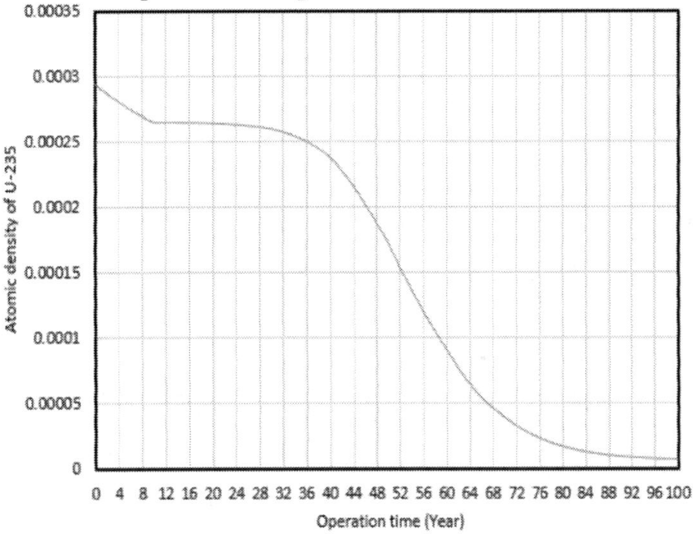

Figure 14. Atomic density of U-235.

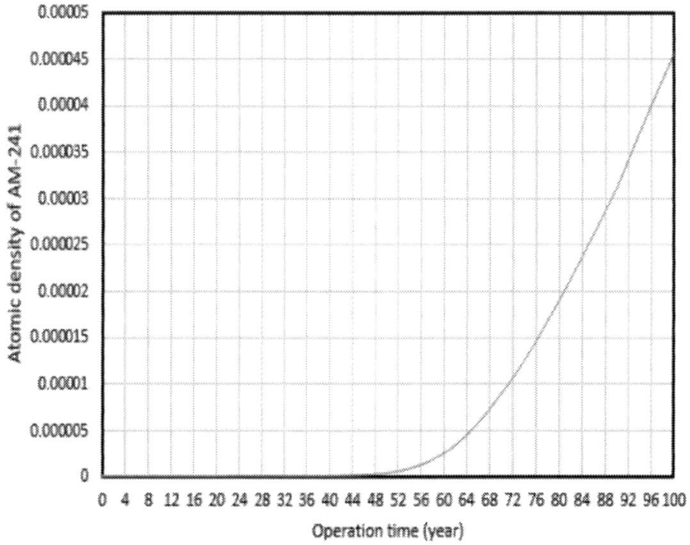

Figure 15. AM-241 atomic density.

Figure 16 presents Pu-240 atomic density versus time. From the beginning, until about 40 years of burnup, Pu-240 accumulation is slow however, after 40 years the accumulation of Pu-240 grows rapidly until the end of life. Figure 17 depicts Pu-242 atomic density versus time. Pu-242 atomic density accumulation is rather slow in the first half of burnup time but accelerates rapidly in the second half of burnup time.

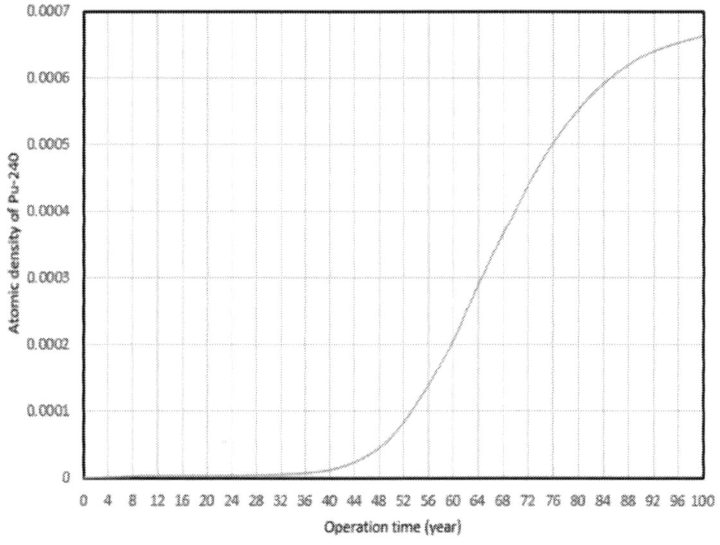

Figure 16. Pu-240 atomic density.

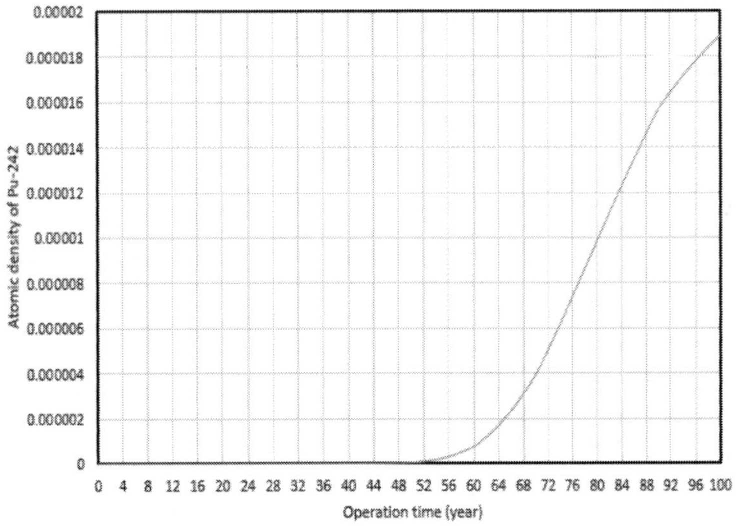

Figure 17. Atomic density of Pu-242.

Figure 18 presents axial relative power density distribution in the beginning of life, in the middle of life and in the end of life. The distribution moved toward fuel regions with higher U-238 and the relative power density tends to be smaller in the end of life. The reflector region covers 70 cm on the top, bottom, and right, hence the relative power density distribution is zero. At the beginning of life, the peak power occurs approximately at 142 cm from zero (9th region). This is due to the first region placed near to the tenth region. On the other hand, at the end of life, the highest relative power appears at about 160 cm (8th region).

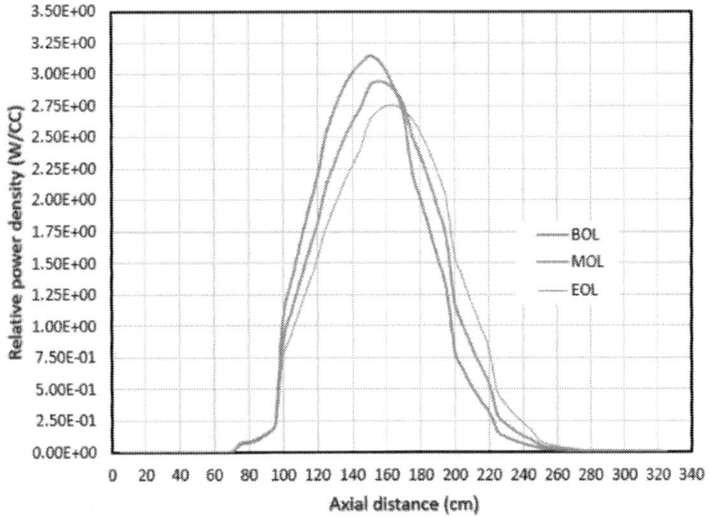

Figure 18. Relative power density.

4. Conclusion

Burnup performance of modified CANDLE shuffling in the axial direction on gas cooled fast reactor with UN-Th fuels has been examined. The neutron aspects such as integral conversion ratio, infinite multiplication factor, burnup level, power densities, atomic densities and effective multiplication factor have been employed to analyze the working principle of reactor design. The thermal power reactor is 3000MWt, and helium gas andTH-9 have been selected as coolant and cladding materials respectively. The effective multiplication factor is greater than unity, indicating that the reactor can operate throughout the burnup period using UN and Th as fuel cycle input. The burnup level at the end of operation is approximately 30.29% HM. The modified CANDLE strategy allows for continuous reactor operation with UN-Th fuels.

References

[1] Hong P, Tran H and Sekimoto H 2016 Energy Burnup performance of small-sized long-life CANDLE high temperature gas-cooled reactors with U – Th – Pa fuel End of Life *Ann. Nucl. ENERGY* **91** 36–47

[2] Syarifah R D, Investigator P, Aula M H, Ardianingrum A, Janah L N and Maulina W 2022 Comparison of Thorium Nitride and Uranium Nitride Fuel on Small Modular Pressurized Water Reactor in Neutronic Analysis Using Srac Code *Eastern-European J. Enterp. Technol.* **2** 21–8

[3] Ariani M, Su'ud Z, Waris A, Khairurrijal, Monado F and Sekimoto H 2012 The feasibility study of small long-life gas cooled fast reactor with mixed Natural Uranium/Thorium as fuel cycle input *AIP Conf. Proc.* 1448 59–64

[4] Shafii M A 2021 Neutronic analysis of sodium-cooled fast reactor design with different fuel types using modified CANDLE shuffling strategy in a radial direction 1–12

[5] Widiawati N, Su'ud Z, Irwanto D, Permana S, Takaki N and Sekimoto H 2021 Enhancing the performance of a long-life modified CANDLE fast reactor by using an enriched 208Pb as coolant *Nucl. Eng. Technol.* 53 423–9

[6] Sekimoto H and Yan M 2007 Design study on small candle reactor *13th Int. Conf. Emerg. Nucl. Energy Syst. 2007, ICENES 2007* 1 489–97

[7] Nguyen H, Nishiyama J and Obara T 2020 Burnup Performance of CANDLE Burning Reactor Using Sodium Coolant *Nucl. Sci. Eng.* 00 1128–42

[8] Peters M 2021 Core Configuration Analysis for Modular Gas- cooled Fast Reactor (GFR) using OpenMC *J. Phys. Conf. Ser.*

[9] Su'ud Z, Miftasani F, Ilham, Sarah A, Ariani M, Sekimoto H, Waris A and Sidik P 2017 Design Study of Small Modified CANDLE based Long Life Gas Cooled Fast Reactors *Energy Procedia* 131 6–14

[10] Su'ud Z and Sekimoto H 2012 Design study of long-life Pb-Bi cooled fast reactor with natural uranium as fuel cycle input using modified CANDLE burn-up scheme

[11] Ariani M, Su'ud Z, Monado F, Waris A, Arif I, Ferhat A and Sekimoto H 2013 Optimization of Small Long Life Gas Cooled Fast Reactors With Natural Uranium as Fuel Cycle Input **261** 307–11

[12] Sekimoto H 2014 Six Requirements for Nuclear Energy System and Candle Reactor 3–7

[13] Ndayiragije J P 2024 Design study of gas-cooled fast reactor with natural uranium as fuel employing modi fi ed CANDLE shu ffl ing strategy in the axial direction *Kerntechnik* 1–11

[14] Ndayiragije J P, Su'ud Z and Waris A 2024 Design study of a 450MW thermal Modified CANDLE fast reactor using helium gas as a coolant *J. Phys. Conf. Ser.*

[15] Ndayiragije J P, Su'ud Z and Waris A 2024 A comparison of neutronic studies on 400 MW thermal fast reactor with modified CANDLE burn-up schemes using helium gas, lead bismuth eutectic and liquid sodium coolants *Int. J. Nucl. Energy Sci. Technol.* 17 66–84

[16] Widiawati N, Su'ud Z, Irwanto D, Permana S, Takaki N and Sekimoto H 2022 Design study of 208Pb-Bi eutectic-cooled reactor with natural uranium as fuel cycle input with radial fuel

shuffling *Ann. Nucl. Energy* 171 109003
[17] Ridhani M R, Aziz N and Basuki K T 2016 Analysis Comparison Npp Thorium Fuel As an Alternative Energy Sources 1 109–15

IOP Conf. Series: Materials Science and Engineering **1330** (2025) 012019 doi:10.1088/1757-899X/1330/1/012019

Comparative Study of Natural Background Radiation Between Mamuju Regency, West Sulawesi and West Java

Zulfahmi[1*], Sidik Permana[1], Adi R. A. Abdullah[1], Imam Ghazali Y[2], Ismail Humolongo[1], Haryo Seno[3]

[1] Doctoral Program of Nuclear Engineering, Bandung Institute of Technology, Bandung, Indonesia
[2] Joint Doctoral Program for Sustainability Research, The University of Electro-communications, Tokyo, Japan.
[3] Indonesia's National Research and Innovation Agency, Bandung, Indonesia

*E-mail: psidik@itb.ac.id

Abstract. Some areas have unique radiation levels, caused by the high density of buildings or deposits of radioactive minerals. Mamuju Regency is the area with High Natural Background Radiation (HNBR) despite the low density of the population, while West Java possessing low-level radiation is the area with the largest population in Indonesia. Measurements of dose levels in these two areas were carried out for two weeks covering nine villages in Mamuju and nine Regencies in West Java. The method used is random sampling to collect data from one meter on the surface, while the tool used was the Ludlum model 19 series 8 Analog Survey meter. The results of radiation measurements were 2.208 ± 0.1420 mSv/year from 180 data in the Mamuju Regency and 0.3086 ± 0.0044 mSv/year in 103 data in West Java. The big difference between the two is due to the presence of radioactive element-carrying minerals in the Mamuju Regency, while in West Java areas there are lack of radioactive elements mineralization.

1. Introduction

Human interaction with radiation has existed since the first-time humans were on earth. The radiation from radioactive substances that humans receive comes from cosmic rays that bombard the earth every day, from rocks and soil where radioactive elements are deposited throughout the continental crust, and from human activity itself.[17,1]

The radiation dose received by humans varies from place to place. Some areas naturally receive radiation with high intensity, these areas are called High Natural Background Radiation Areas (HNBRA), and areas with low levels of radiation are called Low-Level Natural Background Radiation Areas. In the world, several regions are known to have naturally high natural radiation dose values, including Ramsar (Iran), Yangjiang (China), Kerala, Tamil Nadu, Orissa Beach (India), Gauarapari, Pocos de Caldas, Araxa, and Tapira (Brazil).[1] In Indonesia, Mamuju Regency is known to be the area with the highest radiation dose rate in Indonesia, reaching an average of above 300 nSv/hour (0.3 μSv/hour), even in one village, Takandeang village, it reached 2,844 nSv/hour (2.84 μSv/hour),[18] the dose rate values in Mamuju range from 982 to 26,769 nSv/hour.[10] While the average value of radiation dose exposure globally is 2.4 mSv/year (273 .8 nSv/hr or 0.2737 μSv/hour).[17] This means that in several places in the world, the population

IOP Conf. Series: Materials Science and Engineering **1330** (2025) 012019 doi:10.1088/1757-899X/1330/1/012019

receives radiation exposure higher than in normal conditions. Of course, it is a concern for the health impact on humans considering the nature of ionizing radiation which can hurt human health.

The radiation dose received by humans can be known by measuring the external radiation level of gamma radiation. By measuring the level of gamma radiation in an area, we can obtain data on the distribution of gamma radiation dose rates in an environment. Radiation dose rate mapping is carried out for various purposes, for example for environmental gamma radiation monitoring studies in areas where a nuclear power plant is to be built,[6, 16] monitoring radiation due to nuclear reactor accidents,[19] monitoring the effects of environmental changes on changes in radiation,[11] providing a database for the distribution of radiation rates in a regional area such as a country or island.[3,6,11]

Gamma radiation dose rate mapping is carried out by measuring radiation dose above ground level using ground survey, car-borne survey, and air-borne survey methods. A survey meter is a tool used to measure radiation dose rate directly which must be portable so that it can be used to measure in all fields. To measure the dose rate of gamma radiation, this tool is usually called a gamma spectrometer. To detect environmental radiation, a gamma survey meter uses a detector. One of the detectors commonly used in field surveys is the NaI(Tl) scintillation detector, while semiconductor detectors such as the HPGe detector are commonly used in laboratories to measure radionuclide levels from soil samples.

The comparison between these two locations aims to examine the diversity of radiation exposure and its underlying causes. This analysis serves to update the existing database for monitoring environmental quality changes and their subsequent effects, ultimately contributing to the formulation of environmental policies. It is crucial to consider that Java, the most densely populated island in Indonesia, is projected to house nuclear power plants in the future. Conversely, Sulawesi possesses regions characterized as High Natural Background Radiation Areas (HNBRA), consequently exhibiting relatively higher radiation exposure levels. [3,7,9,10,15]

2. Material and Method

2.1 Sampling Site

Mamuju is located in the West Sulawesi region, consisting of lowlands, especially areas on the coast to the mountains with an altitude of 1600 m in the eastern region. The measurement area is ±100 km^2 so data collection was carried out along the road from Taan village in the southern part of Tapalang District to the northwestern part of Ahu Village and in the northern part of Tapalang, Takandeang Village. Covering ten villages or sub-districts plus additions. While West Java is an area with the highest population in Indonesia. Data acquisition was carried out along the road from Bandung City to Cirebon City. Because the location is so wide, the data collection is only devoted to the eastern part of West Java, which includes 10 cities/regencies with an area of ±5500 km^2.

IOP Conf. Series: Materials Science and Engineering **1330** (2025) 012019 doi:10.1088/1757-899X/1330/1/012019

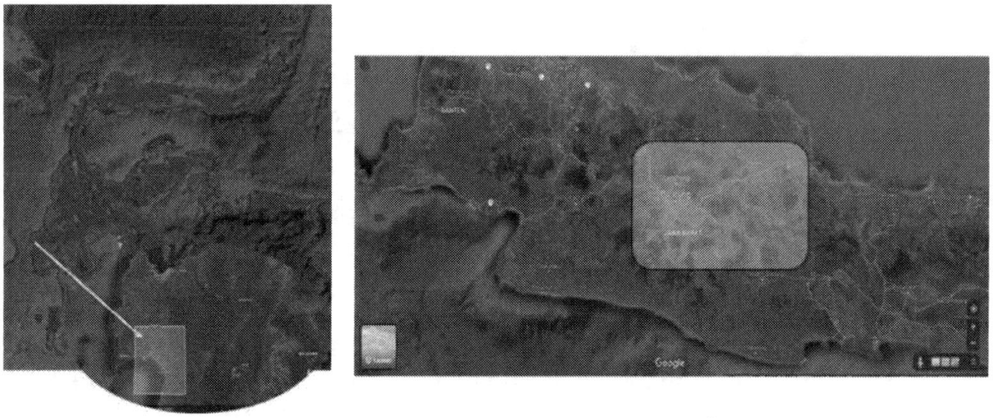

a. b.

Figure 1. Measurement Locations, a. Mamuju Regency b. West Java

2.2 Tools

The Ludlum model 19 series 8 is a scintillation-based radiation survey meter with a detector size of 2.5x2.5 cm NaI (Tl) with a measuring range of 0-25 μ Sv/hour. The sensitivity of this tool (137Cs gamma) is around 175 CPM per 1x10-3 μSv/hour (175 CPM per μR/hour).[12] Because this tool is an analog device that still uses needle observation in every measurement. Measurement uncertainty is determined by the smallest graduation on the instrument's measurement scale for each respective measurement range.

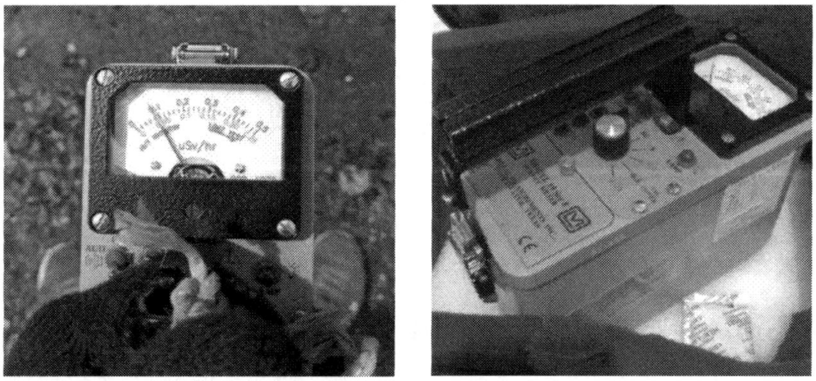

Figure 2. Survey meter gamma radiation Ludlum Model 19 Series 8

To obtain measurement position recording data for each sample measured, a Garmin Etrex 10 branded Global Positioning System tool is used.

2.2 Method

For the Mamuju area, data acquisition was carried out at each point with a grid of 250 m points except for data at Mount Ahu, data collection was carried out as much as possible due to the high value of the dose rate. Measurements were taken along the road from the border of Taan Village to Ahu and Takandeang Villages, plus several control points around the shoreline and measurement point further inland. Measurements for each data were carried out over 1 meter of the ground surface,[2,8] with a systematic random sampling technique that was taken on foot or using a motorcycle.

IOP Conf. Series: Materials Science and Engineering **1330** (2025) 012019 doi:10.1088/1757-899X/1330/1/012019

Meanwhile, in West Java, due to such a large area, the measurement grid is adjusted by 5 to 10 km per sample measuring point. The technique used is the same as before and is measured above 1 meter. During the measurement, the location was reached by car.

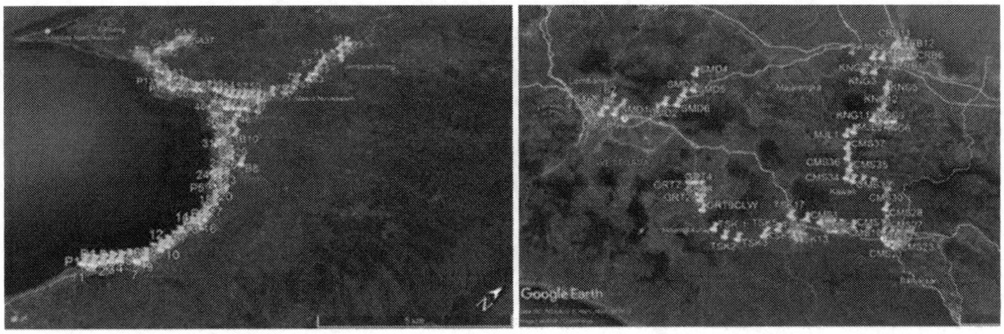

a b

Figure 3. Measurement points a. Mamuju Regency, b. West Java

3. Result and Discussion

3.1 Mamuju

The results of measuring the radiation dose rate in Mamuju is shown in table 1 below:

Table 1. Measurement Result in Mamuju Regency

No	Location	Number of measurement points	Average Gamma Dose Rate (μSv/h)
1	Taan Village	16	0,1708 ± 0,0125
2	Urban Village of Kasambang	19	0,2099 ± 0,0125
3	Urban Village of Galung	15	0,2283 ± 0,0125
4	Tampalang Village	12	0,2510 ± 0,0125
5	Orobatu Village	12	0,2636 ± 0,0125
6	Pasabu Village	19	0,2197 ± 0,0125
7	Ahu Village	5	0,2450 ± 0,0125
8	Ahu Hill	37	0,5699 ± 0,0500
9	Takandeang Village	25	0,2625 ± 0,0125
10	Taan Beach-Ahu	20	0,0981 ± 0,0125
Total		180	Average 0,2519 ± 0,0162

The results of the total measurements show that the average value of the radiation dose rate is 0.2519 ± 0.0125 μSv/hour or 2.208 ± 0.1420 mSv/year including control data in the beach. Since we measure radiation on the land, we can avoid the beach data. The average dose rate on the land is 0.2689 ± 0.0162 μSv/hour or 2,3571± 0.1420 m Sv/hour, with the lowest dose value on the coast of Kasambang Village at 0.0250 ± 0.0125 μSv/hour and the highest in the Ahu Hill area with a value of 0.8000 ±0.05 μSv/hour.

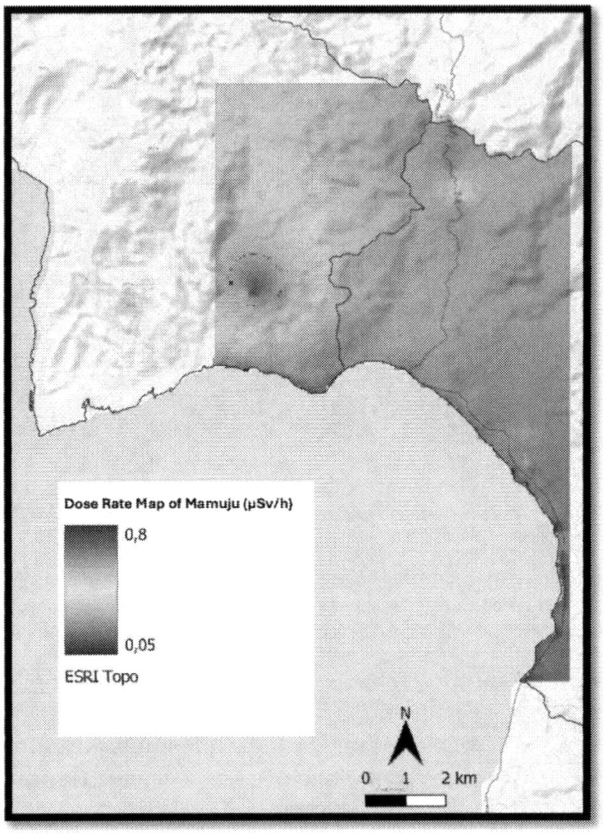

Figure 4. Contour Map of the dose rate (µSv/hour) measured in Tapalang Sub-District and West Tapalang Sub-district, Mamuju Regency, West Sulawesi.

Areas with elevated radiation dose rates are visually represented using a color gradient, with red indicating the highest values. The highest value, 0.8 µSv/hour (800 nSv/hour) in Ahu village is 2 times more than the global average value of radiation exposure received by humans of 2.4 mSv/year (0.273 µSv/hour).[17] This is related to the distribution of radioactive minerals which, based on BATAN's research, were formed in the villages of Takandeang and Ahu villages.[7] In Tapalang District there are indications of uranium and thorium mineralization. The uranium minerals are gummite and autunite, which were identified in the villages of Takandeang and Botteng.[9] In the village of Ahu it was identified that there was uranium mineralization,[4] this is what shows the red color on Mount Ahu, which can be seen on the contour in Figure 4.

In other villages, shown in green, the radiation dose rate is 0.2 – 0.5 µSv/hour, although not as high as in Ahu village and Takandeang village, but still higher than the average island radiation dose rate. Java, namely 0.46 mSv/year (0.052 µSv/hour).[5]

3.2 West Java

The value of the average radiation dose rate in West Java is shown in the following table:

Table 2. Average radiation dose rate at all measurement points in West Java Regencies

No	Location	Number of measurement points	Average gamma dose rate (μSv/h)
1	Bandung	6	0.0325 ± 0,0005
2	Sumedang	6	0.0342 ± 0,0005
3	Garut	8	0.0194± 0,0005
4	Tasikmalaya	16	0.0117± 0,0005
5	Ciamis	24	0.0168± 0,0005
6	Banjar	12	0.0145± 0,0005
7	Majalengka	2	0.0325± 0,0005
8	Cirebon	15	0.0383± 0,0005
9	Kuningan	14	0.0378± 0,0005
Total		103	Average 0,0352± 0,0005

The average dose rate obtained was 0.0352 ± 0.0005 μSv/hour or 0.3086 ± 0.0044 mSv/year.

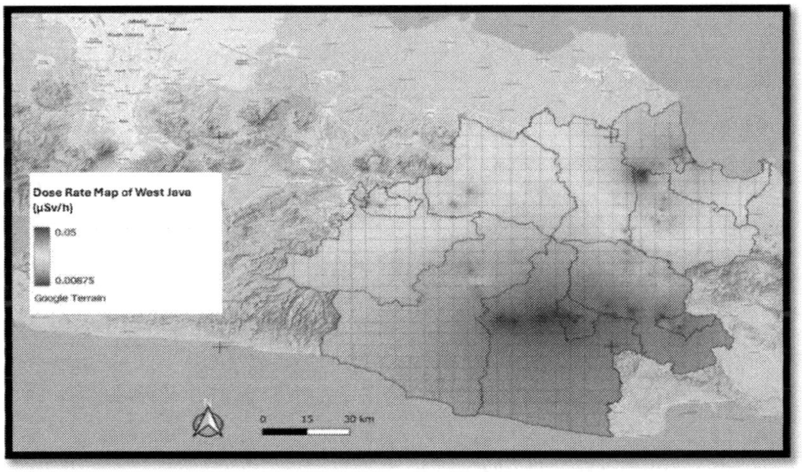

Figure 5. Contours of radiation dose rates at all measuring points in West Java

Measurements in West Java are all carried out on highways and in urban centers. In Bandung, Tasikmalaya City, and the Cirebon area, the influence of the radiation dose rate is caused by buildings and human activities. In areas such as Sumedang, Ciamis, Banjar, and Kuningan measurements are carried out around villages, so the soil type may be a control for radiation. The value of the radiation dose rate in West Java is controlled by the type of soil and buildings around the measuring point.

3.3. Comparison of Dose Rate from Mamuju Regency and West Java

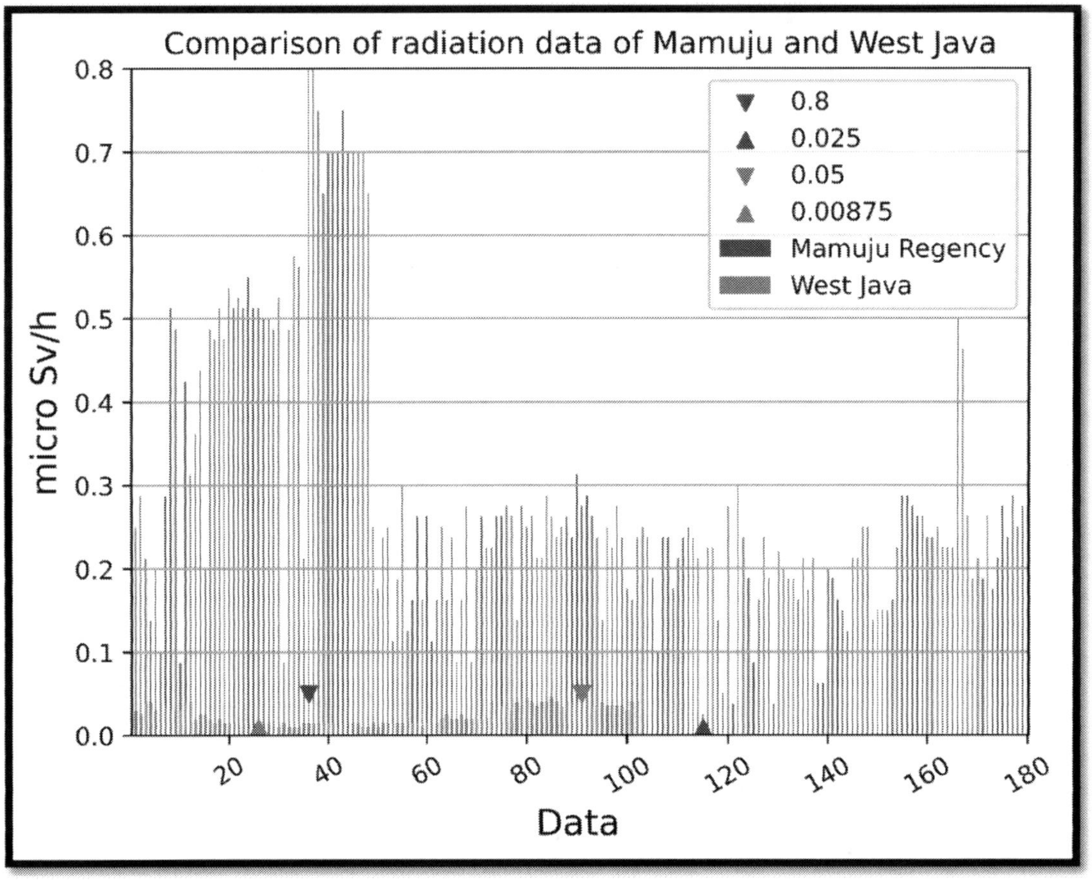

Figure 6. Comparison of the measured radiation dose rate values in Mamuju and West Java Regencies

The average dose rate data for Mamuju Regency is 0.2519 ± 0.0125 μSv/hour or 2.208 ± 0.1420 mSv/year, 8 times the average value of radiation dose rate in West Java 0.0352 ± 0.0005 μSv/hour or 0.3086 ± 0.0044 mSv/year. In previous studies, the average value of the radiation dose rate in Mamuju Regency was 300 nSv/hour or 0.3 μSv/hour,[14] while the dose rate values in Java and West Java were 0.0386 μSv/hour.[5]

In Mamuju, high dose rates were concentrated by radioactive minerals, uranium (ppm eU) up to 2,914.30 ppm, thorium (ppm eTh) up to 15,511 ppm eTh and potassium (%K) up to 98.80% (Sukadana, 2021). Soil types include Entisol, Histosol, Ultisol, and mostly Inceptisol.[13] In contrast to Mamuju Regency, where the control of the dose rate value is due to the concentration of radioactive element mineralization. In West Java, no dose rate anomaly was found so radiation dose rate control was more due to various things such as buildings, human activities, use of fertilizers, and, terrestrial sources, namely rocks and soil with low levels of radioactive elements.

In various soil types in Java, the presence of low radionuclide elements was found. Uranium concentration (U-238) in grumosol, red-yellow podzolic, regosol, latosol, andosol, litosol and regosol soil types up to ≥ 2 ppm. Thorium concentration (TH-232) in regosol, grumosol, latosol, andosol, alluvial, litosol, red-yellow podzolic, red Mediterranean and yellow podzolic concentrations ranged from <1 ppm to ≥ 7 ppm. Potassium concentration (K-40) in red yellow podzolic, alluvial, regosol, grumosol, latosol, andosol soil types up to ≥ 1 %.[5] Because no samples

IOP Conf. Series: Materials Science and Engineering **1330** (2025) 012019 doi:10.1088/1757-899X/1330/1/012019

were taken for radionuclide analysis from various soil types, we only compared them with soil types from the measured areas as shown in table 3 below:

Table 3. Types of Soil in the West Java Region

No	Location	Soil type	Source
1	Bandung City	Alluvial and Andosol	Jabarprov.go.id, 2021
2	Sumedang Regency	Latosol, Grumosol, Andosol, and Regosol.	Sumedang dalam angka, 2015
3	Garut Regency	alluvial, andosol association, litosol association, mediteran association, podsolik association, dan regosol association	garutkab.go.id, 2021
4	Ciamis Regency	Latosol, Podzolik, Aluvial, complex Resina and Grumusol	RPJMD Kabupaten Ciamis Tahun 2014-2019
5	Tasikmalaya Regency	Associations of gray regosol, gray-brown regosol, litosol, and reddish latosol	RPIJM Tasikmalaya
6	Banjar Regency	Organosol soil, glei humus with alluvial and physiographic parent material, red-yellow podzolic complex soil, and laterite. Latosol soil red yellow complex podzolic soil, latosol	RPIJM Banjar, 2021
7	Majalengka Regency	Latosol, Podsolik, Grumosol, Aluvial, Regosol, Mediteran, and their associations	westjavainc.org
8	Kuningan Regency	podzolic soil, regosol dan latosol and alluvial sediment	http://repository.umy.ac.id
9	Cirebon Regency	alluvial, grumosol dan latosol soil	cirebonkota.go.id

The average total radiation exposure in all countries is around 2.4 mSv/year including the average of cosmic radiation, terrestrial gamma radiation, inhalation (mainly radon), and ingestion.[17] The annual dose rate in Mamuju Regency is higher than in West Java, and in one area it can be seen in Figure 7. that it exceeds the world's average radiation dose rate in the Mount Ahu area, although it does not include the results of radiation measurements due to radon and others.

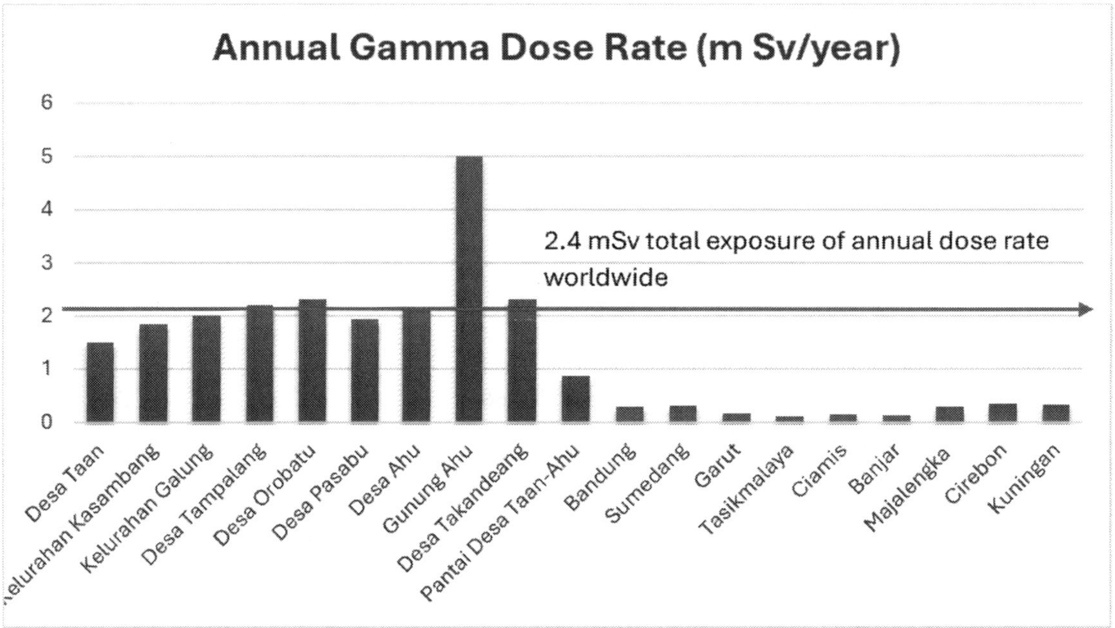

Figure 7. Gamma radiation dose rate in a year at each location due to gamma external exposure

The United Nations Scientific Committee on the Effects of Atomic Radiation (UNSCEAR) reported that the population-weighted average of external exposure dose rates from terrestrial gamma radiation indoors and outdoors is 0,084 μSv/hour and 0,059 μSv/hour. West Java can be categorized as a low radiation exposure area, while Mamuju exhibits high radiation levels and is classified as a High Natural Background Radiation Area (HNBRA). However, several other regions in Indonesia are also known to have relatively high radiation exposure, including the Bangka Belitung Islands, Kalan in West Kalimantan, and areas within North Sumatra.[3,22]

Table 4. Comparison between Present Study of radiation Exposure[23,24,25]

Country	HNBRA	Area characteristic	*Average annual internal dose (mSv)*	*Average annual external dose (mSv)*	*Average annual effective dose (mSv)*
Brazil	Guarapari	*Monazite sands*	–	–	7
China	Yangjiang	*Monazite particles*	4.1	2.1	6.4
India	Kerala	*Monazite sands*	2.4	4.5	6.9
Iran	Ramsar	*Sulfurous hot springs and Ra^{226}*	4.2	6	10.2
Indonesia	Mamuju	*Volcanic rocks and basalt rock type containing Th^{232} and U^{238}*	26.9	5	32
	Bangka Belitung		1,17	0,69	
	West Kalimantan			0,69	
	World average		–	0.48	24

Mamuju Regency warrant attention. However, recent research indicates no significant correlation between residing in a High Natural Background Radiation Area (HNBRA) and increased cancer incidence. Nonetheless, a significant association was found with mortality attributed to non-cancer diseases. This finding necessitates further investigation, considering the multitude of confounding factors. In Mamuju, the indigenous population appears to have developed immune systems adapted to the elevated radiation levels. Nonetheless, this adaptation remains a concern for non-native residents. [20,21]

4. Conclusion

Based on the measurement results for each location, the average radiation dose in the eastern part of West Java was 0.3086 ± 0.0044 mSv/year. This means that the acceptance of radiation exposure by the people of the eastern part of West Java is still below the world's average exposure dose rate of 2.4 mSv/year excluding other radiation exposure. The highest dose rates were measured at two locations in the Kab. Kuningan of 0.0500 ± 0.0005 μSv/hour. This can be proven by mapping by the QGIS application into a contour shape of radiation exposure which shows the northern part of the total measurement area showing red color as a visualization that the dose rate value is higher than the southern part in the eastern part of West Java. For the district, Mamuju dose rate measurements from Taan-Takandeang Village obtained an average radiation dose rate of 0.2853 ± 0.0125 μ Sv/hour with a range of 0.0250 ± 0.0125 μSv/hour which is on the coast to 0.8000 ± 0.0500 μSv/h on Mount Ahu. If referring to the value of exposure to the world's average dose rate, then the measured dose rate in Kab. Mamuju is 2.208 ± 0.1420 mSv/year almost get value excluding other radiation exposure. The radiation dose rate value of Kab. Mamuju is about 8 times higher than the east-central part of West Java. Based on previous research in the district. In Mamuju there are radioactive mineral elements in Ahu village, namely gummite, and autunite. It is recommended that future research expand to encompass a wider geographical area within Indonesia and include estimations of the potential implications associated with varying radiation exposure levels. This comprehensive approach would generate valuable reference data for health-related assessments in Indonesia. Additionally, the utilization of more advanced and diverse instrumentation is encouraged to enhance the accuracy and scope of future investigations.

5. Acknowledgments

The work has been supported by the Bandung Institute of Technology in Research program and the team by supporting and advising through the research and publication of this article.

References

[1] Abubakar Sadiq Aliyu dan Ahmad Termisi Ramli. 2015. *Review. The world's high background natural radiation areas (HBNRAs) revisited: A broad overview of the dosimetric, epidemiological and radiobiological issues*. Radiation Measurements. **73** (2015). 51-59.

[2] Beck. H.L., DeCampo, and J., Gogolak, C. 1972. In Situ Ge (Li) NaI(Tl) Gamma-Ray Spectrometry. Health and Safety Laboratory. US Atomic Energy Commission, HASL258.

[3] Dadong Iskandar, Syarbaini, Sutarman, Bunawas, Kusdiana 2007. *Mapping the Outdoor Gamma Dose Rate in Indonesia. Abstract.*

[4] Frida Rosidatul Mu'awanah, Bambang Priadi, Widodo, I Gde Sukadana, Rian Andriansyah. 2018. *Mobilitas Uranium pada Endapan Sedimen Sungai Aktif di Daerah Mamuju, Sulawesi Barat.* Jurnal Eksplorium Batan. Vol **39** : 95-104.

[5] Gatot Suhariyono, Buchori dan Dadong Iskandar. 2007. *Laju Dosis Radiasi Gamma Lingkungan di Pulau Jawa.* Jurnal BATAN. ISSN 0216-3128.

[6] Gatot Suhariyono, Untara, dan June Mellawati. 2021. *Dose Rate Of Environmental Gamma Radiation In Some Location Of West Kalimantan As a Site Candidate Of Nuclear Power Plant.* Jurnal ResearchGate. Publikasi Online 11 November 2021.

[7] H. Syaful, I.G. Sukadana, and A. Sumaryanto. 2014. *Radiometric Mapping for Naturally Occurring Radioactive Materials (NORM) Assesment in Mamuju, West Sulawesi.* Journal of Atom Indonesia. Vol **40** : 33-39.

[8] IAEA. 2003. *Guidelines for radioelement mapping using gamma-ray spectrometry data.* IAEA Tecdoc 1363. Vienna. Austria.

[9] I Gde Sukadana, Agung Harijoko, Lucas Donny Setijadi. 2015. *Tataan Tektonika Batuan Gunung Api di Komplek Adang, Kabupaten Mamuju, Provinsi Sulawesi Barat.* Eksplorium. Volume **36** No. 1. Mei 2015: 31-34.

[10] I Gde Sukadana, I Wayan Warmada, Agung Harijoko, Frederikus Dian Indrastomo, Heri Syaeful1. 2021. *The Application of Geostatistical Analysis on Radiometric Mapping Data to Recognized the Uranium and Thorium Anomaly in West Sulawesi, Indonesia.* IOP Conf. Series: Earth and Environmental Science 819 (2021) 012030.

[11] Kazumasa Inoue, Masahiro Fukushi, Tan Van Le, Hiroshi Tsuruoka, Shogo Kasahara, and Veerasamy Nimelan. 2020. *Distribution of gamma radiation dose rate related with natural radionuclides in all of Vietnam and radiological risk assessment of the built-up environment.* Nature. Scientific Report. **10**. 12428. https://doi.org/10.1038/s41598-020-69003-0.

[12] Ludlum Manual. 2019. Ludlum Model 19 Micro R Meter. https://ludlums.com/products/all-products/product/model-19#documents.

[13] Masnawi, Achmad. 2015. *Analisis Sektor dan Komoditas Unggulan di Kabupaten Mamuju Sulawesi Barat.* Sekolah Pascasarjana Institut Pertanian Bogor. Bogor

[14] PTBGN-BATAN. 2018. *Laporan Kinerja Pusat Bahan Galian Nuklir Tahun 2017.* Badan Tenaga Nuklir Nasional.

[15] Sukadana dkk, 2021, The Application of Geostatistical Analysis on Radiometric Mapping Data to Recognized the Uranium and Thorium Anomaly in West Sulawesi, Indonesia. IOP Conference Series: Earth and Environmental Science, (2021) 012030 IOP Publishing doi:10.1088/1755-1315/819/1/012030.

[16] Thinova. L., T. Checak, Kluson. J., Troyek. T. 2006. *Use Of Gamma Spectrometry Method for Environmental Monitoring in the Area of NPP.* Journal Of Physics. IOP ebooks. Ser. **41** 071.

[17] UNSCEAR. 2000. *United Nations Scientific Committee on the Effects of Atomic Radiation: Sources, Effects and Risks of ionizing Radiation:Annex B, Exposures from Natural Radiation Sources* vol. I. United Nations Publication, 2000.

[18] UNSCEAR. 2008 *Report to the General Assembly: Sources and Effect of Ionizing Radiation* vol. I. United Nations Publication, 2010.

[19] Yukihisa Sanada a, Kazuya Yoshimura a, Yoshimi Urabe b, Takeyuki Iwai b, Estiner W. Katengeza. 2020. *Distribution map of natural gamma-ray dose rates for studies of the additional exposure dose after the Fukushima Dai-ichi Nuclear Power Station accident.* Journal of Environmental Radioactivity. 223-224. https://doi.org/10.1016/j.jenvrad.2020.106397.

[20] Darlina, Rahardjo, T., Surniyantoro, H. N. E., Rahajeng, N., & Syaifudin, M. 2022. *The Effects of High Level Natural Radiation in Mamuju - Indonesia on the Immune System of Its Residents.* Atom Indonesia, *48*(1), 21–27. https://doi.org/10.17146/AIJ.2022.1118

[21] Kranrod, C., Kudo, H., & Tokonami, S. (2024). *What can we learn from high background radiation area (HBRA) studies in three Asian countries: India, China and Indonesia? Radiological aspects in various HBRAs.* Journal of Radiation Research, *65*(Suppl 1), i32. https://doi.org/10.1093/JRR/RRAE032

[22] Gogoi, P. P., Phukan, S., & Barooah, D. (2024). Terrestrial gamma radiation dose (TGRD) rates and radiological risk assessments in Tiru region of the Naga Schuppen Belt, India. *Atmospheric Pollution Research, 15*(11), 102294. https://doi.org/10.1016/J.APR.2024.102294

[23] Syarbaini, S., Iskandar, D., & Kusdiana, K. (2016). PERKIRAAN DOSIS RADIASI YANG DITERIMA PUBLIK DI PROVINSI KEPULAUAN BANGKA BELITUNG. *Jurnal Ekologi Kesehatan, 14*(4). https://doi.org/10.22435/JEK.V14I4.4710.318-333

[24] Wahyudi, W., Kusdiana, K., Wiyono, M., & Iskandar, D. (2019). ANALISIS DOSIS RADIASI ALAM DARI PAPARAN RADON DAN RADIASI GAMMA DI RUMAH PENDUDUK DI KALIMANTAN BARAT. *GANENDRA Majalah IPTEK Nuklir, 22*(2), 63. https://doi.org/10.17146/GND.2019.22.2.5094

[25] Nugraha, E. D., Hosoda, M., Tamakuma, Y., Kranrod, C., Mellawati, J., Akata, N., & Tokonami, S. (2021). A unique high natural background radiation area in Indonesia: a brief review from the viewpoint of dose assessments. *Journal of Radioanalytical and Nuclear Chemistry, 330*(3), 1437–1444. https://doi.org/10.1007/S10967-021-07908-4/TABLES/2

Analysis of Radionuclide Content in Soil Samples Using the X-Ray Fluorescence Spectroscopy (XRF) Method

Ismail Humolungo[a]*, Sidik permana[a,b], Adi R.A Abdullah[b], Zulfahmi[b], Haryo Seno[c], Alan Maulana[c]

[a]Doctoral Program in Nuclear Engineering Department, Bandung Institue of Technology, Jl. Ganesha No. 10, Bandung, Indonesia.
[b]Physics Department, Bandung Institue of Technology, Jl. Ganesha No. 10, Bandung, Indonesia
[c]Indonesia's National Research and Inovation Agency. Jl. Tamansari No. 71, Bandung, Indonesia.

*Corresponding authors.
E-mail addresses: ihumolungo41@gmail.com

ABSTRACT

The high average value of natural gamma radiation dose rate in Mamuju Regency indicates that the uniqueness is related to the high concentration of radioactive elements of Uranium, Thorium, and Potassium deposited in the rock. In this study, X-Ray Fluorescence tests were carried out with Energy Dispersive X-Ray Fluorescence equipment on 5 soil samples from Mamuju Regency. Measurements were made on powder and *pellet* specimens. The results showed that the average concentration of radioactive elements uranium (U) was 712.17 ppm and 3581.45 for Thorium (Th), while for Potassium (K) was 9127.75 ppm. The thorium element appears to be more dominant with the Th/U ratio reaching 3.99 which is higher than the average Th/U ratio in the Earth's crust of 3.70. A positive correlation between U and Th was obtained with a correlation coefficient value of 0.60, and a positive correlation was also shown by the relationship between U and Th, indicating the enrichment of U along with the enrichment of Th. Analysis of XRF data and dose rate data shows that there are two locations that show a very close relationship between the dose rate and the concentration of radioactive elements in the sample, while in other samples the influence of land use into gardens or highway strips affects the measured dose rate.

***Keywords**: XRF, Dose Rate, Natural Radiation, Soil, Mamuju*

1. Introduction

Natural radioactivity on earth consists of radioactive elements naturally stored in rocks, which then weather into soil, and radioactive elements will stay there in minerals. According to Zhang et al. [1], the radiation detected on a certain piece of land may change due to human disturbance or without disturbance. Measurement of dose rate values with detectors such as Geiger Muller is often done to reduce the amount of radiation in a given area. And the results of the hunt will be diversified by taking rock and soil samples to be tested in the laboratory to determine the content of elements, compounds, or minerals.

A commonly used method is the XRF method, which uses the interaction of X-Rays with matter to determine the chemical composition of a material. This method is widely used in the analysis of radioactive elements in soil or rock as done by Rosiana.I (2020) et al [2], Saputro S.Peta al [3]. Prazeres, C. (201) et al [4], Sain, N.K. et al (2002) [5], and Guembou, JCS, et al (2019) [6]. With the XRF method, the concentration of radioactive elements U, Th and K can be measured so that further analysis can be carried out as needed. As in the field of geology, XRF data analysis can be used to classify the tectonic context of the research area, as done by Sukadana, I.G. et al (2015) [7] and Ghodang, S. (2016) [8].

IOP Conf. Series: Materials Science and Engineering **1330** (2025) 012020 doi:10.1088/1757-899X/1330/1/012020

The first indication of high concentrations of radioactive elements in nature is obtained through radiometric measurements or dose rate mapping. The results of research conducted by Iskandar, D. et al (2007) [9], Syaeful et al [2014] [10]. and A.R.A. Abdullah and S. Permana, 2022 [11] showed that the average natural radiation dose rate in Mamuju Regency, West Sulawesi is high, reaching 0.3 μSv / day. Further research by Sukadana, I.G. et al (2016) [12], Sukadana, I.G. et al (2021) [13], Cakrabuana W., et.al (2021) [14]. suggested that Mamuju Regency is an area that has the potential for uranium and thorium reserves. The results of radiometric or dose rate mapping were confirmed by laboratory sample testing, including the XRF method.

2. Research Methods

This research uses experimental techniques with a focus on testing soil samples using the XRF (X-Ray Fluoresence) method conducted at the Physics Laboratory, Gorontalo State University. It was conducted from April 10 to August 4, 2023.

Table 1. Sampling location in Mamuju Regency

No	Sample code	Location	Sample site characteristic
1	P.B.1	Boteng	Highway peeling
2	P.T.1	Takandeang (Tabangabanga)	Naural (Unprocessed)
3	P.P.1 (54)	Takandeang (Taloba)	Clove graden
4	P.P.1 (19)	Pasabu	Naural (Unprocessed)
5	P.A.1	Ahu	Clove graden

The soil samples used were from research conducted by A.R.A. Abdullah and S. Permana (2002) in Mamuju district, West Sulawesi (Figure 1). 5 soil samples from this site were used as subjects in this study, see Table 1. Based on research [2] and [3], the soil of the research site [12] in Mamuju Regency is classified as a laterite type characterized by high Al content as described by [2] and Godang et al., 2019 [15].

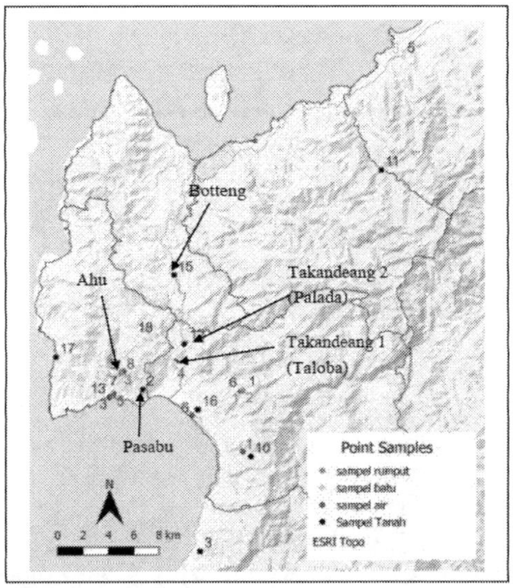

Figure 1. Sampling location (courtesy of [11]).

The samples tested consisted of two specimens, powder, and pellet. Measurements of powder specimens were carried out to obtain information on major and minor elements, while pellet specimens to obtain information on trace elements. XRF measurements were taken using a Bruker Fluorecense Spectroscopy EDXRF S2 Ranger.

Figure 2. Specimens used in the study, powder and pellet form

3. Results and Discussion

3.1 U and Th existence in samples

It is explained in [2] and [15], that the soil from the research area of Mamuju Regency is Laterite type. In more detail, it is explained in [15] and [16] that laterite is a red/brown residual soil that develops in humid, tropical, and subtropical areas with good drainage. Laterite is a near-surface chemical weathering product derived from parent rocks rich in alumnina, iron, and silica. Chemically laterite is characterized by depletion of mobile elements such as Mg, Ca, Ne, K and a little Si and enrichment of mobile elements such as Al, Ti, and Fe. It is explained in [15] that Laterite soils in Mamuju Regency and Mamasa Regency, West Sulawesi, originated from rain-induced weathering processes with a pH of ~6.5 (current pH conditions) originating from the mid to late Miocene era (14 - 5.3 million years ago). The long weathering process caused the parent rock to turn into laterite soil characterized by reduced MgO, CaO, Na2O, K2O, and increased Al2O3. While the parent rock of the soil in Mamuju is phonolite which is an ultrapostasic type basaltic igneous/volcanic rock rich in K2O as described in [2] and [7].

The measurement results show that the major oxide elememtts are Na2O, MgO, Al2O3, SiO2 and CaO. Also detected are some rare earth elements such as CeO2, Pr6O11, Nd2O3, Y2O3, and La2O3, and elements that have high economic value today, namely ZrO2, and BaO. The concentrations of K, U, and Th measured in the XRF test in this study can be seen in table 2.

Table 2. Th, U, and K Concentrations in Measured Sample

No	Sample Kode	Location	Percent (%)			ppm		
			Th	U	K	Th	U	K
1	P.B.1	Boteng	0.235974	0.070295	0.33717	2359.74	702.95	3371.7
2	P.T.1	Takandeang (Tabangabanga)	0.569548	0.086271	0.508846	5695.48	862.71	31371.76
3	P.P.1	Takandeang (Taloba)	0.171439	0	3.137176	1714.39	0	5088.46
4	P.P.1	Pasabu	0.409030	0.142827	0.160083	4090.3	1428.27	4205.99
5	P.A.1	Ahu	0.404733	0.056690	0.420599	4047.33	566.90	1600.83

IOP Conf. Series: Materials Science and Engineering **1330** (2025) 012020 doi:10.1088/1757-899X/1330/1/012020

Referring to [17], the abundance of potassium in the earth's crust is 2.09%. In this study, the potassium concentrations measured by the XRF method from 5 samples belonged to the low-medium level, according to the classification of the results of Syaeful, H. et al (2014a), with the lowest percentage of 0.16% (Pasabu village) and the highest 3.14% (Takandeang village), as shown in table 2.

The U and Th elements contained in the soil of the research samples show a linear correlation as shown in Figure 3 below.

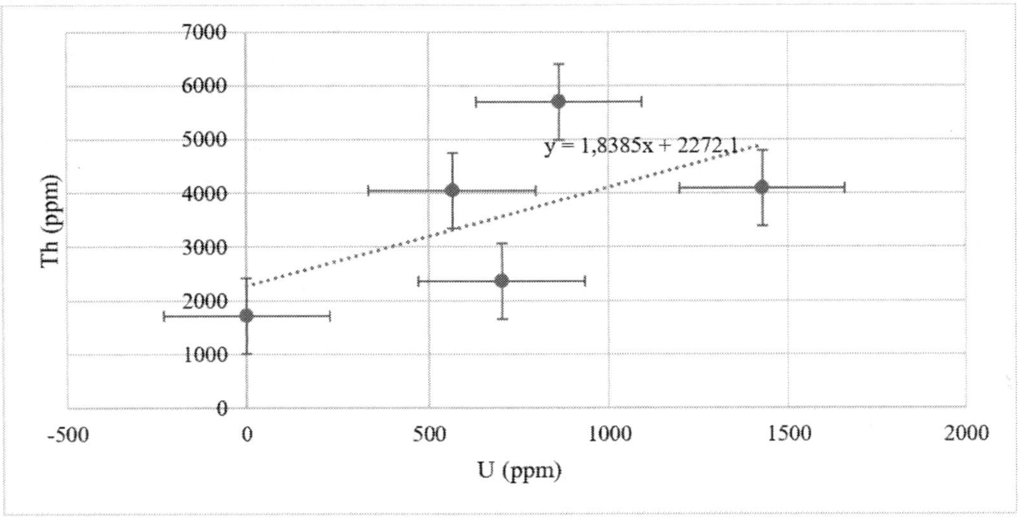

Figure 3. Korelasi positif Th terhadap U

A positive correlation is shown by the relationship between Th and U with a correlation coefficient of 0.60. This implies that uranium enrichment goes hand in hand with thorium enrichment. The enrichment of U and Th that occurs is explained by the following figures 4 and 5.

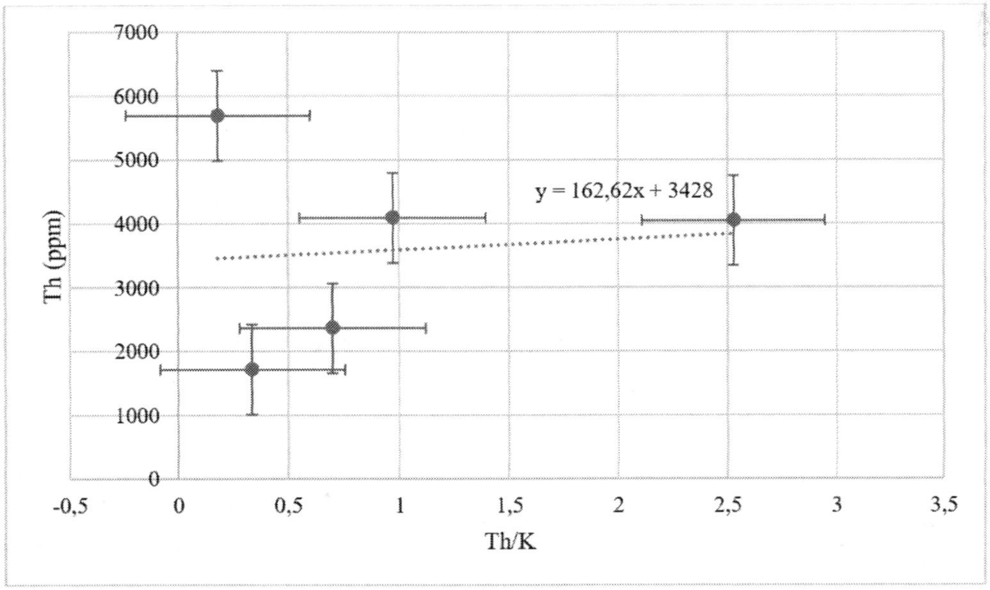

Figure 4. Positive Correlation of Th to Th/K

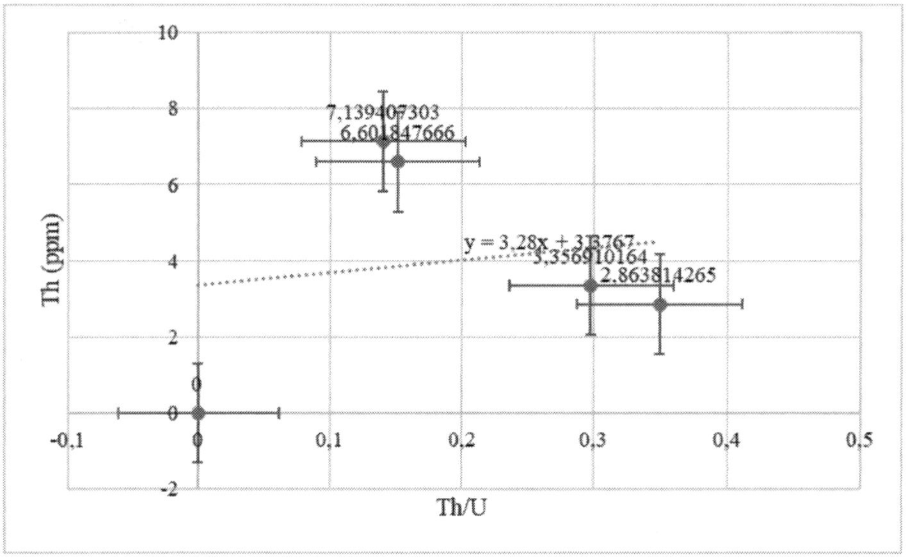

Figure 5. Correlation of Th to Th/U

Figure 5 and 6 above explains the thorium enrichment event. In his research Syaeful, H. et al in [10], explained that the Th/K and Th/U ratio reflects the enrichment of thorium elements on the surface or in geology known as supergene enrichment, this can be seen from the sloping linear regression equation line. Meanwhile, U enrichment cannot be explained due to the lack of available data related to uranium concentration at the sampling location. However, it is theoretically explained that the U enrichment is hydrothermal in nature, characterized by a steep linear regression equation line, as described in [10].

3.2 U and Th Relation with Dose Rate Datas

To verify the high concentration of radioactive elements of uranuim and thorium from XRF measurement data on 5 soil samples from Mamuju Regency, the XRF test results were compared with radiation dose rate data from the research of A.R.A. Abdullah and S. Permana in [11]. The dose rate data is the result of measurements at the place where the five samples were taken, and the measurements were made with the Geiger Muller brand Canberra Mini Radiac. The data is shown in table 3 below.

Table 3. Dose rate of gamma radiation at sampling point (with permission of A.R.A. Abdullah and S. Permana in [11])

No.	Sample Code	Location	Dose Rate (μSv/h)
1	P.B.1	Boteng	1,630
2	P.A.1	Ahu	2,400
3	P.P.1 (54)	Takandeang (Tabangabanga)	1,930
4	P.T.1	Takandeang (Taloba)	1,840
5	P.P.1 (19)	Pasabu	0,820

The relationship between the concentrations of U and Th and the dosage rate measured from each sampling point is shown in Figure 6 below.

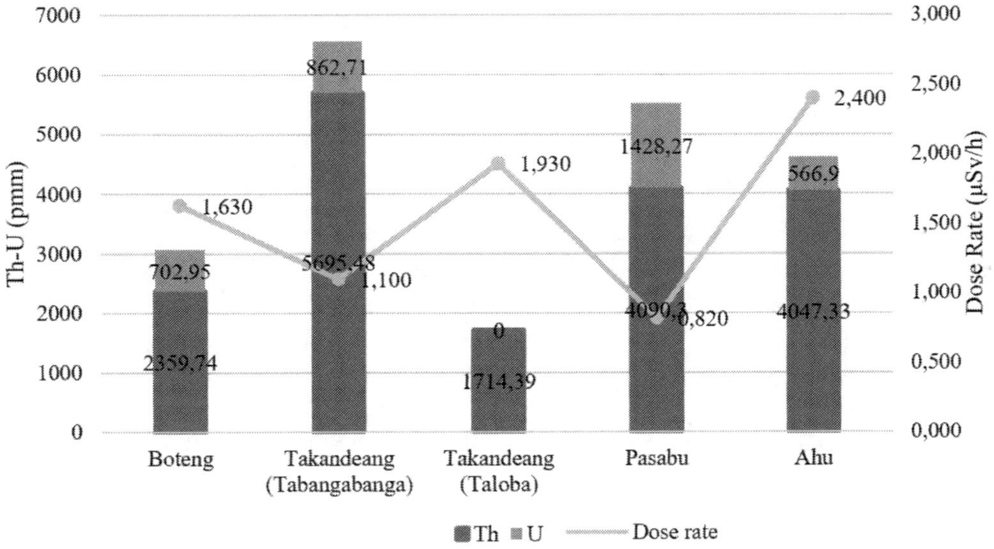

Figure 6. Relationship between U and Th concentrations and radiation dose rate.

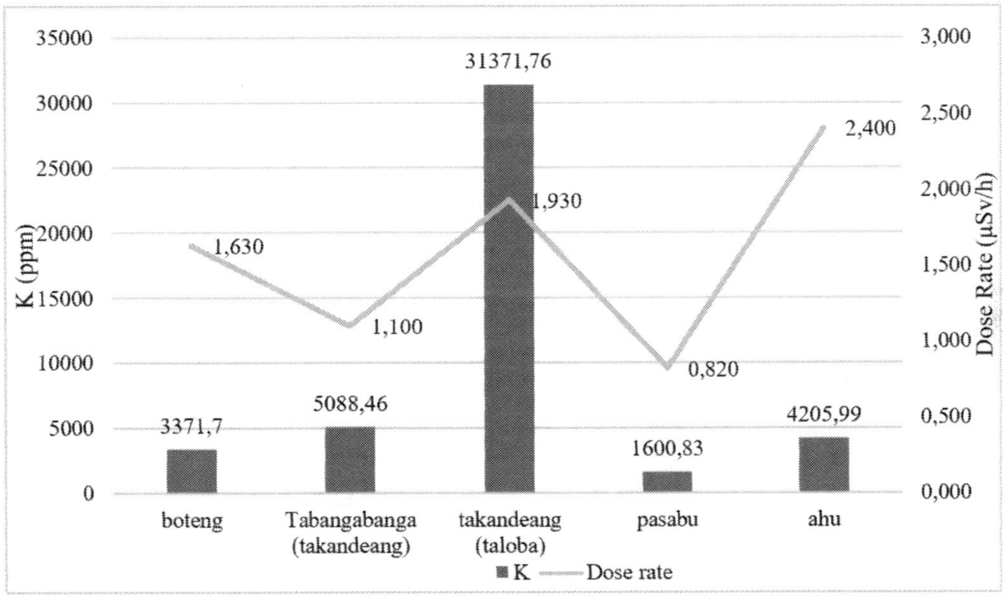

Figure 7. Relationship between K concentration and radiation dose rate

In Ahu village, with the highest dose rate reaching 2,400 μSv/h, the influence of the Th element is dominant. In Pasabu hamlet, the Th concentration is quite high, and the U concentration is the highest among all soil samples, but the measured radiation dose rate is the lowest, which is 0.820 μSv/hour. Then in Tabangabanga hamlet, Takandeang village, the Th concentration was the highest, reaching 5695.48 ppm, but the measured radiation dose rate was only 1,100 μSv/hour, lower than Ahu village and even less than half. In Taloba hamlet, Takandeang village, a high radiation dose rate of 1,930 μSv/hr was measured which was the second highest although the Th concentration was the smallest at 1714.38 ppm and no U element was detected but the K element concentration was the highest, even much higher than the other four samples which reached 31371.76 ppm. The K content in the other samples was below 5100 ppm. In Botteng village, the influence of Th concentrations is evident as the cause of the high radiation dose rate values measured.

From the research of A.R.A. Abdullah and S. Permana in [11], it was revealed that the conditions of the sampling location were different, where there were soil samples taken at the location of the garden, strip (roadside), and natural locations, as described in table 1. The condition of the soil sampling location measured by the XRF method may affect the measured dose rate value and affect the concentration of chemical elements in the soil sample.

In Ahu village and Taloba hamlet, Takandeang village, the location of the clove plantation is a location that has been cultivated and may have been treated with fertilizers that have the potential to increase the content of K, U and Th elements in the soil, as described in M.P. Rahayu's research in [19], which found uranium content in phosphate fertilizers. Particularly in Taloba hamlet, Takandeang village, a very high K content was measured, and it is likely that this K came from fertilizers other than those deposited naturally in rocks which then weathered into soil.

Then in Botteng village, the stripping of soil for road purposes also makes the dose rate value increase because the soil becomes more exposed, so that radioactive elements will emit more radiation. Since the dose rate recording was done near the highway, it is possible that the road-making material mix influenced the measured radiation dose rate (Inoue K, et al (2021). For example, cement as one of the road mixture materials also contains radioactive elements K, U, and Th as in Rasito et al's research in [18].

4. Conclusion

The XRF method can detect the presence of radioactive elements uranium (U), thorium (Th), potassium (K) and rare earth elements in soil samples from Mamuju Regency, West Sulawesi Province which represents an area with unique natural radiation Positive correlation between Th and U indicates the presence of U and Th in linear samples. While K with Th and U, K with Th/K and K with Th/U indicate the enrichment of Th elements on the surface or called supergene enrichmet. Only samples taken in original conditions (not yet processed into gardens or anything else) showed a linear relationship between the concentration of radioactive elements and the measured dose rate. In soil samples from locations that have been processed, in this case into clove gardens and highway strips, the response of dose rate values to U and Th concentrations shows different things. Where potassium was observed. This may be due to the use of fertilizers.

Acknowledgment
This research is supported by the research program of the Ministry of Research, Technology and Higher Education and the Bandung Institute of Technology by providing support and input through research and publication of this scientific article.

5. References

[1]. Zhang, P.C., James, L. Krumhansl, Patrick, V. B. 2002. *Introduction to Properties, Sources, and Characteristics of Soil Radionuclides in Geochemistry of Soil Radionuclides. SSSA Special Publication*. 59. 1–20.

[2]. Rosianna, I., Nugaraha, E.D., Syaeful, H., Putra, S., Hosoda, M., Akata, N., Tokonami, S. 2020. *Natural Radioactivity of Laterite and Volcanic Rock Sample for Radioactive Mineral Exploration in Mamuju, Indonesia*, *Journal of Geosciences*, 10, 376.

[3]. Saputro, S.P., Godang, S., Priadi, B., Basuki, N.I., Himawan, B. 2022. *Geochemical Study of Al–Fe–Ti Enrichment in Rock Weathering: Implications for The Recognizing of Igneous Protolith and The Enrichment of REE in Soil Profile*, *Applied Geochemistry*, 140, 105259.

[4]. C. Prazeres, I. Mártin-Mendez, M. J. Batista, J. Locutura, A. Bel-Lan. 2015. *Distribution of Naturally Radioactive Elements (U, Th and K) in grazing land soils of the Iberian Peninsula Compared with total Natural Radiation. Comunicações Geológicas (2015) 102, Especial I, 73-77.*

[5]. N. K. Saini, P. K. Mukherjee, M. S. Rathi, P. P. Khanna, and K. K. Purohit. 2002. *Trace Element Estimation in Soils: An Appraisal Of ED-XRF Technique Using Group Analysis Scheme. Journal Of Trace and Microprobe Techniques Vol.20, No.4, pp.539–551.*

[6]. Joel Cebastien Shouop Guembou, Maurice Moyo Ndontchueng, Jilbert Eric Mekongtso Nguelem, Gregoire Chene, Steve Kayo, Ousmanou Motapon and David Strivay. *Elemental Quantification and Radioactive Characterization of Soil from Douala Bassa Area: littoral Region of Cameroon using X-and γ-rays Spectrometry. IOP Publishing. Environ. Res. Commun. 1 (2019) 065001. https://doi.org/10.1088/2515-7620/ab1d72*.

[7]. I Gde Sukadana, Agung Harijoko, Lucas Donny Setijadi. 2015. *Tataan Tektonika Batuan Gunung Api di Komplek Adang, Kabupaten Mamuju, Provinsi Sulawesi Barat*. Eksplorium. Volume 36 No. 1. Mei 2015: 31-34.

[8]. Godang Shaban, Fadlin, and Bambang Priadi. 2016. *Geochemical Signatures of Potassic to Sodic Adang Volcanics, Western Sulawesi: Implications for Their Tectonic Setting and Origin.* Indonesian Journal on Geoscience Vol. 3 No. 3 December 2016: 195-214.

[9]. Dadong Iskandar, Syarbaini, Sutarman, Bunawas, Kusdiana. 2007. *Mapping the Outdoor Gamma Dose Rate in Indonesia. Abstract.*

[10]. Syaeful, H., Sukadana, I.G., Sumaryanto, A. 2014: *Radiometric Mapping for Natural Occurring Radioactive Materials (NORM) Assessment in Mamuju*, West Sulawesi.

[11]. Adi R. A. Abdullah dan S. Permana. 2022: *Survey Radiasi Alamiah Kabupaten Mamuju. Laporan Penelitian. Institut Teknologi Bandung*. Tidak Dipublikasikan.

[12]. I Gde Sukadana et al. 2016. *Identification of Mineralization Type and Specific Radioactive Minerals in Mamuju, West Sulawesi*. Journal of East China University of Technology. Vol. 39. Suppl. Dec. 2016.

[13]. I Gde Sukadana, I Wayan Warmada, Agung Harijoko, Frederikus Dian Indrastomo, Heri Syaeful. 2021. *The Application of Geostatistical Analysis on Radiometric Mapping Data to Recognized the Uranium and Thorium Anomaly in West Sulawesi, Indonesia. IOP Conf. Series: Earth and Environmental Science 819 (2021) 012030.*

[14]. W Cakrabuana1, R C Ciputra1 and H Syaeful1. 2021. *Assessing the undiscovered. IOP Conf. Series: Earth and Environmental Science 851. (2021) 012041.*

[15]. Godang, S., Idrus, A., Priadi, B., Basuki, N.I. 2019. *Saprolitization's Characteristics of Rare Earth Elements in Volcanic Regolith on Drill Core 65 in Western Sulawesi, Indonesia. Asian Journal of Physics.* Vol 07. 435-450.

[16]. G. Santha Kumar, P.K. Saini, Rajesh Deoliya, Aman Kumar Mishra, S.K. Negi. 2022. *Characterization of laterite soil and its use in construction applications: A review. Resources, Conservation & Recycling Advances 16 (2022) 200120.*

IOP Conf. Series: Materials Science and Engineering **1330** (2025) 012020 doi:10.1088/1757-899X/1330/1/012020

[17]. William M. Haaynes. 2016. *Abundance of elements in the earth's crust and in the sea, CRC Handbook of Chemistry and Physics, 97th edition (2016–2017), p. 14-17*

[18]. Rasito, Zulfakhri, P.A. Arianta, A. Suherman. 2007. *Konsentrasi Uranium, Thorium, dan Kalium dalam berbagai Produk Semen yang Dipasarkan di Indonesia. Jurnal Sains dan Teknologi Nuklir Indonesia Indonesian Journal of Nuclear Science and Technology Vol. IX, No. 2, Agustus 2008.*

Assessment of Transfer Factors for Natural Radionuclides and Radiocesium from Soil-to-Plant and Plant-to-Cow's Milk on a Cattle Farm in Lembang

Imam Ghazali Yasmint[1,2], Abdul Waris[2], Eko Pudjadi[3]
[1] Joint Doctoral Program for Sustainability Research, The University of Electro-communications, Tokyo, Japan
[2] Department of Physics, Faculty of Mathematics and Natural Sciences, Bandung Institute of Technology, Indonesia
[3] Environmental Laboratory, Subs-section of Environmental Safety, Radioecology Section, Center for Technology of Safety and Radiation Metrology, National Nuclear Energy Agency of Indonesia, Lebak Bulus Raya No.49 Jakarta, Indonesia

E-mail: [a)]imamghazaliyasmint@gmail.com

Abstract. Human populations are exposed to both natural and artificial radiation. This research aims to measure the radioactivity of natural and artificial radionuclides in soil, grass, and cow's milk to determine the possible transfer of radionuclides through the food chain in Lembang, West Java, Indonesia. The samples were obtained from a nearby cattle farm and analyzed with an ORTEC gamma spectrometer with an HPGe detector to determine the radionuclides, including ^{226}Ra, ^{232}Th, ^{40}K, and ^{137}Cs. Analysis indicated that the radioactivity concentration in soil is below the world average. Transfer factor analysis revealed that ^{40}K possesses the highest transfer factors from soil to plant and from plant to milk, validating the results of previous studies. In addition, ^{137}Cs were detected in grass and soil but not in milk samples, indicating no further transmission through the food chain. The findings indicate radionuclide transfer mechanisms and show a low radiation risk level in the study area.

Keyword: *Activity Concentration, Gamma Spectroscopy, Natural Radionuclides, Plant-to-Cow's Milk, Radiocesium, Soil-to-Plant, Transfer Factor.*

1. Introduction

The release of radioactive materials into the environment represents a negative effect of utilizing manmade radioactive substances[1]. Furthermore, natural sources of radiation have existed since the Earth's origin. The level of natural radioactivity in an area is dependent upon the geological composition of its constituent components. Additional natural sources of environmental radioactivity include primordial radionuclides from the uranium and thorium series, radon gas, and cosmic radiation originating from beyond the Earth.[2].

The occurrence of primordial radionuclides in the soil results from radionuclide transport through the weathering of the Earth's crust[2]. Radiocesium is a radioactive isotope of cesium (Cs), generated by nuclear reactions and present in the environment due to nuclear accidents or weapons testing.

Radiocesium possesses a half-life of around 30 years, indicating that half of its radioactivity will diminish within that period of time[3].

Radionuclides can transfer through several pathways, with one significant route being the soil-grass-cow's milk pathway[4]. Radionuclides in the soil will be absorbed by plants via their roots or leaves, a process known as soil-to-plant transfer[5]. Additionally, there is a transfer of radionuclides from grass to cow's milk[6]. These procedures will introduce radionuclides into the food chain, resulting in human radiation exposure through food consumption[7]. The UNSCEAR assessment indicates that natural radiation represents the highest dose absorbed by people[8].

Cows and other animals primarily consume grass, and the presence of radionuclides in the grass indirectly impacts human health through the consumption of meat and milk[9]. Elephant grass (Pennisetum purpureum) is a grass species mostly utilized as livestock fodder, characterized by its high nutritional content[10]. The purpose of this study was to determine the natural radionuclide and radiocesium activity concentrations in soil, grass (*Pennisetum Purpureum*) and cow's milk samples and soil-to-plant and plant-to-cow's milk transfer factor based on the results of activity concentrations obtained.

2. Materials and Methods

2.1. Study area, sampling and preparation

Lembang is a sub-district located in the West Java Province of Indonesia. Lembang (Figure 1) is situated at an elevation ranging from 1,312 to 2,084 meters above sea level. Lembang, situated in the mountains, offers a diverse array of tourist attractions. Furthermore, Lembang emerged as the nucleus of education and research in agriculture and animal husbandry. This establishes Lembang as one of the premier research sites.

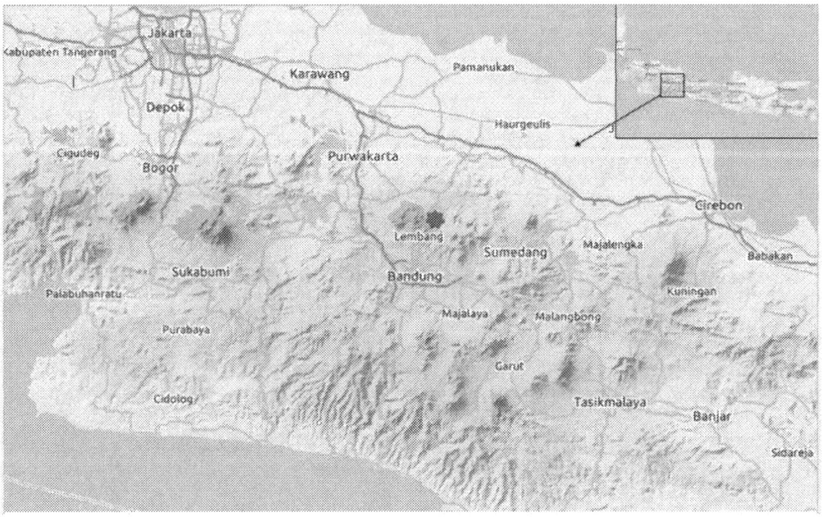

Figure 1. Location of Sampling Site

This study employed an experimental approach by quantifying radionuclides in multiple samples. Five samples were collected from five distinct locations. The specimens consisted of soil and grass. Nine milk samples were collected from three distinct cowsheds. All the grass in these cowsheds was sourced from the same field where the grass sample was obtained. The grass sample was *Pennisetum purpureum*. Sampling was conducted randomly, structured in accordance with the study's

IOP Conf. Series: Materials Science and Engineering **1330** (2025) 012021 doi:10.1088/1757-899X/1330/1/012021

objectives[11]. Soil samples for the analysis of radioactive content are collected from a depth of 5-25 cm. For sedimentation deposition studies, soil must exceed a depth of 25 cm[12]. All components of the grass samples were collected to assess the radionuclide content.

Soil samples were collected from the root zones of the grass at a depth of 10 to 20 cm[13]. The soil and grass sampling locations were identical to acquire the radionuclide transfer factor (TF)[14]. The soil and grass collected were cleansed of any dirt that could be transported during collecting. The soil and grass samples were dried under sunlight for 2-3 days and thereafter transported to the laboratory for preparation.

Soil samples were placed in an oven and dried for several days at 100°C. The dried samples were crushed with a mortar and pestle and subsequently sifted through a 60 mesh sieve[15]. The grass samples were cut into small bits and subsequently placed into a grinder. The samples were crushed with a grass grinder to yield relatively tiny particles. The fairly fine grass samples were processed using a 60 mesh refiner. The milk samples were not processed but were frozen to preserve their freshness.

The soil, grass, and milk samples were placed into 1 L Marinelli-type beakers. The Marinelli-type beakers are specialized containers engineered to accommodate the gamma spectrometer. This container exhibited superior thermal resistance compared to plastic overall. The samples were placed in Marinelli-type beakers until the volume of each sample reached one liter, and then weighed to obtain the final weight of the produced sample. The samples in the Marinelli-type beakers were thereafter closed and sealed with adhesive tape to prevent the escape of radon[14]. The samples were designated with identifiers: soil samples were labeled with 'T' followed by a sample number, grass samples with 'R' followed by a sample number, and milk samples with 'S' followed by numbers. The numerical code for milk samples is segmented, the initial digit represents the cowshed number, while the subsequent digit denotes the sample number. The sealed samples were preserved for 30 days to attain secular equilibrium between the parent and daughter isotopes. After 30 days, the samples were prepared for measurement using a gamma spectrometer[16].

2.2. Activity Concentrations

The activity concentrations were determined using a gamma spectrometer equipped with a High Purity Germanium (HPGe) detector. The sample measuring procedure is illustrated in Figure 2. The utilized detector was a P-type Coaxial HPGe radiation detector, model number GEM60P4. The device exhibits a relative efficiency of 60% and an energy resolution of 1.95 keV at the 1.33 MeV peak and is linked to a 16K multichannel analyzer of DSPEC Pro. The device's results were subsequently analyzed using Computer Software Maestro to present the gamma radiation spectrum. The gamma spectrometer was calibrated utilizing a mixed gamma standard containing five radionuclides: ^{210}Pb, ^{241}Am, ^{133}Ba, ^{137}Cs and ^{60}Co.

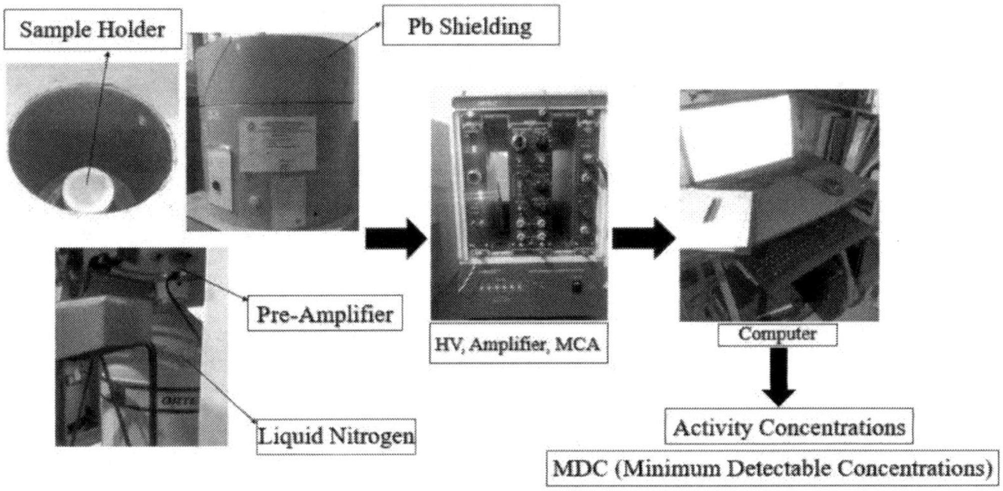

Figure 2. Data Collection Diagram.

Table 1. Gamma energy lines and P_Y of radionuclides used in activity concentrations

Radionuclide	Transition Isotope	Energy (keV)	P_Y
	^{214}Pb	295.207	0.182
	^{214}Pb	351.925	0.351
^{226}Ra	^{214}Bi	609.318	0.446
	^{214}Bi	1120.29	0.147
	^{214}Bi	1764.52	0.151
	^{228}Ac	338.42	0.1126
^{232}Th	^{228}Ac	911.16	0.266
	^{228}Ac	968.97	0.1623
^{40}K		1460.8	0.1067
^{137}Cs		661.657	0.851

Measurements were conducted for 61200 seconds for each sample. The parameters and intensity of gamma energy are presented in Table 1. It was utilized to assess activity concentrations and minimum detectable concentrations (MDC). The evaluation was conducted utilizing the methodology outlined in the Ibikunle report[17]. Uncertainty from the calculation of activity concentrations was derived from multiple factors: uncertainty in gamma yield (~ 0.1%), uncertainty in efficiency calibration (~ 5%) and uncertainty in sample weight (~ 1%)[14].

2.3. Transfer Factors (TF)

The transfer factor is the ratio of radionuclide concentrations in one environmental element compared to other environmental elements containing the same radionuclides[18]. The transfer from soil to grass was determined by calculating the ratio of radioactive concentrations in grass to those in soil, both measured on a dry weight basis[19].

IOP Conf. Series: Materials Science and Engineering **1330** (2025) 012021 doi:10.1088/1757-899X/1330/1/012021

3. Results and Discussion

The activity concentrations for radionuclides of ^{226}Ra, ^{232}Th and ^{40}K in soil, grass (*Pennisetum Purpureum*) and cow's milk are shown in Table 2, Table 3 and Table 4. For radionuclides transfer factor from soil to grass and grass to cow's milk are shown in Figure 3 and Figure 4.

Table 2. Radioactivity Concentration in soil

Sample Label	Type	Radioactivity Concentration in the Sample (Bq kg^{-1})			
		^{226}Ra	^{232}Th	^{40}K	^{137}Cs
T.1	Soil	18.73 ± 0.68	26.39 ± 1.06	75.82 ± 2.12	0.4473 ± 0.068
T.2	Soil	18.98 ± 0.69	26.05 ± 1.04	91.44 ± 2.41	0.3588 ± 0.0694
T.3	Soil	21.93 ± 0.77	31.52 ± 1.20	96.76 ± 2.48	0.2054 ± 0.062
T.4	Soil	20.85 ± 0.73	31.19 ± 1.18	93.07 ± 2.36	0.4485 ± 0.0571
T.5	Soil	22.75 ± 0.79	35.23 ± 1.33	105.20 ± 2.68	0.171 ± 0.0604
Average Concentration		20.65 ± 0.73	30.08 ± 1.16	92.46 ± 2.41	0.33 ± 0.06
Average MDC		0.0026	0.0015	0.0130	0.0004

Table 3. Radioactivity Concentration in grass

Sample Label	Type	Radioactivity Concentration in the Sample (Bq kg^{-1})			
		^{226}Ra	^{232}Th	^{40}K	^{137}Cs
R.1	Grass	1.97 ± 0.36	1.99 ± 0.85	1994.54 ± 30.60	0.1019 ± 0.05
R.2	Grass	1.76 ± 0.28	2.29 ± 0.76	2019.17 ± 31.46	0.1109 ± 0.0504
R.3	Grass	1.38 ± 0.19	2.92 ± 0.78	2119.17 ± 32.78	0.0797 ± 0.0414
R.4	Grass	1.30 ± 0.23	2.28 ± 0.70	1428.23 ± 23.59	0.1634 ± 0.0545
R.5	Grass	1.09 ± 0.21	2.58 ± 0.68	1330.80 ± 22.52	0.1236 ± 0.0309
Average Concentration		1.50 ± 0.25	2.41 ± 0.75	1778.38 ± 28.19	0.12 ± 0.05
Average MDC		0.0100	0.0059	0.0496	0.0014

Table 4. Radioactivity Concentration in cow's milk

Sample Label	Type	Radioactivity Concentration in the Sample (Bq kg^{-1})			
		^{226}Ra	^{232}Th	^{40}K	^{137}Cs
S.1-1	Milk	0.27 ± 0.04	0.15 ± 0.07	65.25 ± 1.68	≤ MDC
S.1-2	Milk	0.24 ± 0.04	0.10 ± 0.07	64.67 ± 1.76	≤ MDC
S.1-3	Milk	0.27 ± 0.04	0.20 ± 0.16	66.67 ± 1.79	≤ MDC
S.2-1	Milk	0.14 ± 0.02	0.20 ± 0.09	61.24 ± 1.67	≤ MDC
S.2-2	Milk	0.08 ± 0.02	0.16 ± 0.08	62.98 ± 1.73	≤ MDC
S.2-3	Milk	0.12 ± 0.02	0.22 ± 0.10	59.94 ± 1.60	≤ MDC
S.3-1	Milk	0.12 ± 0.03	0.19 ± 0.10	55.16 ± 1.59	≤ MDC
S.3-2	Milk	0.15 ± 0.03	0.24 ± 0.13	53.69 ± 1.50	≤ MDC
S.3-3	Milk	0.20 ± 0.04	0.18 ± 0.07	54.90 ± 1.54	≤ MDC
Average Concentration		0.18 ± 0.03	0.18 ± 0.10	60.50 ± 1.65	≤ MDC
Average MDC		0.0023	0.0014	0.0114	0.0003

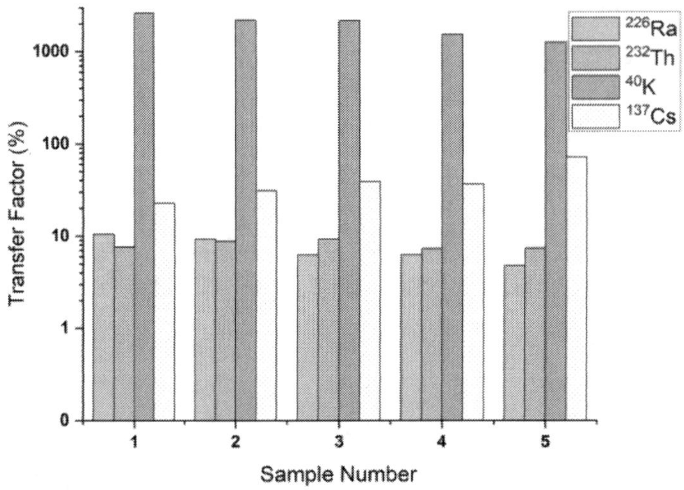

Figure 3. Transfer Factor From Soil to Grass.

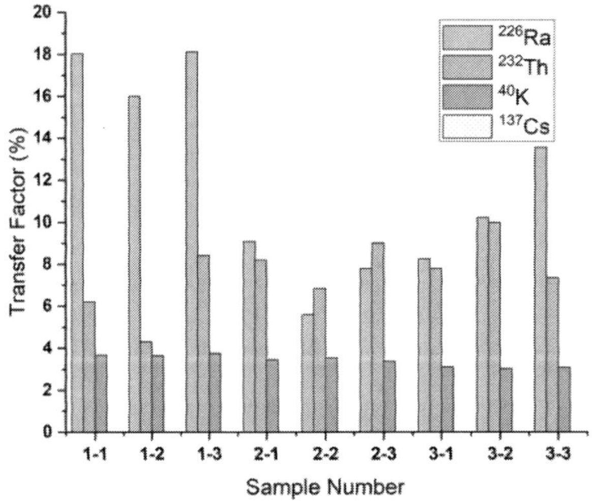

Figure 4. Transfer Factor From Grass to Cow's Milk.

The activity concentrations of ^{226}Ra in soil samples were in the range of 18.73 - 22.75 Bq kg^{-1} with an average of 20.65 Bq kg^{-1}. For grass samples, the activity concentrations of ^{226}Ra were in the range of 1.09 - 1.97 Bq kg^{-1} with an average of 1.50 Bq kg^{-1}. The activity concentrations of ^{232}Th in soil samples were in the range of 26.05 - 35.23 Bq kg^{-1} with an average of 30.08 Bq kg^{-1}. For grass samples, the activity concentrations of ^{232}Th were in the range of 1.99 - 2.92 Bq kg^{-1} with an average of 2.41 Bq kg^{-1}. The activity concentrations of ^{40}K in soil samples were in the range of 75.82 - 105.20 Bq kg^{-1} with an average of 92.46 Bq kg^{-1}. For grass samples, the activity concentrations of ^{40}K were in the range of 1330.80 - 2119.17 Bq kg^{-1} with an average of 1778.38 Bq kg^{-1}. The activity concentrations of ^{137}Cs in soil samples were in the range of 0.17 – 0.44 Bq kg^{-1} with an average of 0.33 Bq kg^{-1}. ^{226}Ra, ^{232}Th, ^{40}K and ^{137}Cs in the soil samples was below the average value of the activity concentration throughout the world of 32 Bq/kg, 45 Bq/kg, 450 Bq/kg and 3.3 Bq/kg[20]. For grass samples, the activity concentrations of ^{137}Cs were in the range of 0.07 – 0.16 Bq kg^{-1} with an average of 0.12 Bq kg^{-1}. In the milk samples, very small amount of ^{226}Ra and ^{232}Th activity concentrations were found. For ^{137}Cs it is not found or its value is below MDC. The activity concentrations of ^{40}K in milk samples were in the range of 53.69 – 66.67 Bq kg^{-1} with an average of 60.50 Bq kg^{-1}.

The Chernobyl fallout will tend to be in the upper layers of the soil since it is localized and relatively new in a deposition. However, nuclear weapons testing has caused ^{137}Cs to be more extensively and deeply distributed in the soil profile. Studies indicate that ^{137}Cs from Chernobyl are mainly stored in the upper 5 cm of soil with limited migration with time. Meanwhile, ^{137}Cs from nuclear tests tend to be deposited deeper because of the impact of extended deposition and soil aging processes[21][22][23]. Based on this, the source of ^{137}Cs in soil samples are likely from a nuclear weapons test. Furthermore, the transfer of ^{137}Cs from soil to vegetation, particularly to grass, can be by root uptake mechanisms and therefore be an important entry route for radionuclides into the food chain. Research has shown that the soil-to-grass transfer factor is similar for ^{137}Cs due to both Chernobyl fallout and nuclear weapon tests[24].

The radionuclides transfer factor from soil to grass of ^{226}Ra were in the 4.8 – 10.5 % range with an average of 7.4 %. For ^{232}Th, they were in the 7.3 – 9.2 % range with an average of 8 %. For ^{40}K, the range of 1265 – 2630 %, with an average of 1965 %. And for ^{137}Cs, they were in the range of 22.7% – 72.2 %, with an average of 40.2 %. There is a very high ^{40}K transfer factor. It means plants can easily absorb ^{40}K. The elevated concentration of ^{40}K transfer factor in plants results from prolonged root absorption processes[25]. Additionally, ^{40}K is an essential nutrient required by plants[25]. The transfer factor values for ^{226}Ra and ^{232}Th are relatively low. The strong affinity of ^{226}Ra for soil particles may account for the low transmission factor of ^{226}Ra[26]. Radium exhibits similar behavior to calcium, hence it will be absorbed alongside calcium during the uptake process by plants[27]. If ^{226}Ra enters the human body, it will imitate calcium and be deposited in bones, resulting in radiation exposure through its decay products[28]. ^{137}Cs exhibit characteristics similar to ^{40}K, facilitating its mobility in soil and enabling plants to readily absorb ^{137}Cs[29]. Moreover, wind-borne radioactive particulates from nuclear experiments can elevate the possibility of ^{137}Cs presence in the grass. The transfer factor values for ^{226}Ra and ^{232}Th were notably low. This is due to the two nuclides being insoluble, leading to their accumulation in the denser regions of the cow's body, resulting in just a minimal quantity being transferred to the milk.

4. Conclusions

Studies on environmental radioactivity have been conducted in Lembang, West Java, Indonesia, focusing on the activity concentrations of ^{226}Ra, ^{232}Th, ^{40}K, and ^{137}Cs, as well as the transfer factors in soil, grass, and cow's milk samples. The average activity concentration in soil is below the global average. The soil-to-plant transfer factors found align with findings from earlier studies, indicating that ^{40}K exhibits the highest transfer factor among radionuclides. This may be due to ^{40}K being one of the essential minerals required by plants. This outcome may serve as a reference for the soil-to-plant transfer factor in West Java, Indonesia, particularly for grass. Additionally, it can be beneficial in assessing direct or indirect radiation exposure in humans. ^{137}Cs was absent in the milk samples, and the detected levels of radium and thorium were minimal. Assuring that the milk produced on the farm is safe for consumption.

5. Acknowledgement

This work was supported by Department of Physics ITB Indonesia, National Nuclear Energy Agency of Indonesia and Department of Food and Animal Husbandry of West Java Province Indonesia. The author would like to thank Wahyudi S.T for his guidance and valuable input in this work. Author also would like to thank the employees of Subs-section of Environmental Safety, Radioecology Section, Center for Technology of Safety and Radiation Metrology, National Nuclear Energy Agency of Indonesia for their nice cooperation during this work.

IOP Conf. Series: Materials Science and Engineering **1330** (2025) 012021 doi:10.1088/1757-899X/1330/1/012021

6. References

[1] Cooper J R, Randle K and Sokhi R S 2003 Radioactive releases in the environment: impact and assessment. *Radioactive releases in the environment: impact and assessment.*

[2] El-Taher A 2013 A Study on Transfer Factors of Radionuclides from Soil to plant *Life Science Journal* **10** 532–9

[3] Rai H and Kawabata M 2020 The Dynamics of Radio-Cesium in Soils and Mechanism of Cesium Uptake Into Higher Plants: Newly Elucidated Mechanism of Cesium Uptake Into Rice Plants *Frontiers in Plant Science*

[4] Karunakara N and Ujwal P 2013 Studies on soil to grass transfer factor (Fv) and grass to milk transfer coefficient (Fm) for cesium in Kaiga region *Journal of Environmental Radioactivity 124*

[5] Haque M and Ferdous J 2017 Transfer of natural radionuclides from soil to plants in Savar Dhaka *Spanish Journal of Soil Science* **7**

[6] Albers B P and Steindl H 2000 Soil-to-plant and plant-to-cow's milk transfer of radiocaesium in alpine pastures: significance of seasonal variability *Chemosphere 41*

[7] Ele Abiama P, Ben-Bolie G H, Amechmachi N, Najib F, El Khoukhi T and Owono Ateba P 2012 Annual intakes of 226Ra, 228Ra and 40K in staple foodstuffs from a high background radiation area in the southwest region of Cameroon *Journal of Environmental Radioactivity* **110** 59–63

[8] United Nations 2000 *Sources and effects of ionizing radiation: United Nations Scientific Committee on the Effects of Atomic Radiation: UNSCEAR 2000 report to the General Assembly, with scientific annexes* (New York: United Nations)

[9] Cui S, Hough R, Fu Q, Qi X, Liu D, Cooper P, Li P and Zhang Z 2019 Concentrations and uptake pathways of polychlorinated biphenyls from soil to grass *Ecotoxicology and Environmental Safety* **182** 109428

[10] Zetina-Córdoba P 2013 Effect of cutting interval of Taiwan grass (Pennisetum purpureum) and partial substitution with duckweed (Lemna sp. and Spirodela sp.) on intake, digestibility and ruminal fermentation of Pelibuey lambs *Livestock Science* 7

[11] International Atomic Energy Agency 2004 *Soil sampling for environmental contaminants.* (Vienna: International Atomic Energy Agency)

[12] Livens F 1990 Measurement of radionuclides in food and the environment. A guidebook *Journal of Environmental Radioactivity* **11** 201–2

[13] Djelic G, Krstic D, Stajic J M, Milenkovic B, Topuzovic M, Nikezic D, Vucic D, Zeremski T, Stankovic M and Kostic D 2016 Transfer factors of natural radionuclides and 137Cs from soil to plants used in traditional medicine in central Serbia *Journal of Environmental Radioactivity* **158–159** 81–8

[14] Saenboonruang K, Phonchanthuek E and Prasandee K 2018 Soil-to-plant transfer factors of natural radionuclides (226 Ra and 40 K) in selected Thai medicinal plants *Journal of Environmental Radioactivity* **184–185** 1–5

[15] Tuo F, Zhang J, Li W, Yao S, Zhou Q and Li Z 2017 Radionuclides in mushrooms and soil-to-mushroom transfer factors in certain areas of China *Journal of Environmental Radioactivity* **180** 59–64

[16] Baeza A, Paniagua J, Rufo M, Guillén J and Sterling A 2001 Seasonal variations in radionuclide transfer in a Mediterranean grazing-land ecosystem *Journal of Environmental Radioactivity* **55** 283–302

[17] Ibikunle S B, Arogunjo A M and Ajayi O S 2019 Characterization of radiation dose and soil-to-plant transfer factor of natural radionuclides in some cities from south-western Nigeria and its effect on man *Scientific African* **3** e00062

[18] International Atomic Energy Agency 2010 *Handbook of parameter values for the prediction of radionuclide transfer in terrestrial and freshwater environments.*

[19] James J P, Dileep B N, Ravi P M, Joshi R M, Ajith T L, Hegde A G and Sarkar P K 2011 Soil to leaf transfer factor for the radionuclides 226Ra, 40K, 137Cs and 90Sr at Kaiga region, India *Journal of Environmental Radioactivity* **102** 1070–7

[20] United Nations 2010 *Sources and effects of ionizing radiation: United Nations Scientific Committee on the Effects of Atomic Radiation: UNSCEAR 2008 report to the General Assembly, with scientific annexes* (New York: United Nations)
[21] Magre A, Boulet B, de Vismes A, Evrard O and Pourcelot L 2023 Identification of the origin of radiocesium released into the environment in areas remote from nuclear accident and military test sites using the $^{135}Cs/^{137}Cs$ isotopic signature *Environmental Pollution* 121606
[22] Almgren S and Isaksson M 2006 Vertical migration studies of ^{137}Cs from nuclear weapons fallout and the Chernobyl accident *Journal of Environmental Radioactivity* 91 90–102
[23] Poręba G and Bluszcz A 2007 Determination of the initial ^{137}Cs fallout on the areas contaminated by Chernobyl fallout *Geochronometria* 26 35–38
[24] Selnaes T and Strand P 1992 Comparison of the uptake of radiocaesium from soil to grass after nuclear weapons tests and the Chernobyl accident *The Analyst* 117 493–496
[25] Dreyer I and Uozumi N 2011 Potassium channels in plant cells: Potassium channels in plants *FEBS Journal* **278** 4293–303
[26] Vandenhove H, Olyslaegers G, Sanzharova N, Shubina O, Reed E, Shang Z and Velasco H 2009 Proposal for new best estimates of the soil-to-plant transfer factor of U, Th, Ra, Pb and Po *Journal of Environmental Radioactivity* **100** 721–32
[27] Chandrashekara K and Somashekarappa H M 2016 Estimation of radionuclides concentration and average annual committed effective dose due to ingestion for some selected medicinal plants of South India *Journal of Radiation Research and Applied Sciences* **9** 68–77
[28] IAEA 2014 *Handbook of Parameter Values for the Prediction of Radionuclide Transfer to Wildlife* (Vienna: INTERNATIONAL ATOMIC ENERGY AGENCY)
[29] Iizuka K et all 2018 Relationship between 137Cs concentration and potassium content in stem wood of Japanese cedar (Cryptomeria japonica) *Journal of Wood Science* **64** 59–64

IOP Conf. Series: Materials Science and Engineering **1330** (2025) 012022 doi:10.1088/1757-899X/1330/1/012022

Evaluation of a PVT-Based Scintillation Detector as a Cost-effective Early Detection Solution in Nuclear Reactor Safety Systems

Imam Ghazali Yasmint[1,2], Cici Wulandari[2,3], Rasito Tursinah[3,4], Amila Amatullah[3], Ismail Humolungo[3], Marisa Variastuti[3] and Zulfahmi[3]

[1]Joint Doctoral Program for Sustainability Research, The University of Electro-communications, Tokyo, Japan

[2]Department of Physics, Faculty of Mathematics and Natural Sciences, Bandung Institute of Technology Ganesha No. 10 Bandung, Indonesia

[3]Department of Nuclear Science and Engineering, Faculty of Mathematics and Natural Sciences, Bandung Institute of Technology Ganesha No. 10 Bandung, Indonesia

[4]Nuclear Energy Research Organization, National Research and Innovation Agency (BRIN), Indonesia

E-mail: [a)]imamghazaliyasmint@gmail.com

Abstract. The safety measures for nuclear reactors are constantly advancing in line with the progress in utilizing nuclear reactors as power sources. Furthermore, the capabilities of early detection systems in nuclear reactors are always evolving to provide the most efficient detectors. The rise of gamma radiation levels serves as an indicator in the safety system of nuclear reactors, and it refers to the gamma radiation detector. An outcome of the advancement of gamma radiation detectors is the creation of detectors that possess exceptional capabilities but come with an expensive cost. Poly Vinyl Toluene (PVT) plastic can function as a gamma detector and is cost-effective. This study aims to investigate the possibility of utilizing PVT-based scintillation detectors as a cost-effective solution for incorporating early detection capabilities into nuclear reactor safety systems. The work involved the characterization of PVT-based scintillation detectors, followed by a comparison with commercially available detectors. This research is concerned with the efficiency of the detector.
Keywords: Efficiency, Gamma Radiation, Plastic Scintillator, PVT, Scintillation Detector

1. Introduction

The status of the world's Nuclear Power Plants (NPP) at the end of 2020 shows that 442 reactors have operated in 32 countries around the world with a total world nuclear power capacity of 392.6 GW(e) and continues to grow with several new reactors under construction. This amount equals 2553.2 terawatt-hours, almost a third of low-carbon electricity and about 10% of world electricity production. The existence of this nuclear power plant, in general, reduces carbon emissions by around two gigatons per year [1]. Early detection in nuclear reactors is included in the general safety requirements in the radiation protection section. The main objective is to ensure the security and safety of the people living around the reactor from radiation leaks [2]. A significant part of early detection is the use of detectors. In this case, the detector is a scintillation detector with thallium-doped sodium iodide (NaI(Tl)). This detector is usually used to monitor the dose rate of environmental radiation in both moving and fixed positions with good scintillation properties [3][4].

One of the drawbacks of NaI(Tl) material is its relatively high price, which requires high costs for its application. The scintillation detector with plastic material is one of the materials that has been widely studied as a nuclear radiation detector option [5]. This material is cheaper than NaI(Tl), although the scintillation light produced is less than NaI(Tl). Some of the most widely used scintillation plastic materials are Polyvinyltoluene (PVT), Polystyrene (PS), and Polymethylmethacrylate (PMMA) [6]. The benefits of PVT-based scintillation detectors are better than their cost and adaptability [7]. PVT is characterized by a shorter time of radioactive decay in comparison to NaI(Tl) and can react quickly for dynamic applications, including security scanning and border inspections. Moreover, its chemical stability and insensitivity against the environment make it an excellent option under hostile conditions [8]. Recent advancements, such as the incorporation of SiPM technology, greatly improved the detection and efficiency of PVT-based systems, which would be usable over a broader energy range from 100 keV to 3000 keV [9].

Apart from their technical advantages, PVT scintillators have also proved to be versatile in most applications. For instance, optical fiber-based systems employing PVT have been engineered for radiation therapy dosimetry, where the scintillator probe is exposed only under extreme conditions while keeping the photodetector safe [10]. The modularity of PVT systems to accommodate modular designs has made them extremely scalable for applications that span from handheld devices to large portal monitoring systems. In this regard, the lack of spectral resolution of PVT has been successfully countered through developments in signal processing technologies, including deconvolution algorithms. These suggest enhanced discrimination following detection towards making the identification between NORMs and Special Nuclear Materials of interest with higher sensitivity[11]. This has significantly reduced false positives in security systems, hence making PVT even more important in modern radiation detection systems. Despite these benefits, PVT detectors have some limitations. The reduced light output compared to NaI(Tl) can limit their effectiveness in applications that require precise spectrometric measurements. Their sensitivity in lower gamma energy ranges may also need to be optimized for certain specialized uses [12].

This study characterized a PVT detector. Characterization is done by determining the detector operating voltage, stability testing, linearity testing, and detector angular response testing. Then, the detector will be compared with other commercial detectors in terms of absolute efficiency.

2. Materials and Methods

The detector is designed in a probe-type configuration. This design was used to ensure the detector is applicable to multiple devices rather than being restricted to a single one. A scintillator and photomultiplier tube (PMT) often comprise probe-type detectors. Furthermore, an end cap typically links the PMT to additional devices with BNC connectors in the PMT part. Figure 1 illustrates the schematic of the gamma radiation detector utilized in this study.

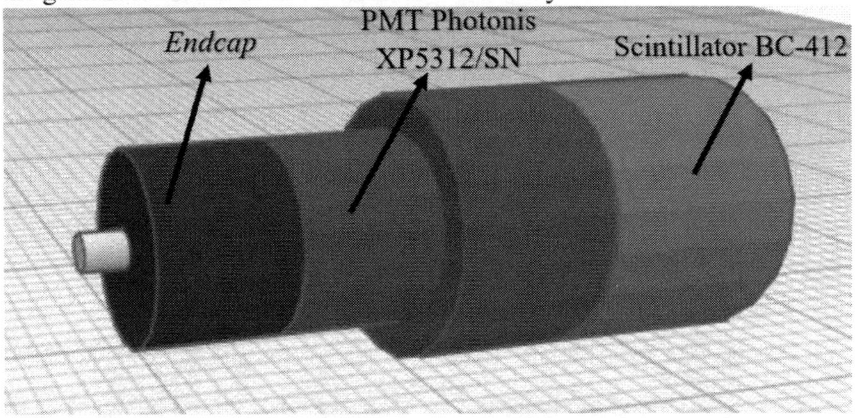

Figure 1. PVT-based Scintillation Detector Schematic Diagram.

The detector utilizes BC-412 PVT material from Saint-Gobain, measuring 3 x 2.25 inches, as a scintillator for the detection of incoming gamma radiation. The rationale for selecting PVT material is its capacity to generate 20% more light than polystyrene [7], with a slightly higher price adjustment. The BC-412 scintillator was selected due to its emission peak at a wavelength of 434 nm, aligning with

IOP Conf. Series: Materials Science and Engineering **1330** (2025) 012022 doi:10.1088/1757-899X/1330/1/012022

the PMT profile employed and suitable for broad applications. The scintillator's dimensions are selected due to their market availability without a precise size need, facilitating the ordering process. This measure also ensures high efficiency.

The PMT XP5312/SN was employed to gather and convert the visible light generated by the scintillation process. The selection of the PMT is due to the PMT XP5312/SN's widespread market availability and its 3-inch diameter, which corresponds with the scintillator's dimension. A diameter of 3 inches is chosen according to what is needed. The measurement of material efficiency indicates that 3 inches of PVT material yields optimal efficiency. An end cap is attached to the base of the PMT, linking the detector to the counter via a BNC-type cable. The scintillator is coated with a reflective layer that reflects the visible light from the PMT. All detector components are covered with Black Polyolefin to enhance their environmental resistance. Figure 2 illustrates the outcome of constructing the detector.

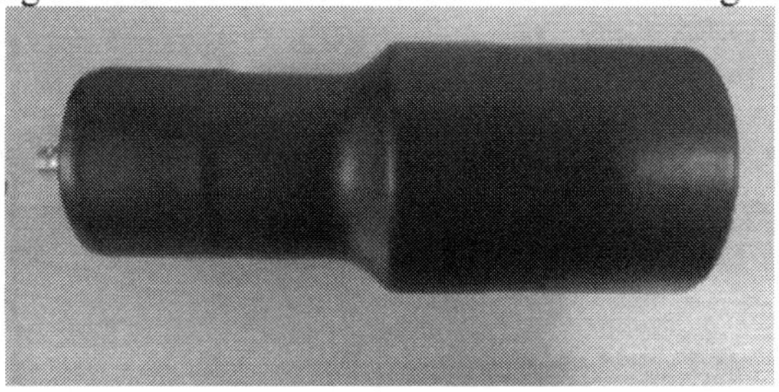

Figure 2. PVT-based Scintillation Detector.

Characterization is achieved by testing to determine the detector's characteristics. The experiments evaluated the detector's operating voltage, stability, linearity, and angular response [13]. A comparison of the fabricated detectors with commercial detectors is also conducted. The detectors utilized for comparison with the BC-412 detector include the Ludlum Module 44-10 detector, a 2 x 2 inch NaI(Tl) detector, and the Technical Associates Model P-6A detector, a Geiger-Muller detector employing argon gas. A counter is required to compare the detectors. The Technical Associates Model PRS-5, a scaler-type counter, is employed to display the count measured by the detector digitally. The comparison experiment utilized gamma sources ^{137}Cs, ^{133}Ba, and lamp mantle for the detectors. Experiments were conducted on a stand to ensure that the experimental conditions, particularly the distance, remained constant. Figure 3 shows the equipment used in this experiment.

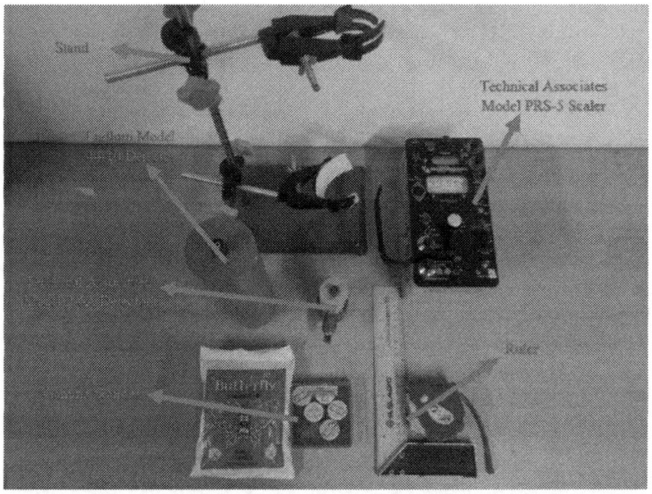

Figure 3. Detector Comparison Experiment Equipment.

3. Results and Discussion

3.1. Detector Operating Voltage Test

Each detector possesses a distinct operating voltage relative to the others. Understanding the detector's operating voltage is crucial for its appropriate adjustment. The operating voltage of a detector often ranges from hundreds to thousands of volts, so it is commonly referred to as high voltage. The operating voltage is determined by measuring the count at each voltage until a plateau (flat) area is reached. This area consistently increases in quantity and often remains stable. The operating voltage of the BC-412 detector was investigated using a Scaler from Technical Associates Model PRS-5. A ^{137}Cs radiation source is employed with a source-detector distance of 10 cm to achieve optimal results.

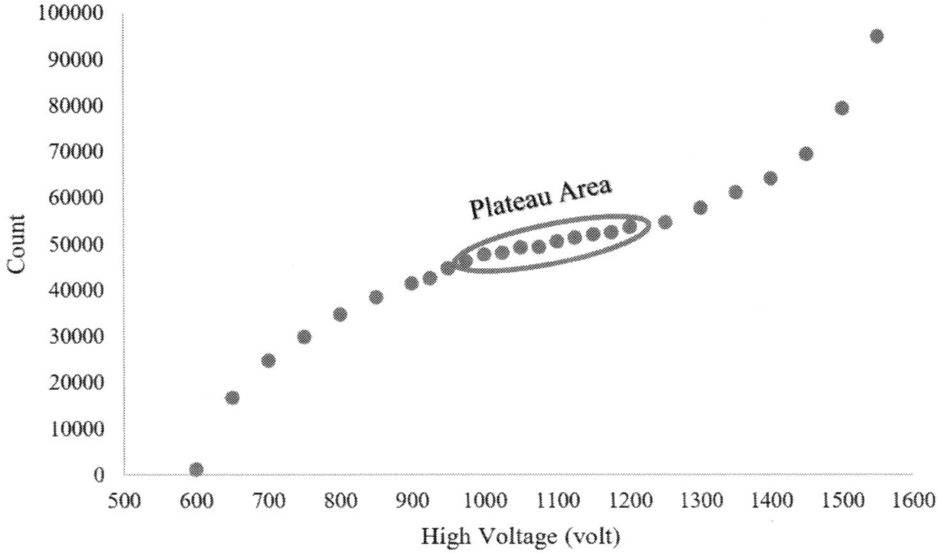

Figure 4. Count Response to High Voltage for BC-412 Detector.

Figure 4 illustrates a graph demonstrating the count's response to elevated voltage. Data retrieval occurs at every 50-volt interval, while in the predicted flat area, it is conducted at a narrower interval of 25 volts. The operating voltage for the BC-412 detector, as indicated in the figure above, ranges from 975 to 1200 volts. The increase in the count and the increase in voltage remain below 10%. The value is often minimal and stable. The plateau enables the detector to function stably and uniformly, facilitating reliable radiation detection across various operating situations. The distinctly identifiable plateau area facilitates calibration, as minor voltage fluctuations within this area minimally affect the overall detection efficiency.

3.2. Detector Stability Test

The emission of radiation is stochastic, consequently, repeated measurements of radiation will similarly exhibit randomness. A method is required to guarantee that measurement results are random due to stochastic radiation sources rather than detector inaccuracies. The prevalent technique employed is the Chi-Square approach. The Chi-Square approach will be employed to assess the stability of the detector [14]. The detector stability test employed a radiation source of ^{137}Cs, exhibiting a measured activity of three µCi, positioned at a distance of 10 cm from the detector. The test lasted about one hour, with data collected every two minutes until 20 data points were acquired. The test result data is subsequently computed for the Chi-Square. Table 1 displays the data from the detector stability test.

Table 1. Detector Stability Test Results.

No.	Count (X_i)	$(X_i - X_{avg})^2$	No.	Count (X_i)	$(X_i - X_{avg})^2$
1	46230	2606.103	11	45957	105008.4
2	46047	54779.4	12	46048	54312.3
3	46542	68094.9	13	46315	1152.602
4	46430	22186.1	14	46463	33105.8
5	46238	1853.303	15	46243	1447.803
6	45842	192764.9	16	46112	28577.9
7	46245	1299.603	17	46153	16396.8
8	46301	398.0025	18	46701	176358
9	46336	3019.502	19	46301	398.0025
10	46379	9594.202	20	46738	208803.3

Based on Table 1, ΣX_i equals 925621, and $\Sigma(X_i - X_{avg})^2$ equals 982157. The Chi-Square value, when inputted into the Chi-Square equation, is 21.22158. The values are subsequently compared with the Chi-Square table. Radiation detectors are considered excellent and stable if the Chi-Square value falls within the probability range of 0.1 to 0.9 [15]. The translated probability value of 0.1 is 11.651, while the probability value of 0.9 is 27.20. The obtained Chi-Square value of 21.22158 falls within the range of 11.651 to 27.20. The results demonstrate that the BC-412 detector is excellent and reliable.

3.3. Detector Linearity Test

If the radiation source detected by the detector is excessively high, there exists a threshold of radiation that the detector is incapable of measuring. This may occur because the radiation has not properly interacted with the detection material before exiting the detector. Furthermore, it may also arise from the detector's dead time, resulting in particular radiation not being registered by the detector. The linearity test of the detector seeks to discover the extent to which the detector maintains linearity in its measurements. The linearity assessment of the detector is often conducted utilizing multiple sources of identical radiation (e.g., ^{137}Cs) with varying doses. As that experiment is impractical, it can be altered by evaluating the source-detector distance function, which assumes that the radiation source dose will vary with distance changes. The effective dose rate for ^{137}Cs at a distance of one meter is 7.789×10^{-5} mSv/h per MBq. This information will act as a reference for evaluating the source-detector distance function. The results of the detector's linearity test are illustrated in Figure 5.

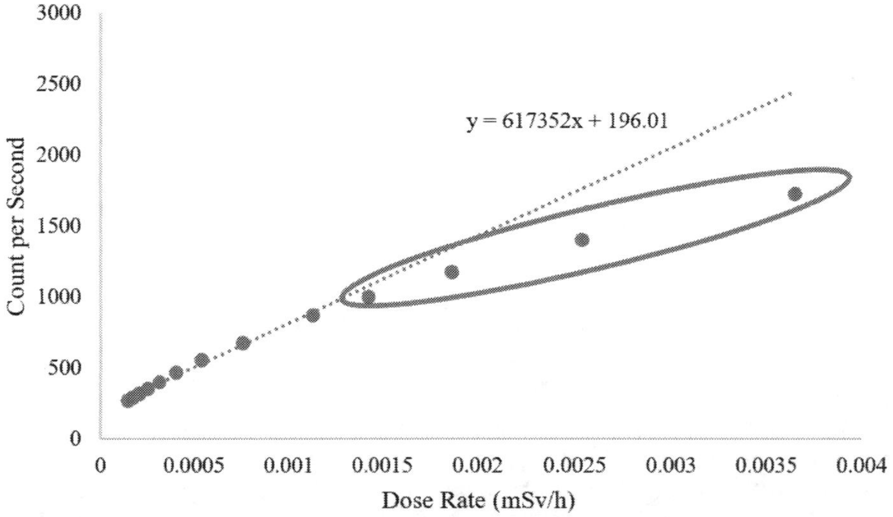

Figure 5. Linearity Detector Test Results.

Figure 5 illustrates that above a certain threshold, increased doses result in the detector's inability to measure linearly. The final indication of the detector's linearity occurs at a dose of 0.00112 mSv/h. The BC-412 detector is capable of measuring doses up to 0.00112 mSv/h.

3.4. Detector Angular Response Test

Despite the dependence of the detector on the source-detector distance, it is also crucial to evaluate the detector's sensitivity to angular variations. This is due to the fact that environmental radiation sources originate from multiple directions. It is essential to understand the response angle profile of the detector to optimize its utilization, particularly in the X-Z plane. The testing procedure is conducted using the source-detector arrangement in a horizontal orientation. The source utilized in this examination is ^{137}Cs. The angle is established by positioning the detector on a 360-degree arc image. Each test is conducted for one minute, with the source position adjusted according to the specified angle variation. Figure 6 displays the outcomes of the detector angular response test.

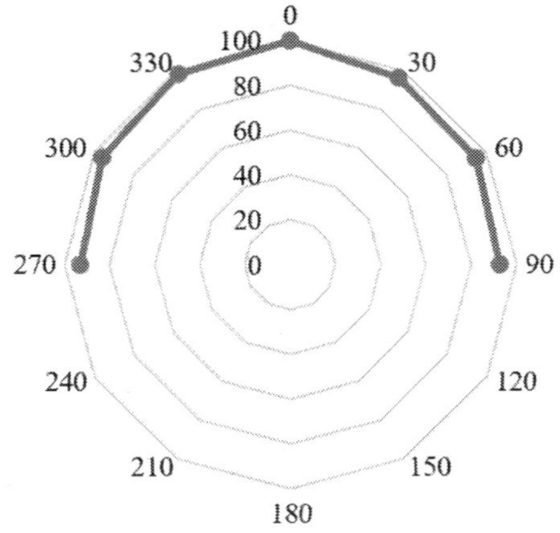

Figure 6. Detector Angular Response Test Results.

Figure 6 demonstrates that the detector angle response test for the X-Z plane yields outstanding results. The minimum angle response value occurs at 90°, representing a decrease of 92.39% from 0°. This indicates that optimal measurements are obtained on the detector's face (0°), whereas measurements taken on the detector's side yield the least results [16].

3.5. Detector Response Comparison

The response of the BC-412 detector was measured using an artificial point gamma source of ^{137}Cs and ^{133}Ba and a natural gamma source of ^{232}Th in the form of a mantle lamp. The results of measuring these sources can be seen in Table 2.

Table 2. Comparison of Detector Responses with Different Sources.

Distance-to-Source Detector (cm)	Detector Geometry	Source	Absolute Efficiency		
			PVT (3" x 2,25")	NaI(Tl) (2" x 2")	GM (0,79" x 1,57")
25	3" x 2.25"	^{137}Cs	0.001426	0.001137	0.000012
		^{133}Ba	0.002414	0.001255	0.000010
		^{232}Th*	0.000429	0.000340	0.000037

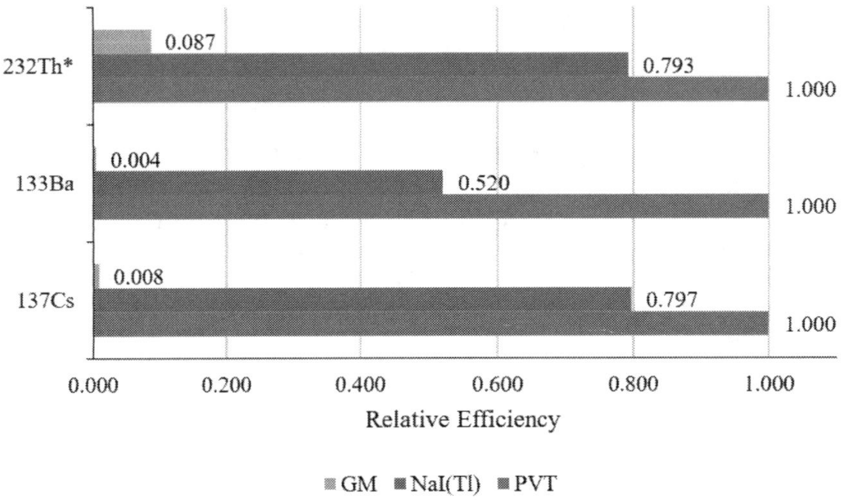

Figure 7. Comparison of Efficiency Relative to PVT.

The variable that is highlighted in relation to detector responses is absolute efficiency. Absolute efficiency is derived from the ratio of the net count to the total gamma[17]. To determine the absolute efficiency of ^{232}Th, the total value of gamma ^{137}Cs is utilized. This is due to the absence of any initial activity from ^{232}Th. The ^{232}Th data is simply utilized for detector comparison, and no definitive conclusions can be established. Table 2 indicates that the BC-412 scintillator plastic detector has the highest absolute efficiency compared to the 2-inch NaI(Tl) and GM detectors. Bar charts and efficiency relative to PVT are employed for illustration in Figure 7 to enhance clarity in comparison. The GM detector exhibits the lowest relative efficiency in comparison to other detectors. This aligns with the attributes of GM detectors, which typically exhibit the lowest efficiency relative to scintillation and semiconductor detectors. The NaI(Tl) detector has lower efficiency compared to the plastic scintillation detector, although possessing superior scintillation properties in terms of light production. The reduced effectiveness of the NaI(Tl) detector is attributed to its smaller scintillator geometry. The scintillator's geometry significantly affects the detector's absolute efficiency[18].

4. Conclusions

A plastic scintillation detector has been designed with PVT BC-412 material. The design process commences with the installation of the reflecting layer, the connection of the scintillator to the photomultiplier tube (PMT), and the assembly of the detector housing. Upon completion of the design process, the detector undergoes testing and characterization. The experiments indicate a flat region for voltage increase, corresponding to the detector's operational voltage range of 975-1200 volts. The increase in detector count in this area remains consistent and is still under 10%. The stability assessment of the detector yielded a Chi-Square value of 21.22158, which remains within the acceptable range for stability. The detector exhibits linearity up to a dose of 0.00112 mSv/h. The optimal measurements for the detector angle response test are obtained on the detector's face. This study demonstrated that the designed scintillation plastic detector had superior absolute efficiency compared to both NaI(Tl) and Geiger-Muller detectors. The produced scintillation detector with PVT material serves as an alternative to NaI(Tl) for early detection capabilities in reactors.

Future research should concentrate on optimizing the material properties of PVT scintillators to enhance light output and sensitivity. Investigations into novel dopants, hybrid materials, and modular designs can maximize performance for particular applications. Evaluating PVT detectors under severe environmental conditions and integrating innovative technologies, such as machine learning techniques, can improve their applicability across many fields.

IOP Conf. Series: Materials Science and Engineering **1330** (2025) 012022 doi:10.1088/1757-899X/1330/1/012022

Acknowledgments
The author would like to acknowledge the Lembaga Pengelola Dana Pendidikan (LPDP), the Ministry of Finance of the Republic of Indonesia, for the educational scholarships provided. The author also would like to acknowledge the Ministry of Education, Culture, Research and Technology of the Republic of Indonesia, the research grant and innovative research grant of the Bandung Institute of Technology.

References
[1] IAEA 2021 Nuclear Technology Review *IAEA Report*
[2] IAEA 2014 Radiation Protection and Safety of Radiation Sources: International Basic Safety Standards *IAEA Safety Standards*
[3] Ji Y-Y, Ho Chung K and Ja Kang M 2020 Assessment of dose rate of detected gamma emitting nuclides using a carborne survey with a large volume NaI(Tl) detector *Progress in Nuclear Energy* **123** 103272
[4] al-Khafaji I H K 2013 A Comparative Study of the Scintillation Detector NaI(Tl) in Two Sizes 2"x2" ,1"x1.5" *Advances in Physics Theories and Applications* **21**
[5] Kim D, Lee S, Park J, Son J, Kim T H, Kim Y H, Pak K and Kim Y K 2020 Performance of 3D printed plastic scintillators for gamma-ray detection *Nuclear Engineering and Technology* **52** 2910–7
[6] Siciliano E R, Ely J H, Kouzes R T, Milbrath B D, Schweppe J E and Stromswold D C 2005 Comparison of PVT and NaI(Tl) scintillators for vehicle portal monitor applications *Nuclear Instruments and Methods in Physics Research Section A: Accelerators, Spectrometers, Detectors and Associated Equipment* **550** 647–74
[7] Pla-Dalmau A, Bross A D and Mellott K L 2001 Low-cost extruded plastic scintillator *Nuclear Instruments and Methods in Physics Research Section A: Accelerators, Spectrometers, Detectors and Associated Equipment* **466** 482–91
[8] Pausch G, Roemer K, Herbach C 2011 Characterization and calibration of large-volume PVT detectors by Backscatter Gating *IEEE Nuclear Science Symposium Conference Record* 2214–9
[9] Meshkian M, Allwork C, Gendotti U, Ellis M, Schotanus P 2017 Applications of innovative SiPM-based PVT scintillator detectors *IEEE Nuclear Science Symposium and Medical Imaging Conference (NSS/MIC)* 1–5
[10] Watanabe K 2022 Applications of scintillators in optical-fiber-based detectors *Jap J Appl Phys* 62
[11] Kouzes R 2011 PVT analysis with a deconvolution algorithm *Symetrica Technical Report*
[12] Liao S, Erasmus R, Jivan H, Pelwan C, Peters G and Sideras-Haddad E 2015 A comparative study of the radiation hardness of plastic scintillators for the upgrade of the Tile Calorimeter of the ATLAS detector *J. Phys.: Conf. Ser.* **645** 012021
[13] Seco J, Clasie B and Partridge M 2014 Review on the characteristics of radiation detectors for dosimetry and imaging *Phys. Med. Biol.* **59** R303–47
[14] Rasito, Suherman A and Sukmabuana P 2014 PENGEMBANGAN SPEKTROMETER GAMMA PORTABEL BERBIAYA MURAH MENGGUNAKAN SOUND CARD USB 10
[15] Abimanyu A 2014 Pengujian Modul Monitor Radiasi Pada Sistem Pemantauan Pengangkutan Zat Radioaktif *JNTETI* **3** 2
[16] Rasito, Ilham P, Oetami R H and Suherman A 2009 SIMULASI KALIBRASI EFISIENSI PADA DETEKTOR HPGE DENGAN METODE MONTE CARLO MCNP5
[17] Tekin H O, Almisned G, Issa S A M, Zakaly H M H, Kilic G, Ene A 2022 Calculation of NaI(Tl) detector efficiency using ^{226}Ra, ^{232}Th, and ^{40}K radioisotopes: Three-phase Monte Carlo simulation study *Open Chemistry* 20:541-549
[18] Kwak S W, Yoo H S, Jang S S, Kim J S, Yoon W K 2008 Spectral Analysis Method of Plastic Scintillator-based Radiation Detector against Nuclear/Radiological Terrorism *Transactions of the Korean Nuclear Society Autumn Meeting*

IOP Conf. Series: Materials Science and Engineering **1330** (2025) 012023 doi:10.1088/1757-899X/1330/1/012023

Correction Factors Evaluation in Calculation of Saturation Activity for TRIGA 2000 In-Core Neutron Flux Measurements

Fahma Roswita[1,3*], Haryo Seno[2], Dikdik Sidik Purnama[2], Nina Widiawati[1], Zaki Su'ud[3], and Nuri Trianti[1]

[1] Research Center for Nuclear Reactor Technology, Research Organization for Nuclear Energy, National Research and Innovation Agency, Building 720 KST B.J. Habibie Serpong, South Tangerang, Indonesia 15314
[2] Research Center for Nuclear Beam Analysis Technology, Organization for Nuclear Energy, National Research and Innovation Agency, Building 720 2nd Floor KST B.J. Habibie Serpong, South Tangerang, Indonesia 15314
[3]Nucelar Science and Engineering Study Program, Faculty of Mathematics and Natural Sciences, Bandung Institute of Technology, Ganesha 10, Bandung, Indonesia 40132

*E-mail: fahma.roswita@brin.go.id

Abstract. Neutron flux measurement for research reactor in-core is typically performed using the neutron activation method. In this study, an ^{197}Au sample The sample in this study used an ^{197}Au was irradiated to measure neutron.The neutron was calculated by determining the saturation activity of the irradiated sample, with correction factors applied for counting efficiency and uncertainty. This study analytically evaluated the influence of each correction factor on the neutron flux measurements in the TRIGA 2000 reactor core. The results indicate significant difference in flux calculations when using $\epsilon_{Overall\ Counting}$ and $\epsilon_{Calibration}$. The averages value of neutron flux using $\epsilon_{Overall\ Counting}$ is only 0.5% of the flux value using $\epsilon_{Calibration}$. The difference is due to the lack of consideration of intrinsic efficiency, which affects the full peak energy efficiency of the detector. As a result, $\epsilon_{Overall\ Counting}$ is also affected by intrinsic efficiency.

1. Introduction

Indonesia has three research reactors, namely the TRIGA 2000 Reactor in Bandung, the Kartini Reactor in Yogyakarta, and G.A. Siwabessy Multipurpose Reactor (RSG) in Serpong, South Tangerang. The oldest research reactor is the TRIGA 2000 reactor which has been in operation for more than 50 years. Because it has been in operation for so long, reactor core management must be implemented very carefully. The main objective of core management is to achieve optimal reactivity along with the most economical use of fuel. The basic principle of reactor core management, namely reshuffling, and refueling, which will change the configuration of the reactor core, is intended to increase excess reactivity (k-eff > 1.0) by considering other safety parameters, such as shutdown margin and power peaking factor both axially and radially.

One of the most important parameters for evaluating research reactor performance is the neutron flux. Accurate neutron flux measurements are essential for optimizing in-core activation methods, calibrating fission power, and verifying the core configuration [1]. Therefore, experimental neutron flux measurement were carried out after changes in the configuration of the reactor core [2]. The purpose of these measurements is to asses of the effect of the new core configuration on the axial distribution of thermal neutron flux and to determine the magnitude of thermal neutron flux. Neutron flux distribution is a key input for various experiments and plays a crucial role in determining the power distribution in the reactor core [3]. The neutron flux values are used to generate flux distribution data for the reactor core, helping identify the most optimal locations for utilizing irradiation facilities.

Additionally, neutron flux measurements are essential for validating simulation codes in research reactors[4]. comparing experimental measurements with simulation calculations support reactor safety and provide necessary corrections for in-core irradiation activities, such as those carried out at the Central Thimble [5] and Rabbit Channel[6].

1.1 Indirect Flux Neutron Measurement

There are two methods for measuring neutron flux: direct and indirect measurement. Direct measurement typically uses neutron detectors, while indirect measurement involves irradiating samples, also known as neutron activation methods [7]. In this experiment, neutron flux measurements were conducted using the indirect method, or neutron activation method, with an ^{197}Au sample of 99% purity. The sample was in the form of a coin with a diameter of 0.5 cm and a thickness of 0.05 cm.

Experiments were conducted using a TRIGA 2000 reactor, operated at a power level of 100 kW. The reactor takes 15 minutes to reach the critical state. Once criticality is achieved, the ^{197}Au sample is inserted into the in-core position. The sample was irradiated for 5 minutes, after which the reactor was shut down to prevent further neutron exposure. The irradiated ^{197}Au sample transforms into ^{198}Au, which emits gamma radiation. This gamma radiation is then measured using a gamma spectrometer. The activation reaction of ^{197}Au can be written as follows:

$$Au^{197}(n,\gamma)Au^{198} \xrightarrow{\gamma} Au^{198} \xrightarrow{t_{1/2}=2.69\,d} Hg^{198} + e^{-} + \bar{\nu} + \gamma\ (E = 411.8\ keV)$$

Neutron flux can be calculated using the following formula [8]:

$$A_{\infty} = \varphi \Sigma_{act} V \quad (1)$$

φ (neutron/cm^2 s) is the neutron flux received by the sample. Σ_{act} (cm^2) is the activation cross section, which depends on the material used, as each material has a different cross section. V(cm^3) is the volume of the sample. A_{∞} (dps) is the saturation activity, defined as the condition where the rate of nuclear production equals the rate of nuclear decay. The saturation activity is determined from the sample counting results using the following equation [8]:

$$A_{\infty} = \frac{\lambda C}{\epsilon(1-e^{-\lambda t_i})e^{-\lambda t_d}(1-e^{-\lambda t_c})} \quad (2)$$

λ = decay constant
C = count (cps)
ϵ = overall counting efficiency (%)
t_i = duration of irradiation (s)

t_d = delay time (s)
t_c = time count (s)

There are two types of samples in this study, Au foil sample and Au foil sample wrapped in cadmium (Cd). The neutron flux received by the Au sample represents the total neutron flux, while the neutron flux received by the Cd-wrapped Au sample represents the epithermal neutron flux. This is because Cd has the ability to absorb thermal neutrons. The results of neutron flux calculations depend on the activity measurements. To calculate the activity of the sample or radioactive source measured by an HPGe detector, the overall counting efficiency is required (as shown in equation 2). This study compares the detector efficiency calibration values with other components of overall counting efficiency, namely geometry efficiency and the inherent relative efficiency of the detector for the thermal neutron flux calculation.

2. Materials and Methods

2.1 HPGe Detector

HPGe (high-purity germanium) detectors are a type of semiconductor detector. Semiconductor detectors have advantages such as the ability to resolve the energy of particles from a polyenergetic energy spectrum [10]. The HPGe detectors work by utilizing the ionization effect that occurs when radiation interacts with the germanium material inside the detector. This interaction generates a voltage that produces an electrical signal, which can be measured. The detector used in this study was a Canberra HPGe, model GC3518 with a relative efficiency of 35% as stated in the datasheet. . This detector has an active diameter of 6.23 cm. Measurements using HPGe detectors depend on the full energy peak efficiency, which is influenced by both geometry and energy conditions. The geometric condition depends on the source and detector geometry, as well as the distance between the source and the detector. The energy is affected by the characteristics of the detector and its intrinsic efficiency.

2.2 Detector Efficiency Calibration

Efficiency calibration is a crucial step for gamma spectrometer equipment. The detector efficiency is determined using standard reference sources and/or Monte Carlo simulations [11]. When using a standard source, the source is measured, and the efficiency of each detected energy peak is calculated. This efficiency data is then used to generate an efficiency curve. The detection efficiency curve depends not only on the detection system but also on the sample shape and sample matrix [12].

In this study, the detector efficiency calibration was performed using a Eu-152 standard source. Eu-152 was selected because it has multiple energy peaks, ranging from low energy at 121.78 keV to high energy at 1408.01 keV. The radioactive ^{198}Au sample has a gamma emission peak at 411.8 keV. This peak is close to one of the energy peaks of Eu-152, which is at 411.12 keV. Therefore, Eu-152 is considered a suitable standard source for calibrating the efficiency of measurements for the ^{198}Au sample. The efficiency for each detected energy peak is calculated, and an energy vs. efficiency curve is generated. The efficiency for each energy peak is calculated using the following equation.

$$A = \frac{cps}{\varepsilon\, I_\gamma} \tag{3}$$

Where A (dps) is the activity in disintegrations per second, cps is the count per second, ε is the efficiency at a specific energy peak (%), and I_γ is the probability of gamma radiation being emitted at that specific energy peak (%).

2.3 Overall Counting Efficiency

Overall counting efficiency (ϵ) represents the efficiency that accounts all the factors affecting a detector's ability to measure the radiation produced by a radioactive source. In other words, overall counting efficiency measures how effectively a detection system (including the detector and measurement geometry) detects gamma radiation from a given source. The value of overall counting efficiency is derived from various components related to measurement geometry, the detector, and the radiation source being measured. In this study, the flux value affected by the activity of the correction using the detector efficiency value and the overall counting efficiency value are compared. Overall counting efficiency can be considered as the total efficiency, with the following calculation formula.

$$\epsilon_{total} = \varepsilon_g \times \varepsilon_r \times \varepsilon_\gamma \tag{4}$$

a. Geometrical Efficiency (ε_g)

Geometric efficiency is the ratio between the number of gamma photons that reach the surface of the detector to the number of gamma photons emitted by the source in all directions. This efficiency depends on factors such as the distance between the source and the detector, the size of the detector, and the direction of the gamma radiation emission. Geometrical efficiency in this research is given by:

$$\varepsilon_g = \frac{\Omega}{4\pi} = \frac{2\pi\left(1 - \frac{d}{\sqrt{d^2 + r^2}}\right)}{4\pi} = \frac{\left(1 - \frac{d}{\sqrt{d^2 + r^2}}\right)}{2\pi} \tag{5}$$

Solid angle (Ω) represents the solid angle (in steradians) subtended by the detector at the source position. The symbol d is the sample distance in cm. The symbol r is the active radius of the HPGe detector in cm.

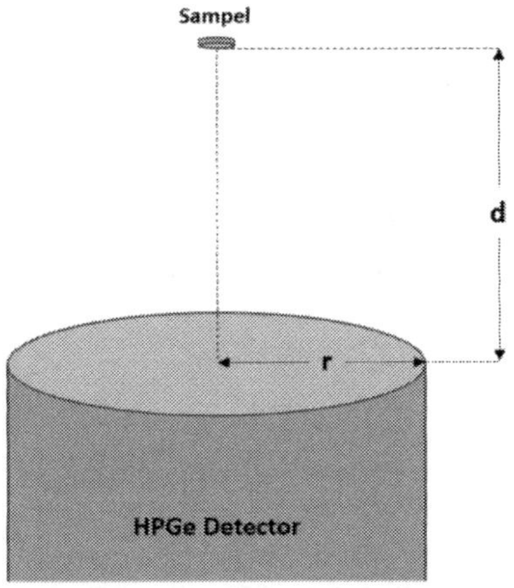

Figure 1. The distance of the sample (d) to the HPGe detector and the active radius of the HPGe detector (r).

b. Relative Efficiency (ε_r)
Relative efficiency is the comparison between the absolute efficiency of a detector and the efficiency of a standard detector. The standard detector used is typically a NaI(Tl) detector of a specified size[10]. In other words, relative efficiency describes the detector's ability to detect a source by comparing it to a detector whit a knowing efficiency for detecting a specific standard source. The relative efficiency value is usually provided in the detector's datasheet.

c. Gamma Emission Probability (ε_γ)
Gamma emission probability referes to the likelihood that a nuclear decay will emit a gamma photon with a specific energy. This probablity depends on the isotope and the energy of the emitted gamma photon. For ^{198}Au, the gamma emission probability at an energy of 411.8 keV is 95.62%.

2.4 Uncertainty Factors

The uncertainty of neutron flux measurement data is caused by several factors, including neutron irradiation position, Au sample weight, counting statistics, counting geometry, and nuclear data[13]. It is also crucial to fully understand the uncertainties pertaining to these measurements, as they could restrict the inferences that can be made from the data and spur additional research to lower the major contributions to the final combined uncertainty. A measurement of the neutron flux spectrum, for instance, at a specific irradiation site might yield valuable data for isotope production optimization, but it might not have enough fidelity to support fuel management activities [4].

It is necessary to calculate a combination of these uncertainties if the measurement result is affected by multiple uncertainty factors. For instance, the uncertainty variables in this study comprise a number of elements, including time, sample mass, and net counting. In general, the equation that follows can be used to calculate the measurement's combined uncertainty factor[14]. The average uncertainty of each counting result evaluated using GENIE 2000 is used to calculate the counting uncertainty ($u(C)$). It is assumed that the time uncertainty ($u(t)$) is 0.5 seconds with t for time in second and the mass uncertainty ($u(m)$) is 0.00005 grams with m for mass in gram.

$$u_{combined} = \frac{u\,(\phi)}{\phi} = \sqrt{\left[\frac{u(C)}{C}\right]^2 + \left[\frac{u(m)}{m}\right]^2 + \left[\frac{u(t)}{t}\right]^2 + \cdots} \quad (6)$$

3. Result and Discussion

Before counting the irradiated ^{198}Au samples, detector efficiency calibration was performed using Eu-152 standard source. The efficiency for each energy peak observed during the counting of the Eu-152 standard source was calculated. The efficiency calculation results at each detected energy peak were used to create a curve, which is displayed in Figure 1. A power trendline was fitted to the data, resulting in the equation $y = 0.0141x^{-0.781}$. The value x was substituted with the value 411.8, representing the energy peak of ^{198}Au, resulting in y = 1.28E-04. This y value was used as the efficiency value ($\epsilon_{Calibration}$) for each activity calculation of the ^{198}Au samples.

IOP Conf. Series: Materials Science and Engineering **1330** (2025) 012023 doi:10.1088/1757-899X/1330/1/012023

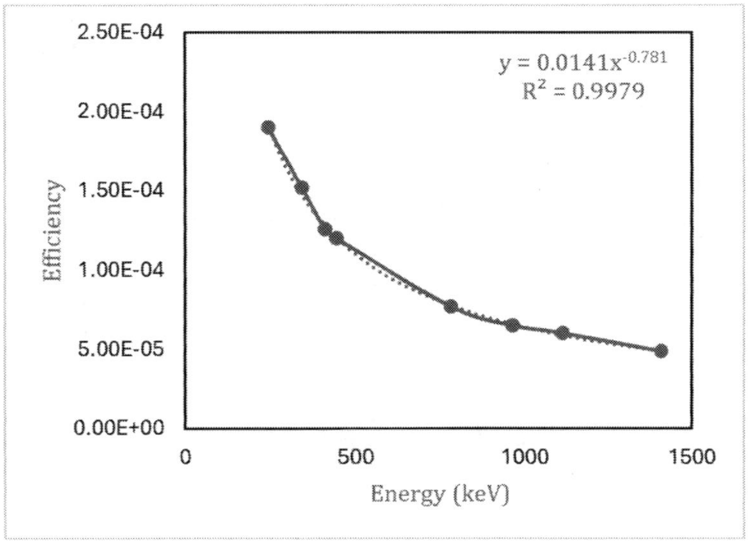

Figure 2. Detector efficiency calibration curve from Eu-152 standard source

In addition to calibrating detector efficiency with a standard source, efficiency—or in this case, overall counting efficiency—can also be obtained from the calculation of total efficiency. This study uses three efficiency components: geometric efficiency, relative efficiency, and gamma emission probability. Geometric efficiency is calculated using equation 10, where the distance between the sample and the detector is 5 cm, and the active radius of the detector is 3.115 cm. The geometric efficiency calculation uses the point source equation because the Au foil used has a very small radius compared to the active radius of the detector. Each different sample shape will have a different geometric efficiency equation. Thus, geometric efficiency is highly dependent on the shape, position, and distance of the source from the detector. From equation 10, a value of ε_g= 7.56E-02 was obtained.

The detector's relative efficiency in this study is assumed to be the same as stated in the detector's datasheet. In other words, the detector is considered to have the same performance as a new detector, without any degradation in function. These two efficiencies are then multiplied by the gamma emission probability of ^{198}Au, which is 95.62%, to obtain the overall counting

Table 1. Comparison of total neutron flux calculations using overall counting efficiency and standard source efficiency calibration

Position (cm)	Activity Average		Total Neutron Flux (neutron/cm² s)	
	$\epsilon_{Overall\ Counting}$	$\epsilon_{Calibration}$	$\epsilon_{Overall\ Counting}$	$\epsilon_{Calibration}$
42.75	5.87×10^{7}	1.16×10^{10}	$(9.65\pm0.03)\times 10^{8}$	$(1.91\pm0.005)\times 10^{11}$
45.75	7.08×10^{7}	1.40×10^{10}	$(1.16\pm0.03)\times 10^{9}$	$(2.29\pm0.006)\times 10^{11}$
48.75	7.72×10^{7}	1.53×10^{10}	$(1.27\pm0.03)\times 10^{9}$	$(2.51\pm0.007)\times 10^{11}$
56.75	2.93×10^{7}	5.79×10^{9}	$(4.82\pm0.04)\times 10^{8}$	$(9.52\pm0.004)\times 10^{10}$
59.75	3.69×10^{7}	7.29×10^{9}	$(6.05\pm0.03)\times 10^{8}$	$(1.20\pm0.003)\times 10^{11}$
62.75	4.26×10^{7}	8.43×10^{9}	$(7.00\pm0.02)\times 10^{8}$	$(1.39\pm0.004)\times 10^{11}$

*The position is measured from the base position of the active reactor core towards the axial top.

IOP Conf. Series: Materials Science and Engineering **1330** (2025) 012023 doi:10.1088/1757-899X/1330/1/012023

efficiency. The overall counting efficiency ($\epsilon_{Overall\ Counting}$) is used to calculate the activity value of each counted ^{198}Au sample.

Table 1 display the calculated results of the total neutron flux received by the uncovered ^{198}Au sample. The neutron flux decreases with increasing position, regardless of whether $\epsilon_{Overall\ Counting}$ or $\epsilon_{Calibration}$ is used. This is because the fuel meat in the fuel rods is typically located in the central axial region. The topmost and bottommost positions of the fuel rod are occupied by the handle tools, namely the top end fitting and bottom end fitting. Therefore, theoretically, the highest neutron flux peak in the reactor core is located in the central axial region. However, there is a significant difference between the total neutron flux using $\epsilon_{Overall\ Counting}$ and that using $\epsilon_{Calibration}$. The total neutron flux with $\epsilon_{Calibration}$ has a higher flux value, averaging in the order of 10^{11} compared to the total neutron $\epsilon_{Overall\ Counting}$ which is in the order of 10^{8}-10^{9}.

Table 2. Comparison of epithermal neutron flux calculations using overall counting efficiency with standard source efficiency calibration

Position (cm)	Activity Average		Epithermal Neutron Flux (neutron/cm^2 s)	
	$\epsilon_{Overall\ Counting}$	$\epsilon_{Calibration}$	$\epsilon_{Overall\ Counting}$	$\epsilon_{Calibration}$
44.25	1.08×10^{7}	2.14×10^{9}	$(1.08\pm0.007)\times 10^{8}$	$(3.56\pm0.013)\times 10^{10}$
47.25	1.32×10^{7}	2.61×10^{9}	$(2.16\pm0.007)\times 10^{8}$	$(4.28\pm0.014)\times 10^{10}$
50.25	1.44×10^{7}	2.85×10^{9}	$(2.38\pm0.008)\times 10^{8}$	$(4.70\pm0.016)\times 10^{10}$
58.25	3.23×10^{6}	6.38×10^{8}	$(5.30\pm0.035)\times 10^{7}$	$(1.05\pm0.007)\times 10^{10}$
61.25	6.47×10^{6}	6.09×10^{9}	$(1.06\pm0.005)\times 10^{8}$	$(9.95\pm0.049)\times 10^{10}$
64.25	3.72×10^{6}	7.35×10^{8}	$(6.10\pm0.038)\times 10^{7}$	$(1.21\pm0.008)\times 10^{10}$

The calculated results of the epithermal neutron flux received by the ^{197}Au sample covered with Cd are shown in Table 2. Similar to the total neutron flux, the epithermal neutron flux decreases with increasing position. The positions of the total and epithermal neutron fluxes are slightly different because the two samples were placed with a small distance between them to avoid neutron absorption interference between the samples. Therefore, the samples were arranged in a staggered position between the Au sample and the Cd-covered Au sample.

A notable difference is also observed between the epithermal neutron flux results using $\epsilon_{Overall\ Counting}$ and those using $\epsilon_{Calibration}$. Specifically, the difference in epithermal neutron flux between the two methods is 99.5%. If the total and epithermal neutron fluxes exhibit differences in value, then the thermal neutron flux will also show similar differences. This can be observed from the data in Table 3. The averages value of neutron flux using $\epsilon_{Overall\ Counting}$ is only 0.5% of the flux value using $\epsilon_{Calibration}$.

IOP Conf. Series: Materials Science and Engineering **1330** (2025) 012023 doi:10.1088/1757-899X/1330/1/012023

Table 3. Comparison of thermal neutron flux calculations using overall counting efficiency with standard source efficiency calibration

Position (cm)	Thermal Neutron Flux (neutron/cm^2 s)	
	$\epsilon_{Overall\ Counting}$	$\epsilon_{Calibration}$
43.5	(7.85±0.036)× 10^8	(1.55±0.007)× 10^{11}
46.5	(9.44±0.038)× 10^8	(1.87±0.008)× 10^{11}
49.5	(1.03±0.004)× 10^9	(2.04±0.008)× 10^{11}
57.5	(4.28±0.044)× 10^8	(8.47±0.087)× 10^{10}
60.5	(5.00±0.033)× 10^8	(2.02±0.014)× 10^{10}
63.5	(6.39±0.045)× 10^8	(1.26±0.009)× 10^{11}

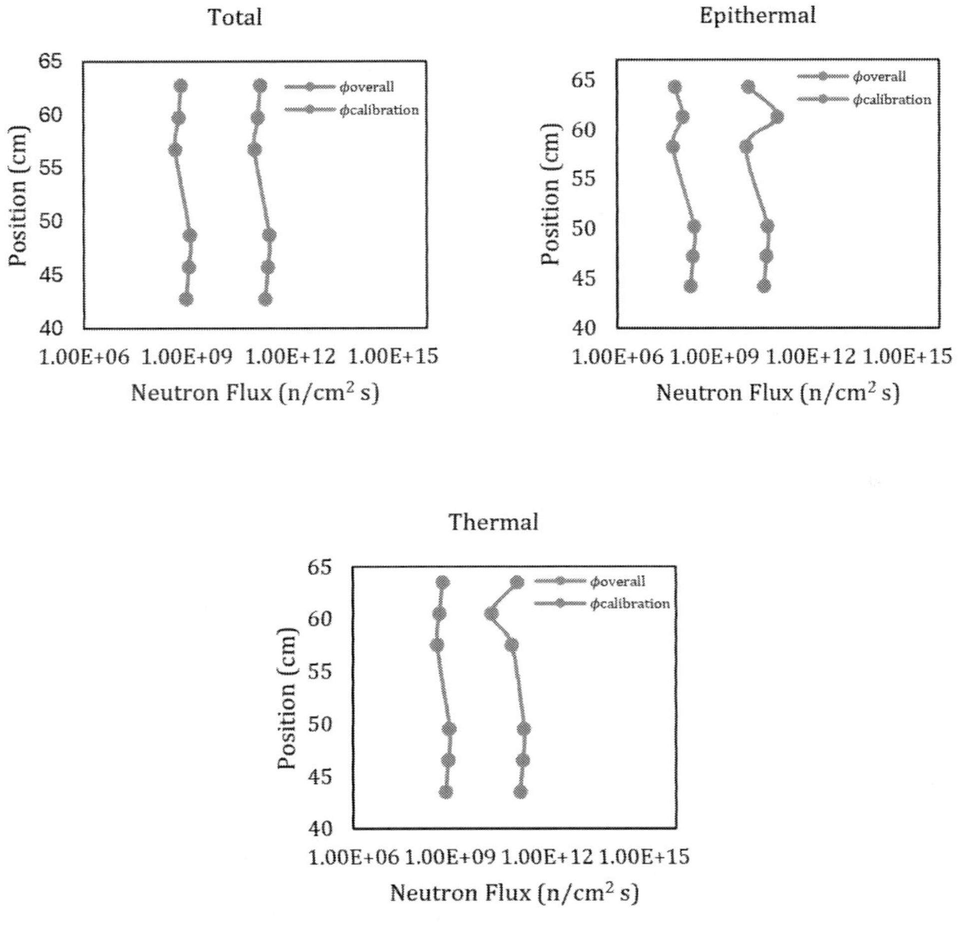

Figure 3. Comparison of total, epithermal, and thermal neutron flux graph

4. Conclusion

There is a very significant difference in flux calculations when using $\epsilon_{Overall\ Counting}$ and $\epsilon_{Calibration}$. The averages value of neutron flux using $\epsilon_{Overall\ Counting}$ is only 0.5% of the flux value using $\epsilon_{Calibration}$. This discrepancy is likely due to the lack of consideration of intrinsic efficiency, which affects the detector's full peak energy efficiency. As a result, the $\epsilon_{Overall\ Counting}$ is also influenced by intrinsic efficiency. Therefore, further investigation into this intrinsic parameter is necessary, either through Monte Carlo simulations or experiments using a standard ^{198}Au source.

References

[1] B. G. Park *et al.*, "Measurement of the neutron flux distribution in a research reactor by neutron activation analysis," *J. Radioanal. Nucl. Chem.*, vol. 330, no. 2, pp. 501–512, 2021, doi: 10.1007/s10967-021-07837-2.

[2] E. Ratnawati, I. Jaka, and H. Ali, "EVALUATION OF THERMAL AND EPITHERMAL NEUTRON FLUX IN THE RABBIT SYSTEM FACILITY RSG GAS TERAS 89," *Nuclear Reactor Management Bulletin*, vol. 13, no. 1, 2016.

[3] M. H. B. Rabir *et al.*, "Measurement and simulation of thermal neutron flux distribution in the RTP core," *IOP Conf. Ser. Mater. Sci. Eng.*, vol. 298, no. 1, 2018, doi: 10.1088/1757-899X/298/1/012029.

[4] E. L. MacConnachie and D. R. Novog, "Measurement, simulation and uncertainty quantification of the neutron flux at the McMaster Nuclear Reactor," *Ann. Nucl. Energy*, vol. 151, Feb. 2021, doi: 10.1016/j.anucene.2020.107879.

[5] F. Roswita, H. Seno, P. Basuki, H. Wafda, A. M. Efendi, and N. Trianti, "Identification of thermal neutron flux in central thimble of TRIGA 2000 reactor with experimental method," *AIP Conf. Proc.*, vol. 2893, no. 1, pp. 4–5, 2023, doi: 10.1063/5.0173580.

[6] D. Chiesa *et al.*, "Measurement and Simulation of The Neutron Flux Distribution in The TRIGA Mark II Reactor Core," *Ann. Nucl. Energy*, vol. 85, pp. 925–936, 2015, doi: 10.1016/j.anucene.2015.07.011.

[7] R. S. A. Kuswandrata, A. Agung, S. Syarip, A. R. Antariksawan, U. S. Hidayat, and A. S. Wicaksono, "Performance Test of SPND Neutron Detector and Neutron Flux Mapping of Kartini Reactor Core," *AIP Conf. Proc.*, vol. 2967, no. 1, 2024, doi: 10.1063/5.0193149.

[8] G. F. Knoll, *Radiation Detection and Measurement*, Fourth Edi. John Wiley & Sons, Inc.

[9] T. P. Hardi and S. Center for Nuclear Industrial Materials Technology - BATAN Puspiptek Serpong Area, "Neutron Flux Measurement and Determination of Cadmium Ratio on Four-Circle Neutron Difractometer / Texture Difractometer (DN2)," 2009.

[10] N. Tsoulfanidis, "MEASUREMENT AND DETECTION OF RADIATION Second Edition," 1995.

[11] S. Sayın, M. Seferinoğlu, E. Yeltepe, B. Çetin, and S. Şentürk Lüle, "Assessment of the detection efficiency calibration of high-resolution gamma-ray spectrometers by EGSnrc and MCNP6.2 Monte Carlo codes," *Radiat. Phys. Chem.*, vol. 203, Feb. 2023, doi: 10.1016/j.radphyschem.2022.110601.

[12] S. Harb, K. S. Din, and A. Abbady, "Study of Efficiency Calibrations of HPGe Detectors for Radioactivity Measurements of Environmental Samples," in *Proceedings of the 3rd Environmental Physics Conference*, 2008, p. 60.

[13] B. G. Park *et al.*, "Measurement of the neutron flux distribution in a research reactor by neutron activation analysis," *J. Radioanal. Nucl. Chem.*, vol. 330, no. 2, pp. 501–512, Nov. 2021, doi: 10.1007/s10967-021-07837-2.

[14] A. Tajani and A. Markowicz, "EDXRF ANALYSIS OF THIN SAMPLES," in *Quantifying uncertainty in nuclear analytical measurements*, International Atomic Energy Agency, 2004, p. 250.

IOP Conf. Series: Materials Science and Engineering **1330** (2025) 012024 doi:10.1088/1757-899X/1330/1/012024

Comparison of Fission and Non-Fission Methods in Mo-99 Radioisotope Production

Dhiya Salma Salsabila[1*], Anis Rohanda[2]

[1] Master Program in Nuclear Science and Technology Department, Faculty of Mathematics and Natural Science, Bandung Institute of Technology, Jalan Ganesha 10 Bandung 40132, Indonesia.
[2] Research Center for Nuclear Reactor Technology, Research Organization for Nuclear Energy, National Research and Innovation Agency, Building No. 80, BJ Habibie Integrated Science Area, South Tangerang 15310, Indonesia.

*E-mail: dhiyasalmas@gmail.com

Abstract.
The utilization of the G.A. Siwabessy Multipurpose Reactor (RSG-GAS) includes the production of radioisotopes. Among the radioisotopes produced by RSG-GAS is Mo-99. The Mo-99 radioisotope is used in the production of Technetium-99m (^{99m}Tc), which is used in diagnostic imaging in the medical field. There are two ways to produce the Mo-99 radioisotope. First, with a fission scheme using Low Enriched Uranium (LEU) in the form of UO_2. Second, with a non-fission scheme by irradiating MoO_3 samples. The aim of this study is to analyze the production results of radioisotopes using the fission scheme compared to the non-fission scheme using the ORIGEN 2.1 program. In this study, an MoO_3 sample weighing 8.753 grams is compared to UO_2 weighing 8.753 grams with the composition: U-235 1.508 grams, U-238 6.206 grams, and O-16 1.039 grams. Then, variations in the irradiation position in the reactor were made, namely at the Central Irradiation Position (CIP) at locations E7 and D6, as well as the Irradiation Position (IP) with reactor power of 5, 15, and 30 MWt. The calculation results show that at CIP-D6 with a power of 30 MWt, the production result of the non-fission scheme is 7.05 Ci. Meanwhile, the production result of the fission scheme under the same conditions is 797.10 Ci. This is due to the better ability of UO_2 to capture neutrons needed in nuclear reactions, as well as having more uranium isotopes that easily undergo neutron fission. Therefore, for Mo-99 production, it is recommended to use the fission scheme.

1. Introduction

The RSG-GAS is a research reactor with a power of 30 MWt. One of the uses of the RSG-GAS is for the production of radioisotopes. Some of the radioisotopes that can be produced by the RSG-GAS include I-131, Mo-99, and Ir-192. These radioisotopes are later used in the medical field, for example, Mo-99 is beneficial for the diagnosis and monitoring of diseases. Due to these benefits, the demand for Mo-99 radioisotopes continues to increase, and various efforts are being made to enhance the efficient production of this radioisotope.

Various studies have been conducted on the production of Mo-99 radioisotopes using different target materials. Research by Budisantoso on the production of Mo-99 using uranyl nitrate solution as the target material in the Kartini Reactor resulted in 3.23 Ci of Mo-99 per 300 cc of solution. [1]. Subsequent research by Kuntjoro on the activity of Mo-99 isotopes in the RSG-GAS reactor demonstrated that irradiating four LEU targets weighing 11.609 grams each produced 300 Ci of Mo-99. [2]. Research by Sutrisno on optimizing LEU targets for Mo-99 production in the RSG-GAS reactor at the CIP position showed that fully filling the CIP with LEU targets weighing 1.2 grams and 3 grams per capsule produced 364 Ci and 911 Ci of Mo-99, respectively. [3].

The two main schemes used in Mo-99 production are the non-fission route, involving the irradiation of MoO_3 samples, and the fission route, involving the irradiation of LEU, specifically uranium dioxide (UO_2). In the fission scheme, research by Surian Pinem on the optimization of radioisotope production in the RSG-GAS reactor indicated that the irradiation positions at E-7 and G-7 are the best core positions for irradiating LEU targets with a maximum mass of 21 grams. [4]. The research was conducted at the Tehran Research Reactor by Ebrahim Abedi, utilizing 1.18 grams of LEU with an irradiation time of 7 days, followed by 1 day of decay process. The neutron flux at the reactor was 1.1E+14, resulting in the production of 290 Ci of Mo-99 [5]. Subsequently, at the Pakistan-1 Research Reactor (PARR-1), 150 Ci of Mo-99 was produced with a neutron flux of 1E+14 [6]. Meanwhile, in the non-fission research scheme conducted by I. Saptiama on the production of molybdenum radionuclide (Mo-99) using natural molybdenum samples (MoO_3), an average Mo-99 yield of 0.81 Ci was achieved with an average target mass of 5 grams [7]. The research by D.S. Sairanbayev successfully optimized irradiation conditions for natural molybdenum (MoO_3) at the WWR-K reactor, resulting in an increase in specific activity of produced Mo-99 from 1.82 Ci/g to 3.31 Ci/g. This was achieved by employing a neutron trap in the form of boron carbide inside the sample capsule [8].

This study aims to compare the production yields of Mo-99 between the two approaches. Previous studies have highlighted the advantages and disadvantages of each approach, yet there remains room for further exploration into the differences in Mo-99 production yields they generate. Factors such as irradiation conditions like position within the reactor and power variations can influence Mo-99 production efficiency. A study is needed to understand the differences in Mo-99 production yields between irradiating MoO_3 samples and LEU using UO_2. ORIGEN2.1 will be employed in this study as the calculation program for radionuclide activity. Variations in sample position and reactor power will be investigated to determine their effects on Mo-99 production yields.

This study is expected to provide valuable insights for the nuclear medicine industry. It aims to establish a foundation for developing more efficient and sustainable production strategies to meet the increasing demand for Mo-99.

2. Methodology

In this study, the activity of Mo-99 was calculated using ORIGEN2.1. Input data included thermal neutron flux values obtained from measurements conducted by the Directorate of Nuclear Facility Management (DPFK) at various points within the RSG-GAS reactor. Two experimental schemes were employed: a non-fission pathway utilizing an 8.753-gram MoO_3 target sample, and a fission scheme employing 8.753 grams of LEU in the form of UO_2, composed of 1.508 grams U-235, 6.206 grams U-238, and 1.039 grams O-16. Irradiation was conducted for 100 hours, followed by a 24-hour cooldown period. In non fission scheme the reactivity of Mo-98 towards neutrons ($\sigma_{n,\gamma} = 0.13\ b$), a high neutron flux and a long duration are required to produce an adequate of Mo-99. Stringers were used to insert the MoO_3 or LEU samples, as illustrated in Figure 1.

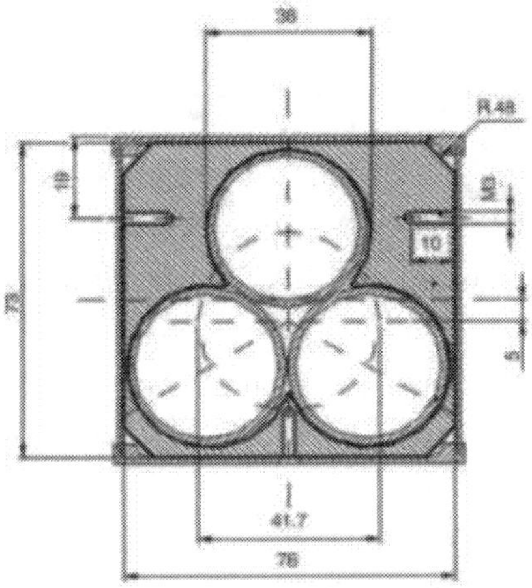

Figure 1. Stringer for irradiation in RSG-GAS [9]

The stringers can only be placed at CIP (E6, D6, E7, and D7) and IP (G7, D9, B6, E4) locations within the core of the RSG-GAS reactor. The maximum weight that can be inserted into each stringer is 100 grams. The core configuration of the RSG-GAS reactor can be seen in Figure 2.

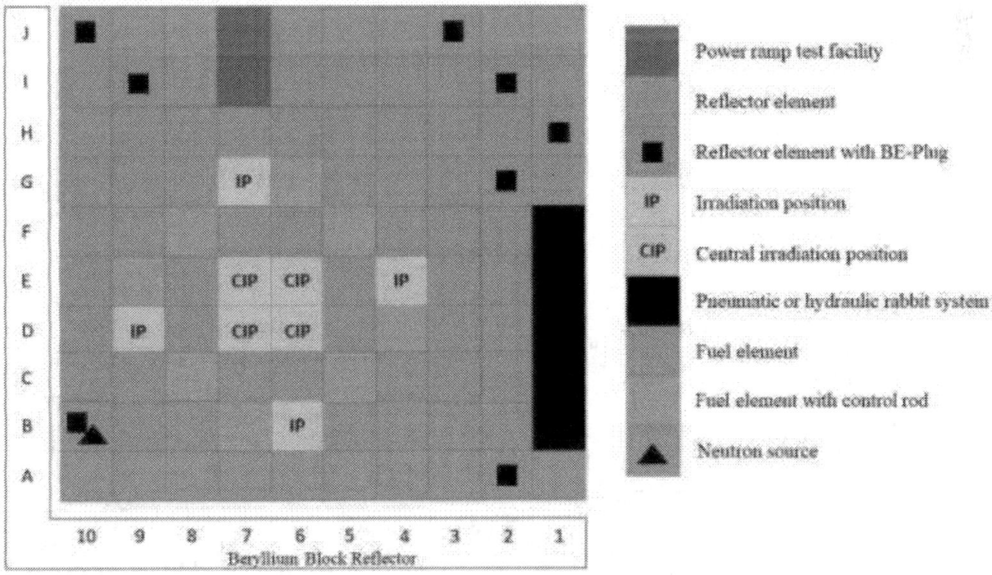

Figure 2. Core configuration of RSG-GAS

In this study, sample placement points were selected at CIP-E7, CIP-D6, and IP, with power variations of 5, 15, and 30 MWt. Neutron flux values are provided in Table 1. The neutron flux values are consistent across all irradiation positions [10].

Table 1. Neutron flux data at various positions in RSG-GAS

Neutron flux data at each position			
Power	CIP-E7	CIP-D6	IP
30 MWt	3.12E+14	3.5E+14	2.36E+14
15 MWt	1.56E+14	1.8E+14	1.18E+14
5 MWt	5.2E+13	5.8E+13	3.93E+13

Next, a comparison was conducted of Mo-99 activity calculations between MoO_3 and UO_2 targets using the Thermal.lib library. Activity calculations were performed at positions CIP-E7, CIP-D6, and IP. ORIGEN2.1, a widely used code for predicting the characteristics of spent reactor fuel and high-level waste, has several limitations when it comes to radioisotope production. One major limitation is its reliance on fixed nuclear data libraries, which may not always match the specific conditions of a given problem. Additionally, ORIGEN2.1 operates as a stand-alone tool and lacks the ability to interact with other modules to obtain problem-specific neutronic data. The software also simplifies modeling, which may not accurately represent complex scenarios involving continuous nuclide feed and chemical removal, essential for certain types of radioisotope production. Furthermore, while ORIGEN2.1 can generate decay emission spectra, it may not provide the detailed spectra needed for specific radioisotope production applications. These limitations can affect the accuracy and applicability of ORIGEN2.1 in certain radioisotope production scenarios.

3. Result and Discussion

Calculation of radioisotope using ORIGEN2.1

Table 2 presents the calculated results of Mo-99 radioisotope using MoO_3 samples at various reactor positions with power variations and their respective neutron fluxes. Irradiation lasted for 100 hours followed by a 24-hour cooling period. At position CIP-D6 with 30 MWt, the highest target value of 80.54 Ci was obtained, reducing to 62.60 Ci after cooling. In contrast, the smallest target value was observed at IP position with 5 MWt, measuring 9.04 Ci and reducing to 7.03 Ci after cooling. This disparity is due to the IP position at 5 MWt having the lowest neutron flux compared to other positions and power levels.

Table 2. Comparison of Mo-99 activity calculation results between non-fission and fission schemes

Position	Power (MWt)	Neutron flux	Non-fission scheme		Fission scheme	
			After irradiation (Ci)	After decay (Ci)	After Irradiation (Ci)	After decay (Ci)
CIP-E7	5	5.20E+13	1.05	0.81	123.50	96.02
	15	1.56E+14	3.14	2.44	365.20	283.80
	30	3.12E+14	6.28	4.88	714.40	555.30

CIP-D6	5	5.83E+13	1.17	0.91	138.40	107.60
	15	1.75E+14	3.53	2.74	408.50	317.60
	30	3.50E+14	7.05	5.48	797.10	619.60
IP	5	3.93E+13	0.79	0.62	93.53	72.70
	15	1.18E+14	2.38	1.85	277.70	215.90
	30	2.36E+14	4.75	3.69	546.20	424.60

Table 2 illustrates the calculated results of Mo-99 radioisotope using UO_2 at various points within the reactor with power variations as previously described. The highest target value is observed at position CIP-D6 with 30 MWt, yielding 797.10 Ci, reducing to 619.60 Ci after cooling. Conversely, the lowest target value is found at IP position with 5 MWt, measuring 93.53 Ci, reducing to 72.70 Ci after cooling.

Figure 3 depicts a comparison of Mo-99 radioisotope production results at 15 MWt. Fission scheme production of Mo-99 yields higher curie values. Black text in the figure represents non-fission scheme production, while red indicates fission scheme production. Location CIP-D6 demonstrates the highest Mo-99 radioisotope production with the fission scheme at 317.60 Ci, whereas the non-fission scheme yields only 2.74 Ci.

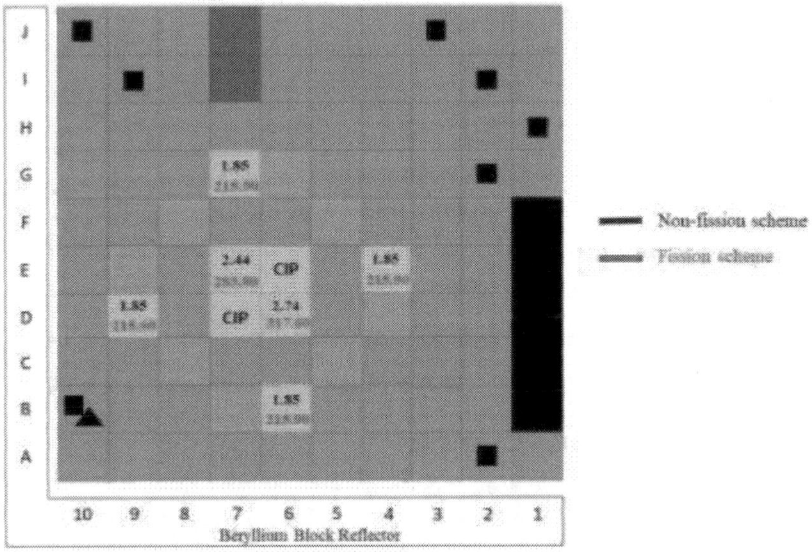

Figure 3. Comparison of Mo-99 production results (in curies) between non-fission and fission schemes at 15 MWt

Upon examination of its reactions, UO_2 undergoes fission reactions. Besides producing Mo-99, fission reactions also generate neutrons that initiate a chain reaction, causing neutrons to interact with other uranium nuclei and perpetuating the reaction continuously. The reaction is [11]

$$^{235}U + n \rightarrow ^{236}U \rightarrow ^{99}Mo + ^{134}Sn + 3n + \text{energy}$$

This chain reaction causes the fission product yield to continuously increase, thereby increasing Mo-99 production along with the ongoing reactions. Meanwhile, in the decay process of MoO_3 into Mo-99, the process occurs when the MoO_3 isotope Mo-98 is exposed to neutron flux within the reactor, a phenomenon known as neutron irradiation. Subsequently, the Mo-98 nucleus captures a neutron through the (n, γ) reaction. The reaction is [11]

$$^{98}Mo + n \rightarrow ^{99}Mo + \gamma$$

The neutron is captured by the Mo-98 nucleus, resulting in the production of Mo-99 and the emission of gamma rays (γ). The (n, γ) reaction does not produce neutrons, thus preventing a chain reaction. This causes the production of Mo-99 to depend on the number of neutrons involved and the availability of Mo-98.

Based on the calculations from Table 1, it is evident that Mo-99 production yields are higher with LEU irradiation using UO_2. This is attributed to UO_2's superior capability to capture and replenish neutrons required for the chain fission reactions that produce Mo-99. Additionally, UO_2 contains more uranium isotopes prone to neutron-induced fission compared to MoO_3.

Recovery and Purification

Another concern in the production of radioisotopes is to maximizing product yield. In fission scheme, after uranium dioxide (UO_2) targets are irradiated, the recovery and purification of Mo-99 involve several chemical processes to isolate Mo-99 from other fission products while ensuring high purity and yield. The irradiated targets are first dissolved using either alkaline or acidic solutions, depending on the target material and processing method. Alkaline dissolution, often used for aluminum-clad targets, employs sodium hydroxide (NaOH) to dissolve the entire target, forming a solution containing sodium molybdate (Na_2MoO_4). Acidic dissolution, typically used for uranium oxide targets, involves nitric acid to selectively dissolve uranium and fission products into a nitrate solution.

Following dissolution, the solution undergoes filtration to remove suspended solids. Mo-99 is then separated from other fission products using ion exchange or solvent extraction techniques. Ion exchange columns, often packed with alumina or other resins, preferentially adsorb molybdate ions (MoO_4^{2-}), which are subsequently eluted using ammonium hydroxide or saline solutions. Advanced methods such as chromatographic purification or sublimation may also be employed to further enhance purity.

The overall recovery yield of Mo-99 typically exceeds 6.15%, with rigorous decontamination steps ensuring that impurities such as uranium, strontium, cesium, and zirconium are removed to meet international pharmacopoeia standards. These processes are optimized for efficiency and speed to minimize losses from radioactive decay during purification [12].

In non fission scheme, after molybdenum trioxide (MoO_3) targets are irradiated, the recovery and purification of molybdenum-99 (Mo-99) involve several chemical processes to isolate and purify the isotope for medical applications. The irradiated targets are dissolved using either acidic or alkaline solutions, depending on the production method. This dissolution process produces a solution containing molybdate ions (MoO_4^{2-}), which then undergoes purification steps such as precipitation, filtration, and ion exchange to remove impurities, including unreacted MoO_3 and other byproducts. Ion exchange columns, typically packed with alumina or other resins, selectively adsorb molybdate ions, which are then eluted using aluminum oxide, ammonium hydroxide or sodium hydroxide solutions. The purified Mo-99 solution is subsequently concentrated and prepared for use in technetium-99m generators. This process achieves high recovery yields, typically exceeding >85% [13].

4. Conclusion

The results of Mo-99 activity calculations using ORIGEN2.1 reveal significant differences based on position points and power levels at RSG-GAS when using MoO_3 and LEU irradiation with UO_2. The findings demonstrate higher Mo-99 production yields with LEU using UO_2. This is attributed to UO_2's superior ability to capture the required neutrons in nuclear reactions, along with its greater abundance of uranium isotopes prone to neutron-induced fission. However, in purification, the product yield of Mo-99 with LEU is 20-50%, while with MoO_3, the yield is up to 80%.

Furthermore, the calculations show that Mo-99 activity values vary depending on the reactor position and power level. At specific positions like CIP-D6 at 30 MWt, Mo-99 production peaks, whereas at others such as IP at 5 MWt, the yields are significantly lower. This underscores the importance of precise position placement and power configuration in enhancing Mo-99 production outcomes. Thus, a thorough understanding of irradiation processes and the selection of appropriate parameters are crucial for optimizing efficiency and Mo-99 production yields.

5. Acknowledgment

The author would like to express sincere gratitude to Prof. Sidik Permana for his kind permission to use the ORIGEN2.1 program in this research. His support are greatly appreciated.

References

[1] B. Trijono Edi and Syarip, "STUDI PRODUKSI RADIOISOTOP Mo-99 DENGAN BAHAN TARGET LARUTAN URANIL NITRAT PADA REAKTOR KARTINI," *GANENDRA*, vol. 5 No. 1, 2002, Accessed: Apr. 12, 2024. [Online]. Available: https://jurnal.batan.go.id/index.php/ganendra/article/view/207/189

[2] S. Kuntjoro, "ANALISIS AKTIVITAS ISOTOP MO-99 DI REAKTOR RSG-GAS," 2016. Accessed: Apr. 12, 2024. [Online]. Available: https://jurnal.batan.go.id/index.php/sigma/article/view/3503

[3] E. Sutrisno, F. Ratnawati, P. Susanti, S. Reaktor, K. Guna -Batan, and G. Puspiptek, "ANALISIS OPTIMASI TARGET 235 U PENGKAYAAN RENDAH (LEU) UNTUK MENDUKUNG PRODUKSI 99 Mo DAN TARGET BATU TOPAZ DI REAKTOR RSG-GAS," 2018. [Online]. Available: http://ganendra.batan.go.id

[4] S. Pinem and T. Surbakti, "Optimization of Radioisotope Production at Rsg-Gas Reactor Using Deterministic Method," 2012. [Online]. Available: https://www.researchgate.net/publication/299738509

[5] E. Abedi, M. Ebrahimkhani, A. Davari, S. M. Mirvakili, M. Tabasi, and M. G. Maragheh, "Neutronic and thermal-hydraulic analysis of fission molybdenum-99 production at Tehran Research Reactor using LEU plate targets," *Applied Radiation and Isotopes*, vol. 118, pp. 160–166, Dec. 2016, doi: 10.1016/j.apradiso.2016.09.011.

[6] A. Mushtaq, M. Iqbal, I. H. Bokhari, T. Mahmood, and A. Muhammad, "Conversion of Molybdenum-99 production process to low enriched uranium: Neutronic and thermal hydraulic analyses of HEU and LEU target plates for irradiation in Pakistan Research Reactor-1," *Nucl Instrum Methods Phys Res B*, vol. 287, pp. 35–45, Sep. 2012, doi: 10.1016/j.nimb.2012.07.002.

[7] I. Saptiama and E. Sarmini, "PEMBUATAN RADIONUKLIDA MOLIBDENUM-99 (99 Mo) HASIL AKTIVASI NEUTRON DARI MOLIBDENUM ALAM UNTUK MEMPEROLEH TEKNESIUM-99m (99m Tc)," 2016.

[8] D. S. Sairanbayev *et al.*, "Optimizing irradiation conditions for natural molybdenum in WWR-K reactor," *Nuclear Engineering and Technology*, 2024, doi: 10.1016/j.net.2024.04.006.

[9] S. Pinem, T. M. Sembiring, and T. Surbakti, "Reactivity insertion accident analysis during uranium foil target irradiation in the rsg-gas reactor core," *Nuclear Technology and Radiation Protection*, vol. 35, no. 3, pp. 201–207, Sep. 2020, doi: 10.2298/NTRP2003201P.

[10] J. Iman, A. Sufmawan, and K. Mustofa, *PENGUKURAN FLUKS NEUTRON TERMAL DI FASILITAS IRADIASI SISTEM RABBIT TERAS 83 REAKTOR RSG-GAS*. 2013.

[11] International Atomic Energy Agency (IAEA), "Photonuclear Production of Radioisotopes," 2024. Accessed: Jun. 17, 2024. [Online]. Available: https://www-pub.iaea.org/MTCD/Publications/PDF/TE-2051web.pdf

[12] S. K. Lee, G. J. Beyer, and J. S. Lee, "Development of Industrial-Scale Fission 99Mo Production Process Using Low Enriched Uranium Target," *Nuclear Engineering and Technology*, vol. 48, no. 3, pp. 613–623, Jun. 2016, doi: 10.1016/j.net.2016.04.006.

[13] Sulaiman, Y. Sugiharto, G. Setiawan, and A. H. Gunawan, "Optimasi pH Alumina dan 99MO Dalam Pembuatan Generator 99Mo.99mTc Berbasis MoO3 Alam," *Urania*, vol. 24, 2018.

Computational Study of Heat Transfer from Clad Surface to Coolant in Pressurized Water Reactor (PWR)

Duwi Hariyanto[1*]

[1] Industrial Technology Faculty, Institut Teknologi Sumatera
Jl. Terusan Ryacudu, Lampung Selatan, Lampung, 35365, Indonesia

*E-mail: duwi.hariyanto@el.itera.ac.id

Abstract. The heat transfer study from the clad surface to the coolant has become a concern in the thermal analysis of nuclear reactors. The study was proposed to model the phenomenon of heat transfer from the clad surface to the coolant in a pressurized water reactor (PWR). Finite element-based modelling was carried out using the computational fluid dynamics (CFD) module in the Fluent software. In the modelling, heat generation was considered only in the fuel pellets, and the coolant flow rate was varied from bottom to top with the help of a pump. The modelling results showed that the lower the coolant flow rate, the higher the temperature on the clad surface. The coolant flow rate of 5,0 m/s can keep the coolant in the liquid phase.

1. INTRODUCTION

Thermal analysis has become a crucial topic, in particular, in the discussion of the transient thermal behaviour of nuclear reactors when the power level or coolant flow rate is subjected to change. Therefore, the transient study is essential in reactor safety and nuclear reactor dynamics modelling. The main objective in thermal analysis is to determine the temperature distribution throughout the core in the certain fission power distribution or coolant flow rate so that the maximum temperature achieved in the core does not exceed the thermal design limits [1]. The reactor thermal power flux distribution had been analyzed for small long-life pressurized water reactor design in paper [2]. The results in the paper showed that the reactor design with Uranium dioxide of 60% fuel with enriched Uranium-235 of 10% -11% -12% and the addition of 0,0125% Gadolinium and 1,0% Protactinium-231 as the burnable poisons could operate for 13 years without refuelling. Also, the distribution of thermal power flux in a pressurized water reactor design with thermal power of 800 MWt had been studied in papers [3-4]. Then, the safety analysis of nuclear reactors based on passive cooling systems had also been carried out in papers [5-8].

Apart from that, one of the exciting topics in the thermal analysis of nuclear reactors is the heat transfer from the clad to the coolant, as reported in related papers. Paper [9] presented the heat transfer modelling from the clad to the coolant and critical heat flux using the advanced Computational Fluid Dynamics (CFD) computer code CFD-Boiling Water Reactor. The paper results showed that the heat transfer modelling from the clad to the coolant provided a good prediction for various heat transfer regimes, including the emergence of critical heat fluxes. Through modelling, paper [10] investigated the heat transfer from the clad to the coolant under pressure water reactor (PWR) conditions and fast transients. The results showed that a fast heating rate could increase the critical heat flux and the critical temperature shift to a higher value. Paper [11] developed a finite difference method for calculating heat transfer from the clad to the coolant. Then, paper [12] used the finite element method to investigate the thermomechanical behaviour of solid and annular nuclear fuel pellets by considering pressurized water reactor (PWR) conditions.

Motivated from the description, modelling was carried out in this paper to study the heat transfer phenomenon from the clad surface to the coolant in a pressurized water reactor. The modelling based on the finite element method was conducted using the computational fluid dynamic (CFD) module in the Fluent 2020 software. This study is expected to be one of the references related to the thermal analysis of light water nuclear reactors.

2. THEORY

The pressurized water reactor (PWR) is a light water reactor that uses thermal neutrons from nuclear fission reactions to generate energy. Most of the energy produced by the fission is quickly stored as heat in the fuel around the fission location. Then, the heat is transferred by thermal conduction across the fuel element, across the gap separating the fuel from the clad, and then across the clad to the clad surface. After that, it is transferred from the clad surface to the coolant by forced convection. The mass flow of the coolant then carries the thermal energy up and out of the reactor core, either as sensible heat (i.e., an increase in coolant temperature) or latent heat (i.e., a thermally induced phase change (boiling)). The illustration is shown in the following figure 1 [1].

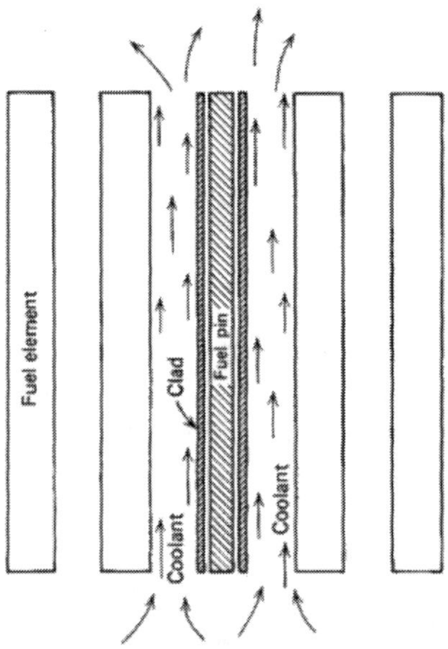

Figure 1. Reactor core cooling illustration [1]

2.1 Heat Conduction Equation

The equation relates to the energy balance in a solid of arbitrary volume. In this case, the energy change rate over time is the difference between the energy gain due to the heat source $q'''(\mathbf{r},t)$ and the energy loss due to heat transfer, which can be written as:

$$\frac{\partial}{\partial t}(\rho c T) = q'''(\mathbf{r},t) - \nabla \cdot \mathbf{q}''(\mathbf{r},t) \qquad (1)$$

where $T(\mathbf{r},t)$ is the solid local temperature, $\rho(\mathbf{r},t)$ and c are the density and specific heat, respectively. $\mathbf{q}''(\mathbf{r},t)$ is a heat flux vector representing the heat flow rate across the surface. Using the diffusion approach, it can be assumed that the heat flux vector $\mathbf{q}''(\mathbf{r},t)$ is proportional to the temperature gradient, as follows:

$$\mathbf{q}''(\mathbf{r},t) = -k\nabla T \qquad (2)$$

Equation (2) is known as Fourier's law of heat conduction. With k as the thermal conductivity, the heat conduction equation can be written as:

$$\frac{\partial}{\partial t}(\rho c T) - \nabla \cdot k\nabla T = q'''(\mathbf{r},t) \qquad (3).$$

2.2 Heat Transfer Equation in Clad

In the case of time-independent heat transfer, constant clad heat conductivity, and no heat production in the clad, the clad temperature drop in the radial direction of equation (2) becomes:

$$-k_C \frac{dT}{dr}\bigg|_{r=r_F} = q'' \qquad (4)$$

Then, the linear power density of the fuel element q' is defined as the power produced per unit length as follows:

$$q' \equiv \pi r_F^2 q''' \qquad (5)$$

Therefore, equation (4) can be written as:

$$-k_C \frac{dT}{dr}\bigg|_{r=r_F} = \frac{q'}{2\pi r_F} = \frac{q''' r_F}{2} \qquad (6)$$

The result of the integration is:

$$T_C - T_S = \frac{q''' r_F^2}{2k_C} \ln \frac{r_F + t_C}{r_F} \qquad (7)$$

where T_C is the clad temperature, T_S is the clad outer surface temperature, r_F is the radius of the fuel element, and t_C is the clad thickness. However, the thickness of the clad is small enough compared to the radius of the fuel element so that by expanding the logarithmic term, equation (7) can be written as:

$$T_C - T_S = \frac{q''' r_F}{2} \frac{t_C}{k_C} = \frac{q'}{2\pi r_F} \frac{t_C}{k_C} \qquad (8).$$

2.3 Heat Transfer Equation from Clad Surface to Coolant

The heat transfer from the clad surface to the coolant is a very efficient process. The heat transfer from the clad surface to the coolant is described by Newton's law for cooling as:

$$q'' = \frac{q'}{2\pi(r_F + t_C)} = h_s(T_s - T_{fl}) \qquad (9)$$

with h_s as convection heat transfer coefficient.

3. METHOD

The computational fluid dynamic (CFD) module in the Fluent software was used to model the heat transfer phenomenon from the clad surface to the coolant. Generally, Fluent software can be used to solve differential equations governing fluid dynamics and heat transfer in fluids based on the finite element method. The basic concept of the finite element method is to divide the mathematical model into simple geometric components. The response of each element can be found from the value of the function on the other elements through the nodes connecting the elements. In the modelling, heat generation was only in Uranium dioxide pellets, and the condition was in a pressurized water reactor (PWR).

The object in the modelling carried out in this paper was the clad and coolant, as shown in figure 2(a). Therefore, the geometry designed as input in the Fluent software can be shown in figure 2(b). Table 1 shows the parameter values used in the modelling. Also, it was considered that the coolant was water flowing from bottom to top due to the force from the pump and the coolant flow rate was varied.

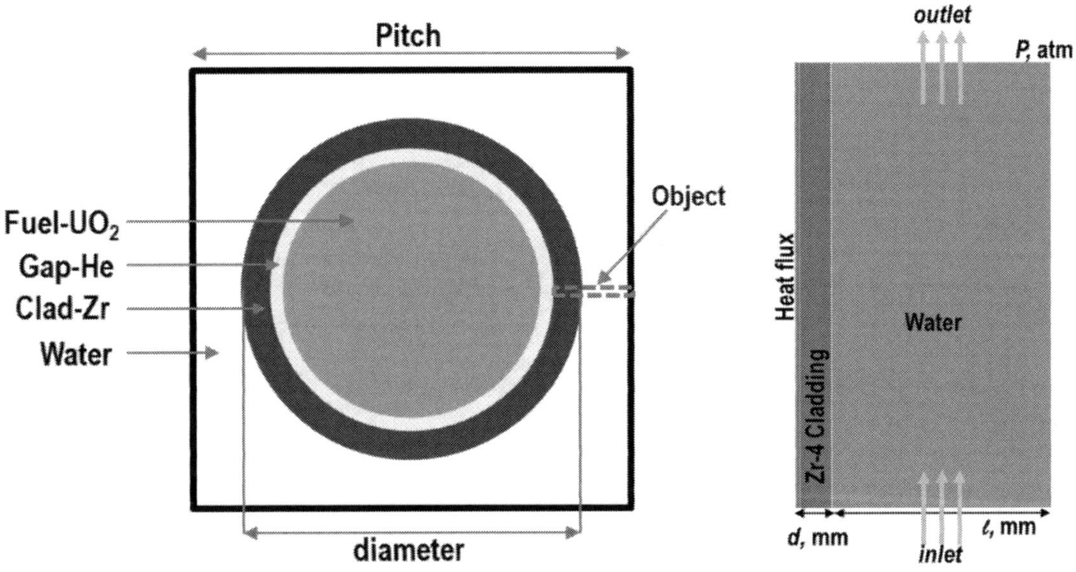

(a) (b)

Figure 2. (a) Standard cross-section of nuclear reactor fuel (b) input geometry of Fluent

Table 1. Specifications of PWR 3441 MWt/1150 MWe [1]

Parameter		
Pitch element	12,500	mm
Diameter element	9,400	mm
Coolant zone width (ℓ)	1,550	mm
Clad thickness (d)	0,572	mm
Object height	13,800	mm
Maximum heat flux	183	W/cm^2
Pressure	155	bar
Inlet temperature	573	K
Outlet temperature	605	K
Maximum fuel temperature	2061	K

4. RESULTS AND DISCUSSION

The heat transfer modelling from the clad surface to the coolant had been carried out using Fluent 2020. The heat flux generated by the fuel made the temperature on the inside of the clad reach 1853 K, which is below the maximum fuel temperature [1]. Then, the heat was transferred to the coolant. The temperature of modelling results from the clad to the coolant at the centre line is shown in figure 3.

Figure 3 shows that the higher the coolant flow rate, the lower the temperature at the boundary between the clad and coolant surfaces. The temperature difference at the boundary between flow rates of 5,0 m/s and 1,0 m/s reached 22%, while between flow rates of 5,0 m/s and 0,01 m/s reached 100%. It showed that when the flow rate decreased, the temperature of the clad surface increased. The temperature distribution contour is shown in figure 4. It can be observed that the temperature distribution from the inside to the clad surface tended to be homogeneous at low flow rates, as shown in figures 4 (a) and (b). The clad temperature contour varied at flow rates of 1,0 m/s and 5,0 m/s, as shown in figures 4 (c) and (d). In this case, from the inside to the clad surface, the temperature decreased. Then, it can be observed in figure 4, the distribution of coolant temperature from inlet to outlet had a significant temperature increase at a coolant flow rate of 0,01 m/s.

Figure 5 shows the water volume fraction results from Fluent software. A value of 0 indicates water in the liquid phase, while 1 indicates water in the vapour phase. It can be observed that at a coolant flow rate of 0,01 m/s, the percentage of water in the vapour phase reached 100% at the top of the clad surface. A sizable percentage of water in the vapour phase can also be observed at a flow rate of 0,1 m/s. At a flow rate of 1,0 m/s, the percentage of water in the vapour phase at the top of the clad surface reached 50%, but overall, the percentage of water in the liquid phase was more dominant.

Meanwhile, the coolant flow rate of 5,0 m/s indicated that water was more dominant in the liquid phase. In the actual case, the PWR design has been provided with a scram signal when the flow rate has reduced to 70% of its initial rate, causing the control rod to descend into the core. It aimed to anticipate a loss of flow accident that can cause the core to overheat [13].

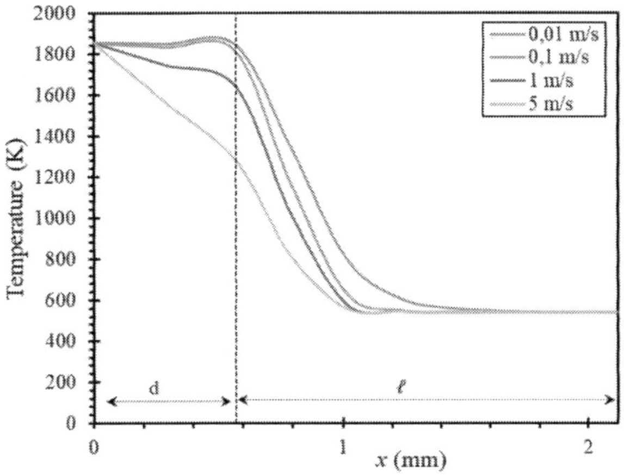

Figure 3. The temperature modelling results along the clad and coolant

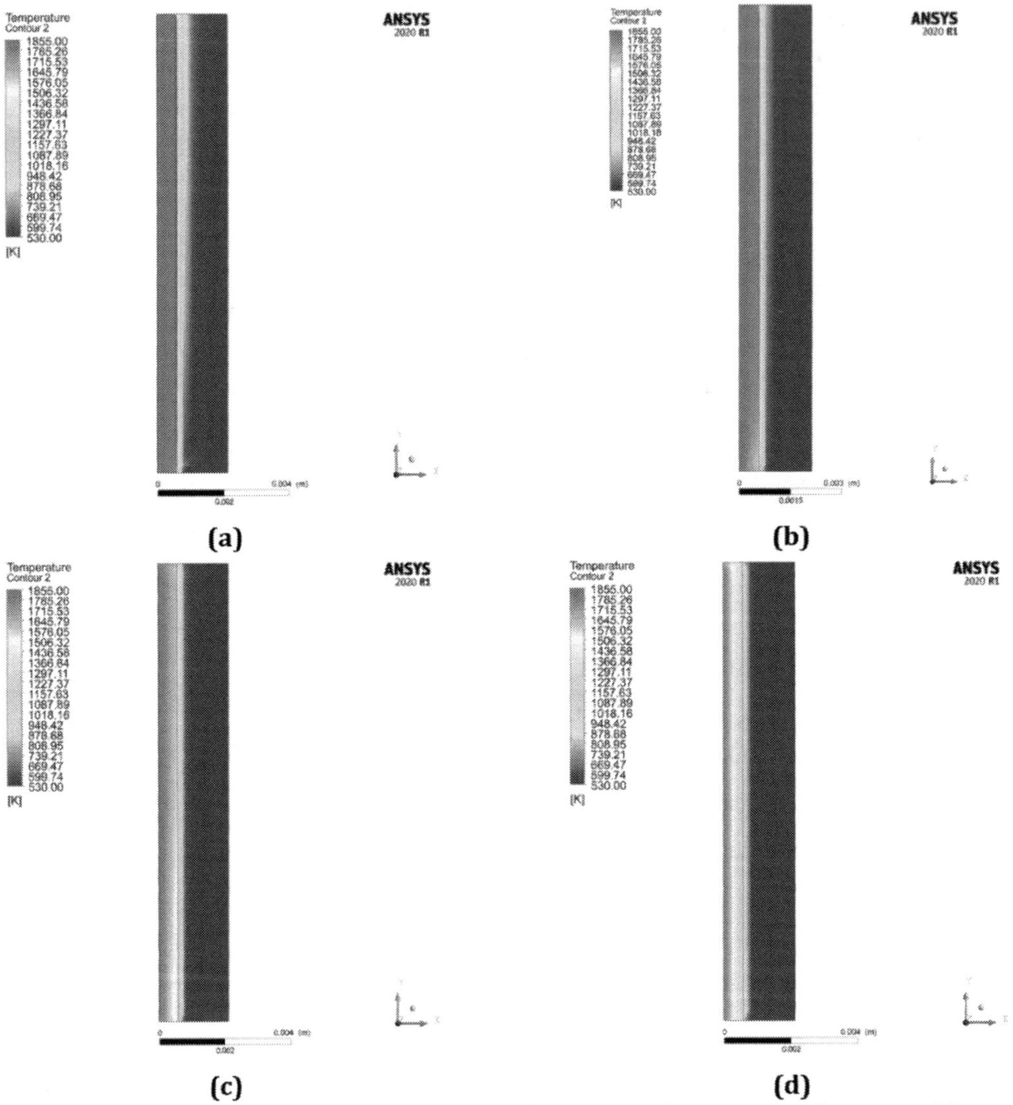

Figure 4. Contour of temperature distribution at flow rates (a) 0,01 m/s (b) 0,1 m/s (c) 1,0 m/s, and (d) 5,0 m/s

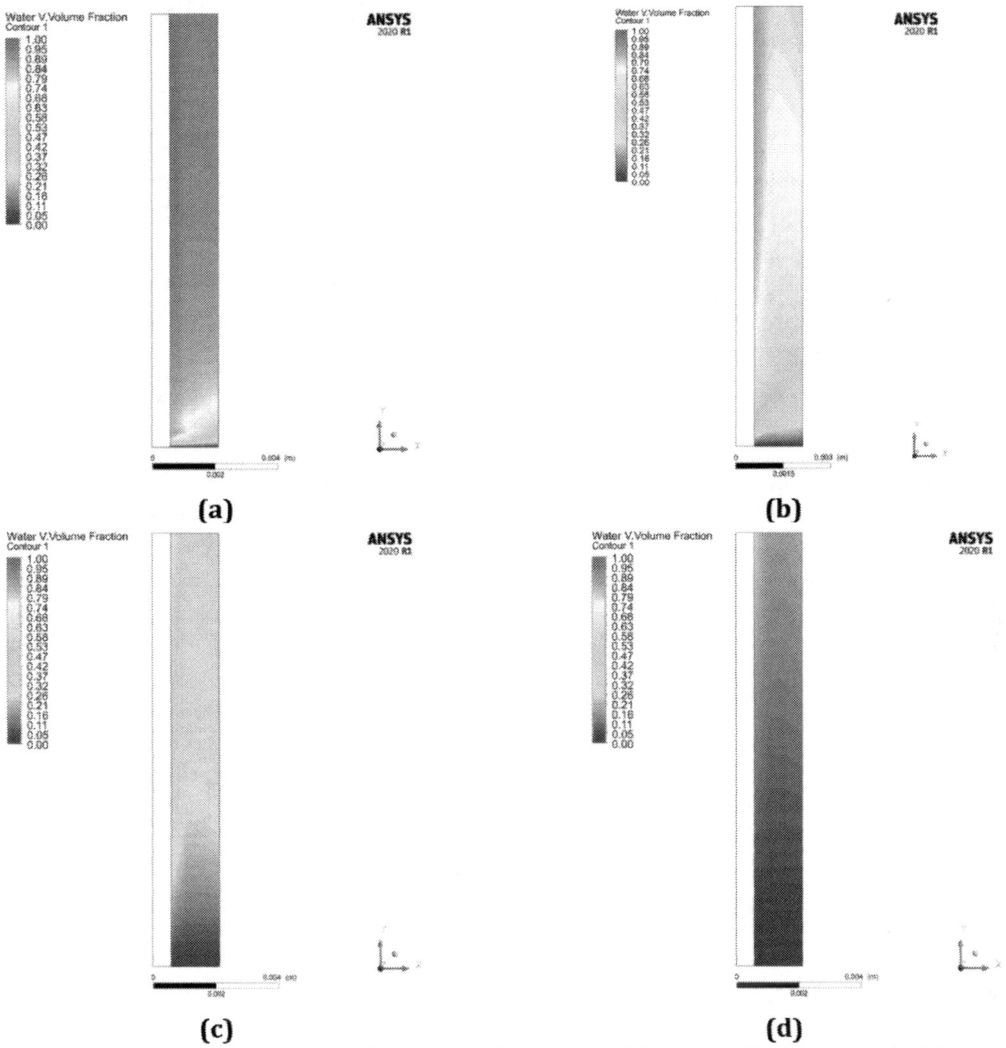

Figure 5. Contour of water volume fraction at flow rates (a) 0,01 m/s (b) 0,1 m/s (c) 1,0 m/s, and (d) 5,0 m/s

5. CONCLUSION

The study of heat transfer from the clad surface to the coolant in a pressurized water reactor using Fluent software provided accurate results in describing physical phenomena. The coolant flow rate significantly influenced the temperature distribution and water phase around the clad surface. Higher coolant flow rates resulted in lower temperatures at the boundary between the clad and coolant, with notable temperature differences of 22% between flow rates of 5.0 m/s and 1.0 m/s, and 100% compared to 0.01 m/s. At low flow rates, such as 0.01 m/s, the temperature distribution from the inside to the clad surface tended to be homogeneous, but the clad surface temperature increased sharply, with almost all the water transitioning to the vapor phase. Conversely, at higher flow rates, water remained predominantly in the liquid phase, enabling more effective cooling. These findings highlight the importance of maintaining optimal coolant flow rates to prevent reactor overheating, as demonstrated in PWR designs equipped with a scram system to mitigate loss-of-flow accidents.

IOP Conf. Series: Materials Science and Engineering **1330** (2025) 012025 doi:10.1088/1757-899X/1330/1/012025

ACKNOWLEDGMENT

The author would like to acknowledge the Decentralization Research Program of the Ministry of Research, Technology, and Higher Education and the ITB Research Innovation and Research International Grant program for their support. We would also like to express our gratitude to the National Research and Innovation Agency (BRIN) Indonesia for providing the ANSYS Fluent license, which was essential for the completion of this research. Furthermore, our heartfelt thanks go to Prof. Dr.Eng. Ir. Sidik Permana from the Faculty of Mathematics and Natural Sciences, Institut Teknologi Bandung, and Dr. Alan Maulana from the Research Center for Nuclear Reactor Technology, National Research and Innovation Agency, Indonesia, for their invaluable insights and contributions throughout this research.

References

[1] J. J. Duderstadt and L. J. Hamilton, *Nuclear Reactor Analysis*. New York: Wiley, 1976.
[2] D. Hariyanto, N. Yuningsih, and S. Permana, Neutronic Analysis of Small Long-Life Pressurized Water Reactor using (Th-U)O_2 Fuels with Gd_2O_3 and Pa-231 as Burnable Poisons, *Indonesian J. Phys.*, **31**(1), 2020.
[3] D. Hariyanto and S. Permana, Three-Dimensional (X-Y-Z) Core Design of Long-Life, *Indonesian J. Phys.*, **31**(1), 2020.
[4] D. Hariyanto and S. Permana, Design Conceptual of 800MWt Long Life Pressurized Water Reactor Using (Th-U)O_2 Fuels with Gd_2O_3 and Pa-231 as Burnable Poisons, *J. Phys. Conf. Ser.*, **1493**, 012006, 2020.
[5] D. Hariyanto and S. Permana, Experimental Investigation of Natural Circulation In A Single-Phase Loop With Different Widths, *International Journal of Electronics and Electrical Engineering*, **8**(2), 7, 2020.
[6] D. Hariyanto, S. Permana, and Suprijadi, Experimental and Simulation Approach of The Loop Geometry Effect on The Natural Circulation System of The Advanced Nuclear Reactor, *Int. J. Energy Res.*, **45**(8), 11892–11903, 2021.
[7] D. Hariyanto and S. Permana, Experimental Study of Thermal Behavior of A Single-Phase Natural Circulation Loop with Vertical Heater and Cooler, *J. Phys. Conf. Ser.*, **1949**, 012024, 2021.
[8] D. Hariyanto, S. Permana, and Suprijadi, Geometry Aspect Review of The Natural Circulation System as Innovation in Passive Cooling of Nuclear Reactor, *Iop Conf. Ser. Earth Environ. Sci.*, **753**, 012015, 2021.
[9] S. Lo, Advances in the Modeling of Clad Heat Transfer and Critical Heat Flux in Boiling Water Reactor Fuel Assemblies, The 12th International Topical Meeting on Nuclear Reactor Thermal Hydraulics (NURETH-12), (Sheraton Station Square, Pittsburgh, Pennsylvania, U.S.A). 2007.
[10] V. Bessiron, Modelling of Clad-to-Coolant Heat Transfer For Ria Applications, *J. Nucl. Sci. Technol.*, **44** (2), 211–221, 2007.
[11] J. Łach and W. Pieczka, Heat Transfer at The Clad-Cooling Fluid Boundary, *Int. J. Heat Mass Transf.*, **25**(10), 1595–1603, 1982.
[12] M. Y. Arafat and D. Datta, A Finite Element Approach to Investigate The Thermomechanical Behavior of Solid and Annular Type Nuclear Fuel Rods, *Mater. Today Proc.*, **47**, 3919–3926, 2021.
[13] B. Foad, S. H. Abdel-Latif, and T. Takeda, Reactivity feedback effect on loss of flow accident in PWR, *Nuclear Engineering and Technology*, **50**(8), 1277–1288, 2018.

IOP Conf. Series: Materials Science and Engineering **1330** (2025) 012026 doi:10.1088/1757-899X/1330/1/012026

Simulation on Neutron Imaging Facility Based on 30 MeV Cyclotron

Dian Adi Prastowo[1*], Fahrurrozi Akbar[1], Andeka Tris Susanto[1], Rissa Damayanti[1], Anas Fahmi Imron[1], Haryo Seno[1], Rasito[1], Ridwan[1], Mujamilah[1], and Abu Khalid Rivai[1]

[1] Research Center for Nuclear Beam Analysis Technology, The National Research and Innovation Agency, South Tangerang, Indonesia

*E-mail: dian.adi.prastowo@brin.go.id

Abstract. A new neutron imaging (NI) facility based on a 30-MeV cyclotron is currently being developed at the National Research and Innovation Agency (BRIN) of Indonesia. This study aims to simulate an optimal neutron delivery system for NI using Monte Carlo method by Particle and Heavy Ion Transport Code System (PHITS) software. The cyclotron accelerates 30 μA protons up to 30 MeV which are directed onto a beryllium target to produce neutrons. After being moderated, neutrons are ejected using a one-meter-long radial beam tube. Our investigation centred on the influence of different reflector and filter materials on the neutron flux. At the start of the beam tube, the overall neutron flux is 3.74 x 10^6 neutrons.cm^{-2}.s^{-1}, of which 42.7% are thermal. Our findings suggest that an aluminium reflector coupled with a 5-cm thick graphite filter is the most effective configuration for minimizing neutron flux loss. This configuration delivers 2.17x10^4 neutrons.cm^{-2}.s^{-1}, of which 69.02% are thermal neutrons. This number is comparable to other thermal neutron imaging facilities worldwide.

1. Introduction

Neutron imaging (NI) is a non-destructive technique with growing applications in materials science, engineering, and security. Traditionally, reactor-based neutron sources have been the primary source for NI, but the development of compact accelerator-based neutron generators now offers new opportunities [1-4]. BRIN, the agency responsible for research and development in Indonesia, also plans to build a 30 MeV cyclotron, in which the construction of NI facilities is one of the planned instruments.

While NI installations are commonly associated with reactor-based neutron sources, there has been increasing interest and development in cyclotron-based NI systems. Cyclotrons offer several advantages, including compactness, flexibility in neutron energy, and potential for on-site operation. However, they typically deliver lower neutron flux, which poses challenges for NI. The acceptable neutron flux for NI experiments is about 10^5 neutrons/(cm^2.s) and higher, but achieving fluxes higher than 10^6 neutrons/(cm^2.s) at sample positions is quite difficult based on the experience of other NI facility developers. Moreover, the spatial resolution of early cyclotron-based NI systems was not very high, typically around 300 – 500 microns [5], compared to less than 100 microns for NI based on synchrotrons or reactors [6-7].

One of the key factors to get maximum results in NI applications is how to obtain a convergent neutron beam on the surface of the test object. For this reason, the design of a collimation system is very important and very decisive for obtaining high-resolution images. Existing NI systems based on 30 MeV cyclotrons often suffer from low neutron flux [8-12] and limited spatial resolution. Comprehensive planning is required starting from the selection of reflector materials, along with a moderator system with controlled temperature, as well as the design of the collimation system configuration along with the materials needed to realize it [13-14].

Therefore, this study addresses these challenges by investigating innovative collimator designs for BRIN's upcoming cyclotron. This paper discusses collimator design, which focuses on variations in reflector and filter materials to achieve the minimum flux for NI at 10^5 neutrons/(cm^2.s). To accomplish this goal, we employed Monte Carlo method to model and optimize collimator performance. By systematically varying materials and dimensions, we aim to significantly increase neutron flux to improve image quality. Successful optimization of the collimator will enable high-resolution NI at BRIN which in turn will contribute to advancements in various research and industrial applications.

2. Methodology

The neutron production and transport in a laboratory setting were simulated using the PHITS Monte Carlo code (version 3.34) developed by the Japan Atomic Energy Agency (JAEA). This code accurately models the transport of various particles, including photons, neutrons, charged particles, and heavy ions, and can replicate nuclear and atomic reactions (Sato, 2020) [15].

A simplified laboratory model was constructed, comprised of a concrete-walled room, a steel proton transport tube, a high-density polyethylene moderator, (ρ = 0.94 g/cm^3), a beryllium target, and a neutron guide tube. A 30 MeV proton beam, 1 cm in radius, was incident on the cylindrical beryllium target (1 cm radius, 0.8 cm thickness). The polyethylene moderator surrounds the proton tube, the target, and the radial beam tube. The proton-beryllium reactions will produce fast neutrons, which are then moderated into thermal neutrons after colliding with high-density polyethylene.

Initially, we followed the BRIN's NI facility design at RSG GA Siwabessy, especially for reflector and filter materials composition, and improved it by employing other reflector and filter materials. To mimic the NI's beam tube at the RSG-GAS reactor, a 1-meter-long beam tube was positioned 15 cm from the target. This tube was filled with atmospheric air and lined with various reflector materials - which are selected for their high coherent scattering length [16-18] - and filter materials, chosen based on their common use in neutron facilities [19-21]. Table 1 outlines the materials and filter thicknesses used in the simulations. All elements were assumed to be in their natural abundance. The top view of the laboratory is depicted in Fig 1 while the longitudinal section of the beam tube is depicted in Fig 2.

The simulations were conducted in two stages. First, proton-beryllium collisions were simulated, and the resulting neutron and photon data were stored in a dump file. This dump file served as the source for the second stage, where neutron transport through the beam tube was modelled. This approach eliminates the need to repeat proton-beryllium collisions for each simulation. Another reason is that simulating proton-beryllium reactions is slow and time-consuming, while simulating neutron transport is relatively faster. Thus, this second stage can be done more efficiently, independently from the first stage.

IOP Conf. Series: Materials Science and Engineering **1330** (2025) 012026 doi:10.1088/1757-899X/1330/1/012026

After each simulation, the neutron flux at the beam tube's end was tallied, along with the neutron energy spectrum. The flux was categorized into thermal (<100 meV), epithermal (100 meV - 10 keV), and fast (>10 keV) neutrons. The optimal reflector and filter materials were determined based on their ability to maximize thermal neutron flux and minimize fast neutron transport, respectively.

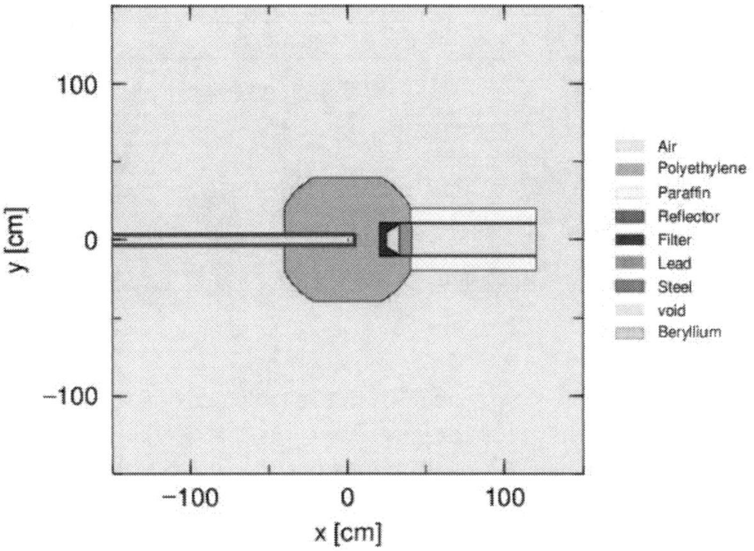

Figure 1. Longitudinal section of the collimator/beam tube.

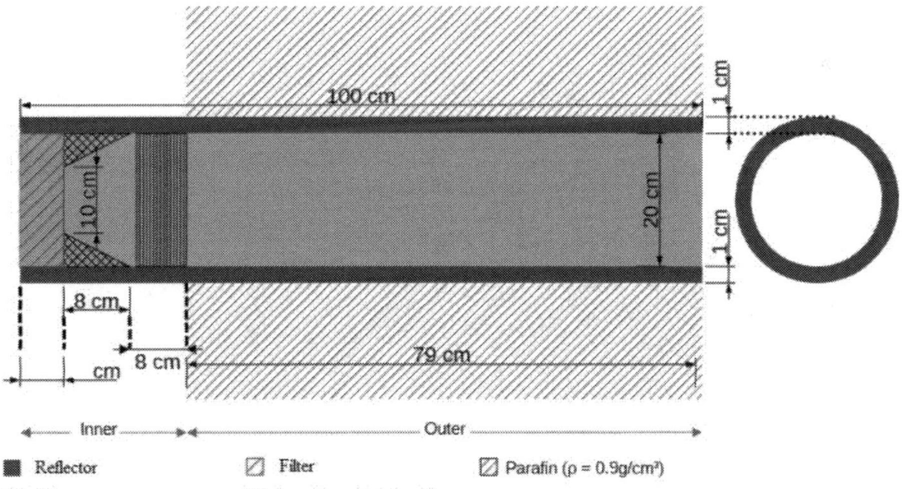

Figure 2. Longitudinal section of the collimator/beam tube.

IOP Conf. Series: Materials Science and Engineering **1330** (2025) 012026 doi:10.1088/1757-899X/1330/1/012026

Table 1. Variables used in the simulation; sample name is composed of reflector material code followed by filter material code and filter thickness code.

Reflector		Filter			
Material Code	Material Name	Material Code	Material Name	Thickness Code	Thickness (cm)
A	Aluminium (Al)	B	Beryllium (Be)	05	5.0
N	Nickel (Ni)	G	Graphite (C)	10	10.0
S	Silica (SO_2)	S	Sapphire (Al_2O_3)	15	15.0
T	Copper (Cu)				

3. Results and Discussion

The neutron and photon energy spectrum from the collision of $^9Be(p,n)^9B$ is shown in Fig 3. These neutrons have various energies, ranging from low-energy neutrons up to high-energy neutrons above 25 MeV. The highest proportion of this spectrum is located at the range of 0.36-0.54 MeV with decreasing proportion towards the higher end of the spectrum. From 30 μA of the proton beam, the overall calculated fluence is 2.57 x 10^{10} neutrons over the solid angle of 4π. This number is higher than most of available portable neutron sources such as DD or DT generators. After the spectrum was captured, it was used for the remainder of the experiment as the source.

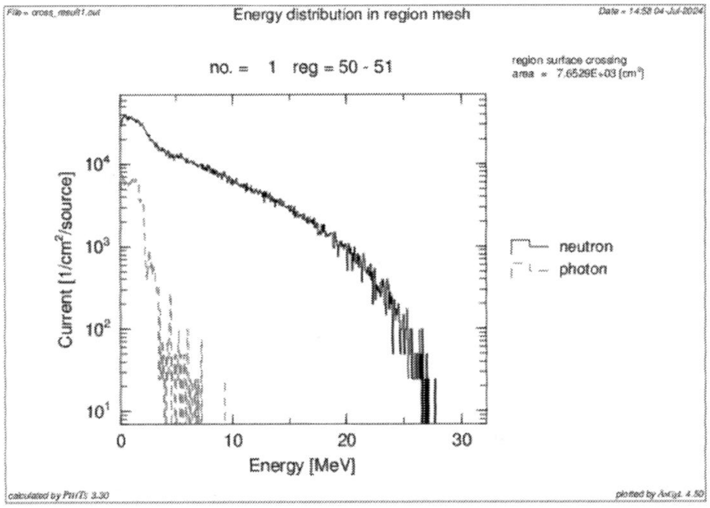

Figure 3. Neutron and photon energy spectrum after $^9Be(p,n)^9B$ reaction at 30 MeV

After traversing through a high-density polyethylene moderator, the neutrons are slowed down and once again the energy spectrums were captured in Fig 4. The average number of neutron currents flowing into the beam tube was 3.74 x 10^6 neutrons/(cm^2.s), of which 42.7% of them, or about 1.60 x 10^6/(cm^2s) were in the thermal range. The thermal peak has appeared but at the same time, there is also a peak at a fast energy range. So, in order to select the best reflector

IOP Conf. Series: Materials Science and Engineering **1330** (2025) 012026 doi:10.1088/1757-899X/1330/1/012026

material, the neutron current exiting the beam tube should not decrease too much from this initial current. Meanwhile, the proportion of thermal neutrons compared to the overall flux should be higher than 42.7% with no peak is epithermal or fast energy range, which indicates than the filter is capable of decreasing the epithermal and fast neutrons from travelling through the beamline.

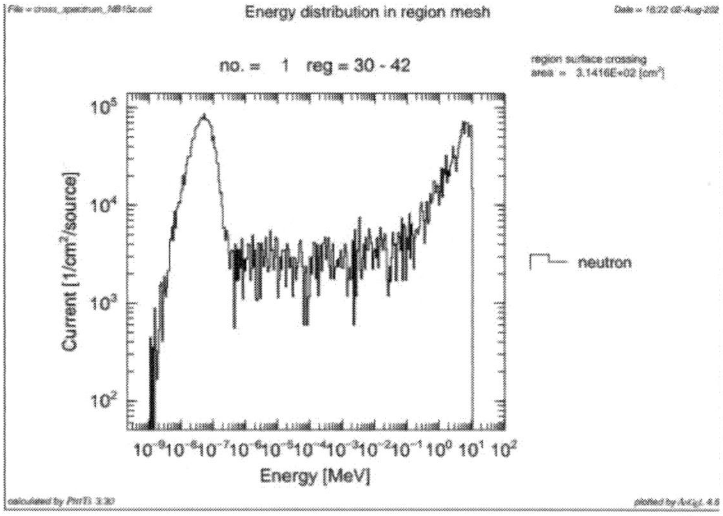

Figure 4. The neutron energy spectrum at the start of the beamline

The first objective of our research was to determine the best reflector material that lines up the entire collimator. This reflector works as follows: every material has its own coherent neutron scattering length (b_{coh}). It means that neutrons which comes below the critical scattering vector (Q_c) will be reflected. The scattering vector (Q) is directly proportional to the grazing angle - the angle between the trajectory of the neutron and the surface of the material - while inversely proportional to the de Broglie wavelength of the neutron. Since the wavelength is inversely proportional to the energy, it means that thermal neutrons will have less Q than fast neutrons. Therefore, thermal neutrons will be reflected up to a higher grazing angle than fast neutrons, which makes them travel farther in the guide tube. Thus, the optimal collimator would not only deliver more neutrons but also selectively prefer the thermal ones. The neutron scattering lengths are available in the software from JENDL 4.0 database and the parameter in the source code is set to enable neutron elastic scattering.

Table 2 summarises the result of the simulation with different collimator materials. The filter is set with 5 cm of graphite. From the table, it can be deduced that aluminium provides the best reflection. This is however contrary to the hypothesis, which states that better reflections are acquired with material with the highest elastic scattering length, which is nickel (10.3 pm compared with 4.13 pm for aluminium) [16]. Moreover, the higher the elastic scattering length the lower the number of neutrons reaching the end of the beam tube. This should be investigated in further research.

Table 2. Results in simulation with different collimator materials

Sample Code	Reflector Material	Filter Material	Thickness (cm)	Thermal Flux ($cm^{-2}s^{-1}$)	Percentage of all neutrons
AG05	Aluminium (Al)	Graphite (C)	5	1.50×10^4	69.02%
NG05	Nickel (Ni)	Graphite (C)	5	1.51×10^3	12.40%
SG05	Silica (SO_2)	Graphite (C)	5	1.24×10^4	66.65%
TG05	Copper (Cu)	Graphite (C)	5	3.35×10^3	26.82%

The next objective of this research is to select the best filter material to sort out thermal neutrons from fast neutrons. The selectivity is calculated by the ratio between neutrons entering the filter and neutrons exiting the filter. Higher selectivity means more neutrons being transmitted than being absorbed. Here, the best filter is graphite, although the difference between graphite and sapphire is not significant. Graphite will filter more fast neutrons while also reducing the thermal neutrons. Meanwhile, sapphire will transmit more thermal neutrons while also leaving more fast neutrons. On the other hand, beryllium will reduce the thermal neutrons more than fast neutrons. Beryllium is actually better to be used in cold neutron facilities, as it would select out neutrons above 5 meV (equivalent to a wavelength of 4 angstroms) [21]. The result of the second experiment is tabulated in Table 3.

Table 3. Results in simulation with different filter materials

Sample Code	Reflector Material	Filter Material	Thickness (cm)	Thermal Selectivity	Epithermal selectivity	Fast Selectivity
AB05	Aluminium (Al)	Beryllium (Be)	5	30.12%	34.27%	35.97%
AG05	Aluminium (Al)	Graphite (C)	5	36.13%	32.95%	31.75%
AS05	Aluminium (Al)	Sapphire (Al_2O_3)	5	34.69%	31.00%	26.30%

The subsequent aim is to determine the effect of filter thickness on the thermal neutron flux at the end of the beam tube. The data is presented in Table 4, which uses the information from previous experiments regarding reflector and filter material, which are aluminium and graphite, respectively. The data shows that a thicker filter will separate the fast neutrons from the thermal neutrons more effectively. However, at the same time, it will decrease the thermal neutron flux as the filter also absorbs thermal neutrons, albeit less significantly.

Overall, the sample AG05 delivers the most thermal neutrons at the rate of $1.50 \times 10^4/(cm^2s)$, which comprises about 69% of all neutrons that reach the sample. This thermal flux and percentage are still lower than the IAEA recommendation which states that thermal neutron radiography facilities should provide neutrons at the rate of $10^5/(cm^2.s)$, of which 90% should be thermal [22-23]. These results, however, are comparable to current radiography facilities built on accelerator sources, such as facilities owned by PKUNIFTY, INEST, or RIKEN [10,24-25].

Table 4. Results in simulation with different filter thickness

Sample Code	Reflector Material	Filter Material	Thickness (cm)	Thermal Flux ($cm^{-2}s^{-1}$)	Percentage of all neutrons
AG05	Aluminium (Al)	Graphite (C)	5	1.50×10^4	69.02%
AG10	Aluminium (Al)	Graphite (C)	10	7.58×10^3	63.64%
AG15	Aluminium (Al)	Graphite (C)	15	6.00×10^3	75.62%

Finally, we also observe the delivered neutron spectrum from the most optimal design, which is from sample AG05. The spectrum shows that the neutrons have been thermalized with one peak at 43 eV which corresponds to the wavelength of 1.38 Å. At this energy, the flux is about 610 neutrons/(cm^2s). This suggests that this facility would not be capable of performing neutron elastic scattering experiments, which require thermal flux circa 10^6/(cm^2s) [26-27].

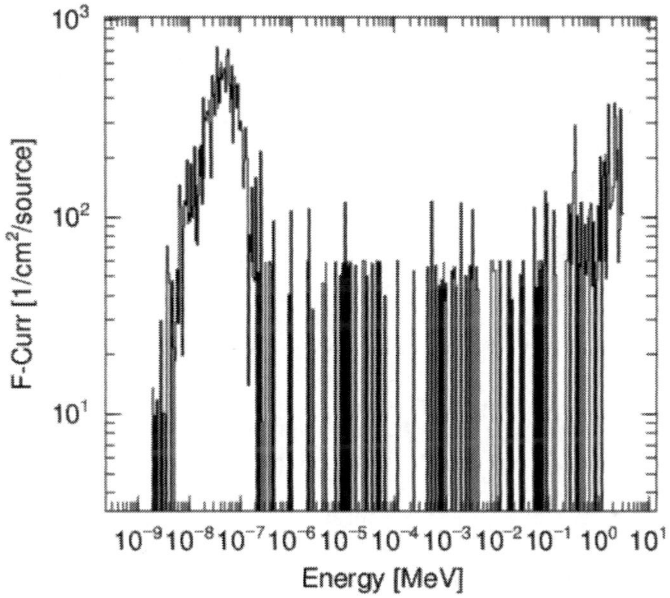

Figure 5. The neutron energy spectrum at the end of the beamline

4. Conclusion

This paper outlines a feasibility study on the building of a neutron imaging facility based on 30-MeV cyclotron that produces 30 mA of proton beam. Our experiments suggest that by utilizing aluminium as the collimator and 5-cm thick of graphite as the filter, we can obtain thermal neutron flux up to 1.5×10^4/(cm^2.s) at the distance of 1 meter from the start of the beam tube. The thermalization ratio is also high, with 69% of neutrons being thermal. Further research should be conducted to increase the thermal flux and thermalization ratio, such as using multilayer super mirror which has been proven to increase thermal neutron delivery.

In the end, this study provides a valuable foundation for the development of advanced neutron facilities for a wide range of applications, especially for neutron imaging purposes. By addressing the identified challenges and exploring future research directions, it is possible to create more efficient and versatile neutron facilities that can contribute to scientific progress and technological advancements.

References

[1] Baxter, D. V *et al.* (2005) 'The Low Energy Neutron Source- Status and Prospects'.
[2] Thulliez, L. *et al.* (2020) 'First steps toward the development of SONATE, a Compact Accelerator driven Neutron Source', *EPJ Web of Conferences*, 239, p. 17011. doi: 10.1051/epjconf/202023917011.
[3] Tran, H. N. *et al.* (2020) 'Neutrons production on the IPHI accelerator for the validation of the design of the compact neutron source SONATE', *EPJ Web of Conferences*, 231, p. 01007. doi: 10.1051/epjconf/202023101007.
[4] Lee, D. W. *et al.* (2024) 'Development status of 30 MeV cyclotron-based neutron source in Korea', *EPJ Web of Conferences*, 298, p. 05004. doi: 10.1051/epjconf/202429805004.
[5] Hiraoka, E. (no date) 'Invited Paper: Neutron Radiography with Cyclotron', pp. 587–592.
[6] Tazawa, S. and Whittemore, W. L. (1993) 'A PROPOSAL FOR AN INTENSE CYCLOTRON-BASED THERMAL NEUTRON SOURCE', *13th International Conference on Cyclotrons and Their Applications*, p. IV10.
[7] Allen, D. A. *et al.* (1994) 'Neutron radiography using a transportable superconducting cyclotron', *Nuclear Inst. and Methods in Physics Research, A*, 353(1–3), pp. 128–133. doi: 10.1016/0168-9002(94)91618-7.
[8] Chia, W.-M., Chen, C.-S. and Yeh, T.-R. (1996) 'NUCLEAR INSTRUMENTS & METHODS IN PHYSICS RESEARCH Conceptual design of the INER cyclotron based neutron radiograph facility', *Nuclear Instruments and Methods in Physics Research A*, 377, pp. 27–31.
[9] Guo, Z. *et al.* (2013) 'Progress of PKUNIFTY - A RFQ accelerator based neutron imaging facility at Peking University', *Physics Procedia*. Elsevier B.V., 43, pp. 79–85. doi: 10.1016/j.phpro.2013.03.010.
[10] Zhang, L. X. *et al.* (2023) 'Resolution analysis of thermal neutron radiography based on accelerator-driven compact neutron source', *Nuclear Science and Techniques*. Springer Nature Singapore, 34(5), pp. 1–13. doi: 10.1007/s41365-023-01227-x.
[11] Morgano, M. *et al.* (2018) 'Unlocking high spatial resolution in neutron imaging through an add-on fibre optics taper', *Optics Express*, 26(2), p. 1809. doi: 10.1364/oe.26.001809.
[12] Lübcke, A. *et al.* (2019) 'Soft X-ray nanoscale imaging using a sub-pixel resolution charge coupled device (CCD) camera', *Review of Scientific Instruments*, 90(4). doi: 10.1063/1.5053593.
[13] Wonglee, S. *et al.* (2018) 'Collimator design for neutron radiography station using Monte Carlo simulation', *Journal of Physics: Conference Series*, 1144(1). doi: 10.1088/1742-6596/1144/1/012095.
[14] Lee, C. H. *et al.* (2021) 'Status of development and planning activities on CANS in Korea', *Journal of Neutron Research*, 23(2–3), pp. 127–141. doi: 10.3233/JNR-210017.
[15] Sato, T. *et al.* (2024) 'Recent improvements of the particle and heavy ion transport code system–PHITS version 3.33', *Journal of Nuclear Science and Technology*. Taylor & Francis, 61(1), pp. 127–135. doi: 10.1080/00223131.2023.2275736.
[16] Cubitt, R. and Fragneto, G. (2002) 'Neutron Reflection: Principles and Examples of Applications (Chapter 2.8.3)', *Scattering*, pp. 1198–1208.
[17] Katyba, G. M. *et al.* (2018) 'Sapphire shaped crystals for waveguiding, sensing and exposure applications', *Progress in Crystal Growth and Characterization of Materials*. Elsevier, 64(4), pp. 133–151. doi: 10.1016/j.pcrysgrow.2018.10.002.
[18] Cousin, F. and Fadda, G. (2020) 'An introduction to neutron reflectometry', *EPJ Web of Conferences*, 236, p. 04001. doi: 10.1051/epjconf/202023604001.
[19] Stamatelatos, I. E. and Messoloras, S. (2000) 'Sapphire filter thickness optimization in neutron scattering instruments', *Review of Scientific Instruments*, 71(1), pp. 70–73. doi: 10.1063/1.1150163.
[20] Adib, M. *et al.* (2005) 'Neutron transmission of single-crystal sapphire filters', *Czechoslovak Journal of Physics*, 55(5), pp. 563–578. doi: 10.1007/s10582-005-0061-7.
[21] Groitl, F. *et al.* (2016) 'A combined radial collimator and cooled beryllium filter for neutron scattering', *Nuclear Instruments and Methods in Physics Research, Section A: Accelerators, Spectrometers, Detectors and Associated Equipment*, 819, pp. 99–103. doi: 10.1016/j.nima.2016.02.056.

[22] 'Report of a Consultancy Meeting: Nondestructive and Analytical Techniques Using Neutrons, 08CT14309, International Atomic Energy Agency (IAEA)' (2019).

[23] Choopan Dastjerdi, M. H. *et al.* (2019) 'A neutron radiography beamline relying on the Isfahan Miniature Neutron Source Reactor', *Nuclear Instruments and Methods in Physics Research, Section A: Accelerators, Spectrometers, Detectors and Associated Equipment.* Elsevier Ltd, 928(February), pp. 20–25. doi: 10.1016/j.nima.2019.03.008.

[24] Guo, Z. *et al.* (2012) 'Neutron radiography with compact accelerator at Peking University: Problems and solutions', *Physics Procedia*, 26(2010), pp. 70–78. doi: 10.1016/j.phpro.2012.03.011.

[25] Kiyanagi, Y. (2018) 'Neutron imaging at compact accelerator-driven neutron sources in Japan', *Journal of Imaging*, 4(4), pp. 17–18. doi: 10.3390/jimaging4040055.

[26] IAEA (2021) 'Compact Accelerator Based Neutron Sources', *Iaea-Tecdoc-1981.*

[27] Putra, E. G. R. *et al.* (2007) 'Performance of the 36 m small-angle neutron scattering spectrometer at BATAN, Serpong, Indonesia', *Journal of Applied Crystallography*, 40(SUPPL. 1), pp. 447–452. doi: 10.1107/S002188980700605X.

AUTHOR INDEX